Birkhäuser

Applied and Numerical Harmonic Analysis

Patricia Alonso Ruiz • Michael Hinz •
Kasso A. Okoudjou • Luke G. Rogers •
Alexander Teplyaev
Editors

From Classical Analysis to Analysis on Fractals

A Tribute to Robert Strichartz, Volume 1

Birkhäuser

Editors
Patricia Alonso Ruiz
Department of Mathematics
Texas A&M University
College Station, TX, USA

Michael Hinz
Fakultät für Mathematik
Universität Bielefeld
Bielefeld, Germany

Kasso A. Okoudjou
Department of Mathematics
Tufts University
Medford, MA, USA

Luke G. Rogers
Department of Mathematics
University of Connecticut
Storrs, CT, USA

Alexander Teplyaev
Department of Mathematics
University of Connecticut
Storrs, CT, USA

ISSN 2296-5009 ISSN 2296-5017 (electronic)
Applied and Numerical Harmonic Analysis
ISBN 978-3-031-37799-0 ISBN 978-3-031-37800-3 (eBook)
https://doi.org/10.1007/978-3-031-37800-3

Mathematics Subject Classification: 42-XX, 46-XX, 58-XX, 28A80

This book is published under the imprint Birkhäuser, www.birkhauser-science.com by the registered company Springer Nature Switzerland AG
The registered company address is: Gewerbestrasse 11, 6330 Cham, Switzerland

Paper in this product is recyclable.

Dedicated to the memory of Robert S. Strichartz.

ANHA Series Preface

The Applied and Numerical Harmonic Analysis (ANHA) book series aims to provide the engineering, mathematical, and scientific communities with significant developments in harmonic analysis, ranging from abstract harmonic analysis to basic applications. The title of the series reflects the importance of applications and numerical implementation, but richness and relevance of applications and implementation depend fundamentally on the structure and depth of theoretical underpinnings. Thus, from our point of view, the interleaving of theory and applications and their creative symbiotic evolution is axiomatic.

Harmonic analysis is a wellspring of ideas and applicability that has flourished, developed, and deepened over time within many disciplines and by means of creative cross-fertilization with diverse areas. The intricate and fundamental relationship between harmonic analysis and fields such as signal processing, partial differential equations (PDEs), and image processing is reflected in our state-of-the-art ANHA series.

Our vision of modern harmonic analysis includes a broad array of mathematical areas, e.g., wavelet theory, Banach algebras, classical Fourier analysis, time-frequency analysis, deep learning, and fractal geometry, as well as the diverse topics that impinge on them.

For example, wavelet theory can be considered an appropriate tool to deal with some basic problems in digital signal processing, speech and image processing, geophysics, pattern recognition, biomedical engineering, and turbulence. These areas implement the latest technology from sampling methods on surfaces to fast algorithms and computer vision methods. The underlying mathematics of wavelet theory depends not only on classical Fourier analysis, but also on ideas from abstract harmonic analysis, including von Neumann algebras and the affine group. This leads to a study of the Heisenberg group and its relationship to Gabor systems, and of the metaplectic group for a meaningful interaction of signal decomposition methods.

The unifying influence of wavelet theory in the aforementioned topics illustrates the justification for providing a means for centralizing and disseminating information from the broader, but still focused, area of harmonic analysis. This will be a key role of ANHA. We intend to publish with the scope and interaction that such a host of issues demands.

Along with our commitment to publish mathematically significant works at the frontiers of harmonic analysis, we have a comparably strong commitment to publish major advances in the following applicable topics in which harmonic analysis plays a substantial role:

*Analytic Number theory * Antenna Theory * Artificial Intelligence * Biomedical Signal Processing * Classical Fourier Analysis * Coding Theory * Communications Theory * Compressed Sensing * Crystallography and Quasi-Crystals * Data Mining * Data Science * Deep Learning * Digital Signal Processing * Dimension Reduction and Classification * Fast Algorithms * Frame Theory and Applications * Gabor Theory and Applications * Geophysics * Image Processing * Machine Learning * Manifold Learning * Numerical Partial Differential Equations * Neural Networks * Phaseless Reconstruction * Prediction Theory * Quantum Information Theory * Radar Applications * Sampling Theory (Uniform and Non-uniform) and Applications * Spectral Estimation * Speech Processing * Statistical Signal Processing * Super-resolution * Time Series * Time-Frequency and Time-Scale Analysis * Tomography * Turbulence * Uncertainty Principles *Waveform Design * Wavelet Theory and Applications

The above point of view for the ANHA book series is inspired by the history of Fourier analysis itself, whose tentacles reach into so many fields.

In the last two centuries, Fourier analysis has had a major impact on the development of mathematics, on the understanding of many engineering and scientific phenomena, and on the solution of some of the most important problems in mathematics and the sciences. Historically, Fourier series were developed in the analysis of some of the classical PDEs of mathematical physics; these series were used to solve such equations. In order to understand Fourier series and the kinds of solutions they could represent, some of the most basic notions of analysis were defined, e.g., the concept of "function." Since the coefficients of Fourier series are integrals, it is no surprise that Riemann integrals were conceived to deal with uniqueness properties of trigonometric series. Cantor's set theory was also developed because of such uniqueness questions.

A basic problem in Fourier analysis is to show how complicated phenomena, such as sound waves, can be described in terms of elementary harmonics. There are two aspects of this problem: first, to find, or even define properly, the harmonics or spectrum of a given phenomenon, e.g., the spectroscopy problem in optics; second, to determine which phenomena can be constructed from given classes of harmonics, as done, for example, by the mechanical synthesizers in tidal analysis.

Fourier analysis is also the natural setting for many other problems in engineering, mathematics, and the sciences. For example, Wiener's Tauberian theorem in Fourier analysis not only characterizes the behavior of the prime numbers, but is a fundamental tool for analyzing the ideal structures of Banach algebras. It also provides the proper notion of spectrum for phenomena such as white light. This latter process leads to the Fourier analysis associated with correlation functions in filtering and prediction problems. These problems, in turn, deal naturally with Hardy spaces in complex analysis, as well as inspiring Wiener to consider communications engineering in terms of feedback and stability, his cybernetics. This latter theory develops concepts to understand complex systems such as learning and cognition and neural networks, and it is arguably a precursor of deep learning and its spectacular interactions with data science and AI.

Nowadays, some of the theory of PDEs has given way to the study of Fourier integral operators. Problems in antenna theory are studied in terms of unimodular trigonometric polynomials. Applications of Fourier analysis abound in signal processing, whether with the fast Fourier transform (FFT), or filter design, or the adaptive modeling inherent in time-frequency-scale methods such as wavelet theory.

The coherent states of mathematical physics are translated and modulated Fourier transforms, and these are used, in conjunction with the uncertainty principle, for dealing with signal reconstruction in communications theory. We are back to the raison d'etre of the ANHA series!

College Park, MD, USA — John Benedetto
College Park, MD, USA — Wojciech Czaja
Medford, MA, USA — Kasso Okoudjou

Preface

This book is the first of a two-volume tribute to Robert Strichartz, the outstanding mathematician whose work enriched various areas of mathematics, and to Robert Strichartz the advisor, mentor, and friend of many in and outside the profession. After Bob's passing on December 19, 2021, we invited colleagues and friends to contribute articles reflecting Bob's mathematical interests. Their generous responses have provided material for two volumes. This first 12-chapter volume is divided into 4 parts, one being a single introductory chapter and the other three containing contributed papers loosely grouped according to broad areas in which Bob made major mathematical advances:

I. Introduction to This Volume
II. Functional and Harmonic Analysis on Euclidean Spaces
III. Analysis on Manifolds
IV. Analysis on Fractals

In the introduction, we attempt a summary of Bob's broad and varied mathematical work and some of its connections to the contributed papers in this volume. We also note his profound contributions as a mathematical mentor, including to generations of undergraduate students during the 30 years he devoted to his Research Experience for Undergraduates program at Cornell University.

We offer our thanks to the many people whose papers appear in this volume; their work is a fitting tribute to Bob's mathematical legacy. Contributors to this volume are: Sebastian Andres, Louisa F. Barnsley, Michael F. Barnsley, David Croydon, Alexander Grigor'yan, John E. Herr, Masanori Hino, Xiaoqi Huang, Yuling Jiao, Palle E.T. Jorgensen, Takashi Kumagai, Yanming Lai, Kathryn O'Connor, Effie

Papageorgiou, Laurent Saloff-Coste, Christopher D. Sogge, Michael E. Taylor, Sundaram Thangavelu, Yang Wang, Eric S. Weber, Haizhao Yang, Yunfei Yang, and Hong-Wei Zhang.

College Station, TX, USA — Patricia Alonso Ruiz
Bielefeld, Germany — Michael Hinz
Medford, MA, USA — Kasso Okoudjou
Storrs, CT, USA — Luke Rogers
Storrs, CT, USA — Alexander Teplyaev
April 2023

Contents

Contributors

Sebastian Andres Department of Mathematics, The University of Manchester, Manchester, UK

Louisa F. Barnsley Australian National University, Canberra, ACT, Australia

Michael F. Barnsley Australian National University, Canberra, ACT, Australia

David Croydon Research Institute for Mathematical Sciences, Kyoto University, Kyoto, Japan

Alexander Grigor'yan Fakultät für Mathematik, Universität Bielefeld, Bielefeld, Germany

John E. Herr Department of Mathematical Sciences, Butler University, Indianapolis, IN, USA

Masanori Hino Department of Mathematics, Kyoto University, Kyoto, Japan

Xiaoqi Huang Department of Mathematics, University of Maryland, College Park, MD, USA

Yuling Jiao School of Mathematics and Statistics, and Hubei Key Laboratory of Computational Science, Wuhan University, Wuhan, P.R. China

Palle E. T. Jorgensen Department of Mathematics, University of Iowa, Iowa City, IA, USA

Takashi Kumagai Department of Mathematics, Waseda University, Tokyo, Japan

Yanming Lai Department of Mathematics, The Hong Kong University of Science and Technology, Kowloon, Hong Kong

Kathryn O'Connor Department of Mathematics, Cornell University, Ithaca, NY, USA

Effie Papageorgiou Department of Mathematics and Applied Mathematics, University of Crete, Crete, Greece

Laurent Saloff-Coste Department of Mathematics, Cornell University, Ithaca, NY, USA

Christopher D. Sogge Department of Mathematics, Johns Hopkins University, Baltimore, MD, USA

Michael E. Taylor Department of Mathematics, University of North Carolina, Chapel Hill, NC, USA

Sundaram Thangavelu Department of Mathematics, Indian Institute of Science, Bangalore, India

Yang Wang Department of Mathematics, The Hong Kong University of Science and Technology, Kowloon, Hong Kong

Eric S. Weber Department of Mathematics, Iowa State University, Ames, IA, USA

Haizhao Yang Department of Mathematics, University of Maryland, College Park, MD, USA

Yunfei Yang Department of Mathematics, The Hong Kong University of Science and Technology, Kowloon, Hong Kong

Hong-Wei Zhang Department of Mathematics, Analysis, Logic and Discrete Mathematics Ghent University, Ghent, Belgium

Part I
Introduction to This Volume

Chapter 1
From Strichartz Estimates to Differential Equations on Fractals

Patricia Alonso Ruiz, Michael Hinz, Kasso A. Okoudjou, Luke G. Rogers, and Alexander Teplyaev

1.1 Functional and Harmonic Analysis on Euclidean Spaces

1.1.1 Harmonic Analysis and Sobolev Spaces

One of Bob's first published papers, based on his dissertation work with Eli Stein, was entitled "*Multipliers on fractional Sobolev spaces*." In it, he began a lifelong study of spaces of smooth and non-smooth functions, including Sobolev, Besov and Hardy spaces, and the space of functions of bounded mean oscillation (BMO) [63, 65, 69, 71, 83, 115, 118]. He was particularly interested in trace and extension operations to small sets and in the situation where a function (or a distribution) or its Fourier transform is restricted to lie in a subset with some specified structure. Among his notable results in this area (other than the Strichartz estimates) are his generalization of Trudinger's inequality [118], his characterization of BMO via convolutions with functions having zero integral [69], and his investigation of Riesz transforms of Hardy functions [83].

P. A. Ruiz
Department of Mathematics, Texas A&M University, College Station, TX, USA
e-mail: paruiz@tamu.edu

M. Hinz
Fakultät für Mathematik, Universität Bielefeld, Bielefeld, Germany
e-mail: mhinz@math.uni-bielefeld.de

K. A. Okoudjou
Department of Mathematics, Tufts University, Medford, MA, USA
e-mail: kasso.okoudjou@tufts.edu

L. G. Rogers · A. Teplyaev (✉)
Department of Mathematics, University of Connecticut, Storrs, CT, USA
e-mail: luke.rogers@uconn.edu; teplyaev@uconn.edu

P. Alonso Ruiz et al. (eds.), *From Classical Analysis to Analysis on Fractals*, Applied and Numerical Harmonic Analysis, https://doi.org/10.1007/978-3-031-37800-3_1

1.1.2 Strichartz Estimates

Many mathematicians are familiar with Bob's name because of the famous Strichartz estimates that are ubiquitous in the study of dispersive PDE. Bob's original estimates [68] give a mixed Sobolev norm estimate of the solution of a wave-type equation by recognizing a classical formula for the solution as an inverse Fourier transform of a tempered distribution supported on a quadratic surface and then using duality to connect this to the Fourier restriction problem previously studied by Stein and Tomas.

When asked, Bob would modestly discount this profound contribution, saying that the main ideas were in the prior work of Segal [56], and he had merely connected it to Stein's unpublished extension of the existing Fourier restriction result of Tomas [123]. In [68], he even said that "none of the methods used in this paper are new" and in personal conversation referred to his results as the simplest case, which he had obtained "in one weekend." In reality, Bob had previously studied harmonic analytic estimates for the wave equation in some detail [62, 66], and his recognition that Segal's use of the (intrinsically one-dimensional) approach of Carleson–Sjölin could be replaced by estimates for the Fourier restriction operator that were applicable in all space dimensions was a substantial breakthrough. The ease with which he came to this result was a reflection of how deeply immersed he was in the theory and applications of Fourier restriction that were then a rapidly developing area of harmonic analysis and shows connections to several of his previous works, particularly that on convolutions having spherical singularities [61].

In our view, his paper introducing what are now called Strichartz estimates is also characteristic of Bob's mathematics; it was typical of him to notice unexpected connections between areas, write down the simplest and most fundamental case of these connections, and then move on to other questions that interested him. His ability to distill the essence of a problem and solve it as an example of what might be possible was part of what let him produce influential mathematics in many areas.

Of course, Strichartz estimates have been dramatically generalized since Bob's original paper. They have been adapted to more general dispersive equations [37], abstracted to more adaptable forms and sharpened to consider endpoint cases [38], and applied in extraordinary generality to both linear and nonlinear PDE [5, 58, 121]. For a new approach to his original results, using techniques that place this squarely within the modern bilinear theory, see the work of Muscalu and Oliviera in Chap. 2 of this volume.

1.1.3 Fourier Analysis and Self-similarity

Bob's earliest papers about fractal sets and measures are those in which he investigated the asymptotic behavior of Fourier transforms of fractal measures [78, 81, 85, 88]. From this point in his career, self-similar sets, measures, and functions took on a central role in his research. Much of this work was in "analysis on fractals," which is discussed in Sect. 1.3 below, but several other significant results on self-similar sets deserve specific mention. One is the introduction of the notion of a fractal blowup and more generally a fractafold [96], which is a fractal-based analog of a manifold and has, following Bob's suggestion in [98], been a central part of the developing study of geometric and analytic structures on fractal sets. Chapter 10 in the present work, contributed by Louisa and Michael Barnsley, investigates the structure of fractal blowups and their connection to tilings of Euclidean spaces. Another is his contributions to the study of orthogonal harmonic function systems for the L^2 space of a self-affine measure [97, 102, 108], in which he became interested by reading the fundamental paper of Jorgensen and Pedersen [29]. The latter theory has developed into a broad set of connections between harmonic analysis and operator theory. Herr, Jorgensen, and Weber give new approach to this topic in Chap. 9, making the substantial step to consider spectral measures with support in two rather than one spatial dimension.

1.1.4 Applications of Harmonic Analysis: Radon Transform, Wavelets, and Distributions

Bob had great enthusiasm for the applications of mathematics, not only because of their utility to non-mathematicians but because of the interesting mathematics that can arise in practical problems as well. An area in which this is particularly visible is his work on the Radon transform. The Radon transform took on major practical significance with the introduction of computed X-ray tomography machines in the early 1970s. Bob saw the study of L^p estimates for this transform as an important question, both with potential practical significance for understanding how the size and smoothness of a signal affects the same properties for the transform and as a test case for assessing the tools of harmonic analysis. His paper [70] is foundational in this regard, in part because it introduced such estimates for the Radon transform in non-Euclidean spaces. He subsequently published an introduction to the mathematical aspects of the Radon transform [72] that won the Lester R. Ford Award for mathematical exposition from the Mathematical Association of America. His contributions to the mathematics of Radon transforms continued with [74, 82, 84, 91].

Another of Bob's notable expository papers is "*How to make wavelets*" [87], a beautiful introduction to an area in which Bob also made theoretical contributions [30, 86, 89, 90, 95, 103]. Bob was a strong believer in the idea that applicable mathematics should be explained in a manner simple enough that students could experiment with it and contended that important tools like Fourier analysis, wavelets, and the theory of distributions should be taught to undergraduates as a set of computational tools much like those encountered in a first calculus course, a proposition that he took sufficiently seriously that he wrote a textbook [92] for this purpose.

For a lovely exposition of the essential features of the instability of water waves, presented by Huy Q. Nguyen and Walter A. Strauss in an elementary style that Bob would certainly have applauded, see Chap. 3. Chapter 4 discusses another development that would have interested Bob, estimates for the application of deep learning to elliptic PDE. Elliptic PDEs were the source of many of his research problems, and deep learning is precisely the sort of mathematics he appreciated: conceptually simple, algorithmically accessible to non-mathematicians, and providing fascinating mathematical structures to investigate.

1.1.5 The Way of Analysis and the Guide to Distribution Theory and Fourier Transform

In his two well-known upper level undergraduate textbooks, [92, 107] and [93, 105], Bob presented a comprehensive view of classical analysis, including Lebesgue integration and harmonic analysis. These books are intimately related to many of Bob's papers in which he made interesting connections from modern mathematics to classical results of Lévy (*Inside the Lévy dragon* [7]), Besicovitch (*Besicovitch meets Wiener-Fourier expansions and fractal measures* [78]), Zygmund (*The school of Antoni Zygmund* [12], joint with Ronald Coifman), Marcinkiewicz (*A multilinear version of the Marcinkiewicz interpolation theorem* [60]). All this work is closely connected to a long series of Bob's popular *American Mathematical Monthly* papers, including [101] *Evaluating integrals using self-similarity*. A detailed analysis involving harmonic functions and eigenfunctions provided in Chap. 5 fits nicely in the type of mathematics that Bob enjoyed.

1.2 Analysis on Manifolds

1.2.1 Riemannian Geometry

Bob began to expand his investigations beyond harmonic analysis on Euclidean spaces quite early in his career. Impressive examples of his work at that time include results on Hardy spaces on manifolds [63], pseudodifferential operators and multipliers on Lie groups [64, 67], harmonic analysis on hyperboloids [66], and Radon transforms on hypersurfaces [70]. Not much later he made a pioneering contribution to the analysis of the Laplacian on complete non-compact Riemannian manifolds [73]. There he used a combination of analytic techniques and, in particular, a spectral theoretic perspective, to prove results on essential self-adjointness, heat semigroups and heat kernels, potentials, functional inequalities, and Riesz transforms, under minimal assumptions. These results complemented earlier work by Th. Aubin, P. Chernoff, H.P. McKean, and S.-T. Yau. The influence of these results and of Bob's other work on Riemannian geometry is reflected in many papers and books, in particular [14, 21, 43, 122]. As Bob's attention expanded to further areas, Riemannian manifolds remained a subject of interest [85], and he continued to consider related geometric ideas until the very end of his career [117].

Within this volume, Bob's exploratory spirit and spectral theoretic point of view are masterly exemplified by Chap. 6. An excellent illustration of the behavior of heat kernels under basic geometric assumptions is contributed by Chap. 8.

1.2.2 Sub-Riemannian Geometry

With [75, 79] and [79], Bob continued his study of analytic tools within a broader geometric framework and built on this to provide one of the most fundamental articles on the entire subject of sub-Riemannian geometry. In this work, he extended earlier studies by R.W. Brockett, C. Carathéodory, W.L. Chow, R. Hermann, A. Koranyi, J. Mitchell, T.J.S. Taylor, and others to give a self-contained account of Christoffel symbols, curves and geodesics, completeness, sub-Laplacians, and related topics in the sub-Riemannian context. In parallel, he provided an in-depth study of Lie bracket relations in [76], with close links to much earlier work of K.-T. Chen and E.B. Dynkin and the Campbell–Baker–Hausdorff–Dynkin formula. His point of view on all these topics was very conceptual, and he was ahead of his time when he emphasized the importance of the sub-Riemannian perspective [77]. In addition, he studied its links to other topics, such as integral transforms [84] and spectral theory [116]. Bob introduced the name "sub-Riemannian geometry" and had a profound influence on extensive subsequent work of many authors, including [22, 23, 25, 45]. The harmonic analytic and sub-Riemannian aspects of Bob's work are beautifully connected by Thangavelu's introduction of a scalar-valued Fourier transform on the Heisenberg group in Chap. 7.

1.2.3 Harmonic Analysis as Spectral Theory of Laplacians

In his paper [80], Bob systematized the classical ideas connecting the spectral theory of self-adjoint operators, representation theory, and harmonic analysis. These ideas are further developed in many of Bob's papers, including [116] *Spectral asymptotics on compact Heisenberg manifolds*, [114] *Average error for spectral asymptotics on surfaces*, [113] *Another way to look at spectral asymptotics on spheres*, and [94] *Estimates for sums of eigenvalues for domains in homogeneous spaces*. In particular, Bob often mentioned that the classical Weyl's asymptotics for manifolds and domains with symmetries should be refined using the representation theory of the symmetry group, specifically by giving a corresponding Weyl-type asymptotic for every irreducible representation of the group. Especially interesting is Bob's recent paper [112] in which Bob introduced an "*asymptotic splitting law*" that describes the proportions of the spectrum that transforms according to the irreducible representations of a finite group that acts as a symmetry group of the Laplacian. Also, in [112], Bob introduced the notion of *spectral mass* for Laplacians with continuous spectrum and formulated a number of conjectures. In a sense, [112] outlines and begins to develop the essential elements of a Weyl asymptotic theory for Laplacians with continuous spectrum. Very interesting recent developments related to Bob's ideas are presented in Chap. 7.

1.3 Analysis on Fractals

1.3.1 Differential Equations on Fractals

Motivated by questions in statistical physics, [1, 17, 19, 24, 50, 55], a new subject, now known in the community as "analysis on fractals," was founded in the late 80s by S. Goldstein, S. Kusuoka, M.T. Barlow, E. Perkins, J. Kigami, R.F. Bass, and T. Lindstrøm. Their foundational articles [3, 8, 20, 31, 32, 39, 40, 42] used techniques from linear algebra, probability theory, and regularity theory for partial differential equations. True to his curiosity and mathematically open mind, Bob was attracted to this research. He noticed that classical harmonic analysis was being developed in the context of metric measure spaces (usually a space equipped with a metric and a doubling measure) and that this included Calderon–Zygmund theory and function spaces, but that differential equations had not yet been studied in fractal setting. What about derivatives of functions defined on intrinsically non-smooth spaces?

In the late 80s and early 90s, Jun Kigami published a series of papers proposing a purely analytic approach to treat the heat equation on the Sierpinski gasket [31–35]. In this topic, Bob found a new direction to apply his vast knowledge of classical harmonic analytic techniques for solving PDEs. To learn first-hand about the developments, he invited Jun Kigami to give a graduate course at Cornell University in Fall 1997. Bob's first articles on the intrinsic analysis on fractals appeared in the second half of the 90s. He addressed topics like wavelets and splines [95], structure and quantitative behavior of solutions to partial differential equation [16], maximum principles on fractals, essential self-adjointness and Liouville theorems on fractal blowups [100], and domain questions for the Laplacian [4]. In [104], he adopted a calculus point of view toward first-order operators and Taylor approximations; in [99], he studied isoperimetric estimates on fractals. These contributions significantly broadened the scope of the intrinsic analysis on fractals in terms of perspectives, methods, and results. Following his natural way of thinking new mathematical paths through examples, Bob started to collect important ideas in [98], a development that would later culminate in one of the very few books on the subject: *Differential equations on fractals* [109], which he called a "soft introduction" [98] to Jun's advanced textbook [36]. During subsequent years, Bob continued to add topics to the discussion, for instance, p-Laplacians [26, 120], links to quantum mechanics [48, 49, 110], distribution theory [52, 54], and pseudodifferential operators [28]. This work provided an unceasing source of ideas for research projects to work on with undergraduate students during his long-established Research Experience for Undergraduates Program at Cornell University. In addition to his existing collaborators, he always welcomed other junior researchers and graduate students to join these projects, creating diverse research groups that included mathematicians at different career stages.

Spectral theory and the behavior of solutions to partial differential equations were two important perspectives in Bob's work in analysis on fractals. Within this volume, Chaps. 11 and 12 provide marvelous expositions on spectral and heat kernel phenomena that are specific to fractals. The following papers of Bob are particularly relevant and important to this topic: [4, 16, 26, 28, 48, 49, 52, 54, 95, 99, 100, 104, 110, 120].

1.3.2 Numerical Analysis on Fractals

Bob spent a huge amount of time and energy on undergraduate research through the Cornell REU and SPUR programs which he founded. As a part of these programs, Bob and more than one hundred undergraduate collaborators obtained many interesting numerical results. A partial list of papers they wrote includes [6, 9, 11, 15, 16, 18, 27, 30, 44, 46, 47, 53, 57, 119]. Among the contributions found here, the introduction of *splines on fractals* was particularly important and fruitful, with many theoretical and potential numerical applications. In one of the first of numerics papers [16], Bob conjectured that waves on fractals have infinite propagation speed. This remarkable conjecture was later proved by one of Bob's undergraduate collaborators, Yin Tat Lee [41].

1.3.3 Intrinsic Analysis on Fractafolds

In his 1999 introductory paper [98] "*Differential equations on fractals*" in the *Notices of the American Mathematical Society*, Bob introduced a new notion of a *fractafold*, a space that is locally modeled on a specified fractal. Thus, a fractafold is the fractal equivalent of a manifold. According to Bob, a "fractafold" is to a fractal what a manifold is to a Euclidean half-space. Bob has a long series of papers based on this fundamental idea: [2, 10, 13, 28, 51, 59, 106, 111]. In particular, Bob proved the existence of isospectral compact fractafolds and studied spectral analysis and intrinsically periodic and almost periodic functions on infinite fractafolds.

1.4 Mentorship

An aspect of Bob's career that made him so impactful was his mentoring work. He was always happy to talk about mathematics with young faculty, and he advised many postdocs, including two of us, and supervised nine Ph.D. students and their theses:

1. Peter Knopf, 1977: *Weak-Type Multipliers.*
2. Nicholas Miller, 1979: *Pseudodifferential Operators and Weighted Sobolev Spaces.*
3. Alberto Setti, 1991: *Eigenvalue and Heat Kernel Estimates for the Weighted Laplacian on a Riemannian Manifold.*
4. Gengqiang Zhou, 1994: *Finiteness and Compactness for the Family of Isospectral Riemannian Manifolds.*
5. Alexander Teplyaev, 1998: *Spectral Analysis on Infinite Sierpinski Gaskets.*
6. Jason Anema, 2012: *Counting Spanning Trees on Fractal Graphs.*
7. Joe Chen, 2013: *Topics in mathematical physics on Sierpinski carpets.*
8. Baris Evren Ugurcan, 2014: *L^p-Estimates And Polyharmonic Boundary Value Problems On The Sierpinski Gasket And Gaussian Free Fields On High Dimensional Sierpinski Carpet Graphs.*
9. Shiping Cao, 2022: *Topics in scaling limits on some Sierpinski carpet type fractals.*

Yet Bob's mentoring went far beyond this. One of his most profound legacies for our profession comes from the enormous amount of time and energy he spent advising undergraduate students through the REU programs he established and led at Cornell for more than 30 years. Through these programs, Bob published papers with more than one hundred undergraduate students, many of whom went on to further study in mathematics and related fields, won awards and fellowships, and are now well-established mathematicians. His deep influence on mathematics will continue through their work.

Acknowledgments P.A.R. was supported in part by the NSF grants DMS 1855349, 1951577, and 2140664. M.H. was supported in part by the DFG IRTG 2235: "Searching for the regular in the irregular: Analysis of singular and random systems." K.O. was supported in part by the NSF grant DMS-2205771. L.R. was supported in part by the NSF grant DMS 1950543. A.T. was supported in part by the NSF grant DMS 1613025 and the Simons Foundation. The authors thank Maria Gordina for helpful discussions related to Bob's impact on Riemannian and sub-Riemannian geometries.

References

1. S. Alexander and R. Orbach, *Density of states on fractals: fractons*, J. Physique Lett. **43** (1982), L623–L631.
2. P. Alonso Ruiz, Y. Chen, H. Gu, R. S. Strichartz, and Z. Zhou, *Analysis on hybrid fractals*, Commun. Pure Appl. Anal. **19** (2020), no. 1, 47–84. MR 4025934

3. M.T. Barlow and R.F. Bass, *The construction of Brownian motion on the Sierpinski carpet*, Ann. Inst. Henri Poinc. **25** (1989), 225–257.
4. Oren Ben-Bassat, Robert S. Strichartz, and Alexander Teplyaev, *What is not in the domain of the Laplacian on Sierpinski gasket type fractals*, J. Funct. Anal. **166** (1999), no. 2, 197–217. MR 1707752
5. Hajer Bahouri, Jean-Yves Chemin, and Raphaël Danchin, *Fourier analysis and nonlinear partial differential equations*, Grundlehren der mathematischen Wissenschaften [Fundamental Principles of Mathematical Sciences], vol. 343, Springer, Heidelberg, 2011. MR 2768550
6. Tyrus Berry, Steven M. Heilman, and Robert S. Strichartz, *Outer approximation of the spectrum of a fractal Laplacian*, Experiment. Math. **18** (2009), no. 4, 449–480. MR 2583544
7. Scott Bailey, Theodore Kim, and Robert S. Strichartz, *Inside the Lévy dragon*, Amer. Math. Monthly **109** (2002), no. 8, 689–703. MR 1927621
8. Martin T. Barlow and Edwin A. Perkins, *Brownian motion on the Sierpiński gasket*, Probab. Theory Related Fields **79** (1988), no. 4, 543–623. MR 966175
9. Brian Bockelman and Robert S. Strichartz, *Partial differential equations on products of Sierpinski gaskets*, Indiana Univ. Math. J. **56** (2007), no. 3, 1361–1375. MR 2333476
10. Shiping Cao, Anthony Coniglio, Xueyan Niu, Richard H. Rand, and Robert S. Strichartz, *The Mathieu differential equation and generalizations to infinite fractafolds*, Commun. Pure Appl. Anal. **19** (2020), no. 3, 1795–1845. MR 4064051
11. Kevin Coletta, Kealey Dias, and Robert S. Strichartz, *Numerical analysis on the Sierpinski gasket, with applications to Schrödinger equations, wave equation, and Gibbs' phenomenon*, Fractals **12** (2004), no. 4, 413–449. MR 2109985
12. Ronald R. Coifman and Robert S. Strichartz, *The school of Antoni Zygmund*, A century of mathematics in America, Part III, Hist. Math., vol. 3, Amer. Math. Soc., Providence, RI, 1989, With the collaboration of Gina Graziosi and Julia Hallquist, pp. 343–368. MR 1025352
13. Ying Ying Chan and Robert S. Strichartz, *Homeomorphisms of fractafolds*, Fund. Math. **209** (2010), no. 2, 177–191. MR 2660562
14. E. B. Davies, *Heat kernels and spectral theory*, Cambridge Tracts in Mathematics, vol. 92, Cambridge University Press, Cambridge, 1989. MR 990239
15. Ron Dror, Suman Ganguli, and Robert S. Strichartz, *A search for best constants in the Hardy-Littlewood maximal theorem*, J. Fourier Anal. Appl. **2** (1996), no. 5, 473–486. MR 1412064
16. Kyallee Dalrymple, Robert S. Strichartz, and Jade P. Vinson, *Fractal differential equations on the Sierpinski gasket*, J. Fourier Anal. Appl. **5** (1999), no. 2-3, 203–284. MR 1683211
17. François Englert, J-M Frère, Marianne Rooman, and Ph Spindel, *Metric space-time as fixed point of the renormalization group equations on fractal structures*, Nuclear Physics B **280** (1987), 147–180.
18. Taryn C. Flock and Robert S. Strichartz, *Laplacians on a family of quadratic Julia sets I*, Trans. Amer. Math. Soc. **364** (2012), no. 8, 3915–3965. MR 2912440
19. Y Gefen, A Aharony, and B B Mandelbrot, *Phase transitions on fractals. iii. infinitely ramified lattices*, Journal of Physics A: Mathematical and General **17** (1984), no. 6, 1277.
20. S. Goldstein, *Random walks and diffusions on fractals*, Percolation theory and ergodic theory of infinite particle systems, IMA Math. Appl., vol. 8, Springer, 1987, pp. 121–129.
21. Alexander Grigor'yan, *Heat kernels on weighted manifolds and applications*, The ubiquitous heat kernel, Contemp. Math., vol. 398, Amer. Math. Soc., Providence, RI, 2006, pp. 93–191. MR 2218016
22. Mikhael Gromov, *Carnot-Carathéodory spaces seen from within*, Sub-Riemannian geometry, Progr. Math., vol. 144, Birkhäuser, Basel, 1996, pp. 79–323. MR 1421823
23. Piotr Hajł asz and Pekka Koskela, *Sobolev met Poincaré*, Mem. Amer. Math. Soc. **145** (2000), no. 688, x+101. MR 1683160
24. S. Havlin and D. Ben-Avraham, *Diffusion in disordered media*, Adv. Phys. **36** (1987), 695–798.
25. Juha Heinonen, *Lectures on analysis on metric spaces*, Universitext, Springer-Verlag, New York, 2001. MR 1800917

26. P. Edward Herman, Roberto Peirone, and Robert S. Strichartz, *p-energy and p-harmonic functions on Sierpinski gasket type fractals*, Potential Anal. **20** (2004), no. 2, 125–148. MR 2032945
27. Steven M. Heilman and Robert S. Strichartz, *Localized eigenfunctions: here you see them, there you don't*, Notices Amer. Math. Soc. **57** (2010), no. 5, 624–629. MR 2664041
28. Marius Ionescu, Luke G. Rogers, and Robert S. Strichartz, *Pseudo-differential operators on fractals and other metric measure spaces*, Rev. Mat. Iberoam. **29** (2013), no. 4, 1159–1190. MR 3148599
29. Palle E. T. Jorgensen and Steen Pedersen, *Dense analytic subspaces in fractal L^2-spaces*, J. Anal. Math. **75** (1998), 185–228. MR 1655831
30. Prem Janardhan, David Rosenblum, and Robert S. Strichartz, *Numerical experiments in Fourier asymptotics of Cantor measures and wavelets*, Experiment. Math. **1** (1992), no. 4, 249–273. MR 1257285
31. J. Kigami, *A harmonic calculus on the Sierpinski spaces*, Japan J. Appl. Math. **6** (1989), 259–290.
32. ______, *Harmonic calculus on p.c.f. self-similar sets*, Trans. Amer. Math. Soc. **335** (1993), 721–755.
33. ______, *Harmonic metric and Dirichlet form on the Sierpinski gasket*, Asymptotic problems in probability theory: stochastic models and diffusions on fractals, Pitman Research Notes in Math., vol. 283, Longman, 1993, pp. 201–218.
34. ______, *Effective resistances for harmonic structures on p.c.f. self-similar sets*, Math. Proc. Cambridge Phil. Soc. **115** (1994), 291–303.
35. ______, *Laplacians on self-similar sets (analysis on fractals)*, Amer. Math. Soc. Transl. **161** (1994), 75–93.
36. Jun Kigami, *Analysis on fractals*, Cambridge Tracts in Mathematics, vol. 143, Cambridge University Press, Cambridge, 2001. MR 1840042
37. Carlos E. Kenig, Gustavo Ponce, and Luis Vega, *Oscillatory integrals and regularity of dispersive equations*, Indiana Univ. Math. J. **40** (1991), no. 1, 33–69. MR 1101221
38. Markus Keel and Terence Tao, *Endpoint Strichartz estimates*, Amer. J. Math. **120** (1998), no. 5, 955–980. MR 1646048
39. S. Kusuoka, *A diffusion process on a fractal*, Probabilistic Methods on Mathematical Physics, Proc. of Taniguchi International Symp. (Katata & Kyoto, 1985) (Tokyo), Kinokuniya, 1987, pp. 251–274.
40. ______, *Dirichlet forms on fractals and products of random matrices*, Publ. Res. Inst. Math. Sci. **25** (1989), 659–680.
41. Yin Tat Lee, *Infinite propagation speed for wave solutions on some post-critically finite fractals*, Analysis, probability and mathematical physics on fractals, Fractals Dyn. Math. Sci. Arts Theory Appl., vol. 5, World Sci. Publ., Hackensack, NJ, [2020] ©2020, pp. 503–519. MR 4472260
42. T. Lindstrøm, *Brownian motion on nested fractals*, Mem. Amer. Math. Soc. **420** (1990).
43. Peter Li and Richard Schoen, *L^p and mean value properties of subharmonic functions on Riemannian manifolds*, Acta Math. **153** (1984), no. 3–4, 279–301. MR 766266
44. John-Peter Lund, Robert S. Strichartz, and Jade P. Vinson, *Cauchy transforms of self-similar measures*, Experiment. Math. **7** (1998), no. 3, 177–190. MR 1676691
45. Richard Montgomery, *A tour of sub-Riemannian geometries, their geodesics and applications*, Mathematical Surveys and Monographs, vol. 91, American Mathematical Society, Providence, RI, 2002. MR 1867362
46. Jens Malmquist and Robert S. Strichartz, *Numerical integration for fractal measures*, J. Fractal Geom. **5** (2018), no. 2, 165–226. MR 3813212
47. Yiran Mao, Robert S. Strichartz, Levente Szabo, and Wing Hong Wong, *Analysis on the Projective Octagasket*, Analysis, probability and mathematical physics on fractals, Fractals Dyn. Math. Sci. Arts Theory Appl., vol. 5, World Sci. Publ., Hackensack, NJ, [2020] ©2020, pp. 297–336. MR 4472253

48. Kasso A. Okoudjou and Robert S. Strichartz, *Weak uncertainty principles on fractals*, J. Fourier Anal. Appl. **11** (2005), no. 3, 315–331. MR 2167172
49. ———, *Asymptotics of eigenvalue clusters for Schrödinger operators on the Sierpiński gasket*, Proc. Amer. Math. Soc. **135** (2007), no. 8, 2453–2459. MR 2302566
50. R. Rammal, *Spectrum of harmonic excitations on fractals*, J. Physique **45** (1984), no. 2, 191–206.
51. Huo-Jun Ruan and Robert S. Strichartz, *Covering maps and periodic functions on higher dimensional Sierpinski gaskets*, Canad. J. Math. **61** (2009), no. 5, 1151–1181. MR 2554236
52. Luke G. Rogers and Robert S. Strichartz, *Distribution theory on P.C.F. fractals*, J. Anal. Math. **112** (2010), 137–191. MR 2762999
53. Robert J. Ravier and Robert S. Strichartz, *Sampling theory with average values on the Sierpinski gasket*, Constr. Approx. **44** (2016), no. 2, 159–194. MR 3543997
54. Luke G. Rogers, Robert S. Strichartz, and Alexander Teplyaev, *Smooth bumps, a Borel theorem and partitions of smooth functions on P.C.F. fractals*, Trans. Amer. Math. Soc. **361** (2009), no. 4, 1765–1790. MR 2465816
55. Rammal Rammal and Gérard Toulouse, *Random walks on fractal structures and percolation clusters*, Journal de Physique Letters **44** (1983), no. 1, 13–22.
56. Irving Segal, *Space-time decay for solutions of wave equations*, Advances in Math. **22** (1976), no. 3, 305–311. MR 492892
57. Chun-Yin Siu and Robert S Strichartz, *Geometry and Laplacian on discrete magic carpets*, to appear in the Journal of Fractal Geometry, arXiv:1902.03408 (2023).
58. Gigliola Staffilani and Daniel Tataru, *Strichartz estimates for a Schrödinger operator with nonsmooth coefficients*, Comm. Partial Differential Equations **27** (2002), no. 7-8, 1337–1372. MR 1924470
59. Robert S. Strichartz and Alexander Teplyaev, *Spectral analysis on infinite Sierpiński fractafolds*, J. Anal. Math. **116** (2012), 255–297. MR 2892621
60. Robert S. Strichartz, *A multilinear version of the Marcinkiewicz interpolation theorem*, Proc. Amer. Math. Soc. **21** (1969), 441–444. MR 0238070
61. Robert S. Strichartz, *Convolutions with kernels having singularities on a sphere*, Trans. Amer. Math. Soc. **148** (1970), 461–471. MR 256219
62. ———, *A priori estimates for the wave equation and some applications*, J. Functional Analysis **5** (1970), 218–235. MR 0257581
63. ———, *The Hardy space H^1 on manifolds and submanifolds*, Canadian J. Math. **24** (1972), 915–925. MR 317037
64. ———, *Invariant pseudo-differential operators on a Lie group*, Ann. Scuola Norm. Sup. Pisa Cl. Sci. (3) **26** (1972), 587–611. MR 420739
65. ———, *Multipliers for spherical harmonic expansions*, Trans. Amer. Math. Soc. **167** (1972), 115–124. MR 306823
66. ———, *Harmonic analysis on hyperboloids*, J. Functional Analysis **12** (1973), 341–383. MR 0352884
67. ———, *Multiplier transformations on compact Lie groups and algebras*, Trans. Amer. Math. Soc. **193** (1974), 99–110. MR 357688
68. ———, *Restrictions of Fourier transforms to quadratic surfaces and decay of solutions of wave equations*, Duke Math. J. **44** (1977), no. 3, 705–714. MR 512086
69. ———, *Bounded mean oscillation and Sobolev spaces*, Indiana Univ. Math. J. **29** (1980), no. 4, 539–558. MR 578205
70. ———, *L^p estimates for Radon transforms in Euclidean and non-Euclidean spaces*, Duke Math. J. **48** (1981), no. 4, 699–727. MR 782573
71. ———, *Traces of BMO-Sobolev spaces*, Proc. Amer. Math. Soc. **83** (1981), no. 3, 509–513. MR 627680
72. ———, *Radon inversion—variations on a theme*, Amer. Math. Monthly **89** (1982), no. 6, 377–384, 420–423. MR 660917
73. ———, *Analysis of the Laplacian on the complete Riemannian manifold*, J. Functional Analysis **52** (1983), no. 1, 48–79. MR 705991

74. ———, *Improved Sobolev inequalities*, Trans. Amer. Math. Soc. **279** (1983), no. 1, 397–409. MR 704623
75. ———, *Sub-Riemannian geometry*, J. Differential Geom. **24** (1986), no. 2, 221–263. MR 862049
76. ———, *The Campbell-Baker-Hausdorff-Dynkin formula and solutions of differential equations*, J. Funct. Anal. **72** (1987), no. 2, 320–345. MR 886816
77. ———, *Realms of mathematics: elliptic, hyperbolic, parabolic, sub-elliptic*, Math. Intelligencer **9** (1987), no. 3, 56–64. MR 895772
78. ———, *Besicovitch meets Wiener-Fourier expansions and fractal measures*, Bull. Amer. Math. Soc. (N.S.) **20** (1989), no. 1, 55–59. MR 948764
79. ———, *Corrections to: "Sub-Riemannian geometry" [J. Differential Geom.* ***24*** *(1986), no. 2, 221–263; MR0862049 (88b:53055)]*, J. Differential Geom. **30** (1989), no. 2, 595–596. MR 1010174
80. ———, *Harmonic analysis as spectral theory of Laplacians*, J. Funct. Anal. **87** (1989), no. 1, 51–148. MR 1025883
81. ———, *Fourier asymptotics of fractal measures*, J. Funct. Anal. **89** (1990), no. 1, 154–187. MR 1040961
82. ———, *Harmonic analysis on constant curvature surfaces with point singularities*, J. Funct. Anal. **91** (1990), no. 1, 37–116. MR 1054114
83. ———, *H^p Sobolev spaces*, Colloq. Math. **60/61** (1990), no. 1, 129–139. MR 1096364
84. ———, *L^p harmonic analysis and Radon transforms on the Heisenberg group*, J. Funct. Anal. **96** (1991), no. 2, 350–406. MR 1101262
85. ———, *Spectral asymptotics of fractal measures on Riemannian manifolds*, J. Funct. Anal. **102** (1991), no. 1, 176–205. MR 1138842
86. ———, *Wavelet expansions of fractal measures*, J. Geom. Anal. **1** (1991), no. 3, 269–289. MR 1120682
87. ———, *How to make wavelets*, Amer. Math. Monthly **100** (1993), no. 6, 539–556. MR 1225202
88. ———, *Self-similar measures and their Fourier transforms. II*, Trans. Amer. Math. Soc. **336** (1993), no. 1, 335–361. MR 1081941
89. ———, *Wavelets and self-affine tilings*, Constr. Approx. **9** (1993), no. 2-3, 327–346. MR 1215776
90. ———, *Construction of orthonormal wavelets*, Wavelets: mathematics and applications, Stud. Adv. Math., CRC, Boca Raton, FL, 1994, pp. 23–50. MR 1247513
91. ———, *A fractal Radon inversion problem*, J. Anal. Math. **64** (1994), 219–240. MR 1303513
92. ———, *A guide to distribution theory and Fourier transforms*, Studies in Advanced Mathematics, CRC Press, Boca Raton, FL, 1994. MR 1276724
93. Robert S Strichartz, *The way of analysis*, Jones & Bartlett Learning, 1995.
94. Robert S. Strichartz, *Estimates for sums of eigenvalues for domains in homogeneous spaces*, J. Funct. Anal. **137** (1996), no. 1, 152–190. MR 1383015
95. ———, *Piecewise linear wavelets on Sierpinski gasket type fractals*, J. Fourier Anal. Appl. **3** (1997), no. 4, 387–416. MR 1468371
96. ———, *Fractals in the large*, Canad. J. Math. **50** (1998), no. 3, 638–657. MR 1629847
97. ———, *Remarks on: "Dense analytic subspaces in fractal L^2-spaces" [J. Anal. Math.* ***75*** *(1998), 185–228; MR1655831 (2000a:46045)] by P. E. T. Jorgensen and S. Pedersen*, J. Anal. Math. **75** (1998), 229–231. MR 1655832
98. ———, *Analysis on fractals*, Notices Amer. Math. Soc. **46** (1999), no. 10, 1199–1208. MR 1715511
99. ———, *Isoperimetric estimates on Sierpinski gasket type fractals*, Trans. Amer. Math. Soc. **351** (1999), no. 5, 1705–1752. MR 1433127
100. ———, *Some properties of Laplacians on fractals*, J. Funct. Anal. **164** (1999), no. 2, 181–208. MR 1695571
101. ———, *Evaluating integrals using self-similarity*, Amer. Math. Monthly **107** (2000), no. 4, 316–326. MR 1763057

102. ______, *Mock Fourier series and transforms associated with certain Cantor measures*, J. Anal. Math. **81** (2000), 209–238. MR 1785282
103. ______, *The shape of the error in wavelet approximation and piecewise linear interpolation*, Math. Res. Lett. **7** (2000), no. 2-3, 317–327. MR 1764325
104. ______, *Taylor approximations on Sierpinski gasket type fractals*, J. Funct. Anal. **174** (2000), no. 1, 76–127. MR 1761364
105. Robert S Strichartz, *The way of analysis*, Jones & Bartlett Learning, 2000.
106. Robert S. Strichartz, *Fractafolds based on the Sierpiński gasket and their spectra*, Trans. Amer. Math. Soc. **355** (2003), no. 10, 4019–4043. MR 1990573
107. ______, *A guide to distribution theory and Fourier transforms*, World Scientific Publishing Co., Inc., River Edge, NJ, 2003, Reprint of the 1994 original [CRC, Boca Raton; MR1276724 (95f:42001)]. MR 2000535
108. ______, *Convergence of mock Fourier series*, J. Anal. Math. **99** (2006), 333–353. MR 2279556
109. ______, *Differential equations on fractals*, Princeton University Press, Princeton, NJ, 2006, A tutorial. MR 2246975
110. ______, *A fractal quantum mechanical model with Coulomb potential*, Commun. Pure Appl. Anal. **8** (2009), no. 2, 743–755. MR 2461574
111. ______, *Periodic and almost periodic functions on infinite Sierpinski gaskets*, Canad. J. Math. **61** (2009), no. 5, 1182–1200. MR 2554237
112. ______, *Spectral asymptotics revisited*, J. Fourier Anal. Appl. **18** (2012), no. 3, 626–659. MR 2921088
113. ______, *Another way to look at spectral asymptotics on spheres*, J. Fourier Anal. Appl. **21** (2015), no. 2, 401–404. MR 3319539
114. ______, *Average error for spectral asymptotics on surfaces*, Commun. Pure Appl. Anal. **15** (2016), no. 1, 9–39. MR 3437531
115. ______, *"Graph paper" trace characterizations of functions of finite energy*, J. Anal. Math. **128** (2016), 239–260. MR 3481175
116. ______, *Spectral asymptotics on compact Heisenberg manifolds*, J. Geom. Anal. **26** (2016), no. 3, 2450–2458. MR 3511483
117. ______, *Defining curvature as a measure via Gauss-Bonnet on certain singular surfaces*, J. Geom. Anal. **30** (2020), no. 1, 153–160. MR 4058509
118. ______, *A note on Trudinger's extension of Sobolev's inequalities*, Indiana Univ. Math. J. **21** (1971/72), 841–842. MR 293389
119. Robert S. Strichartz and Michael Usher, *Splines on fractals*, Math. Proc. Cambridge Philos. Soc. **129** (2000), no. 2, 331–360. MR 1765920
120. Robert S. Strichartz and Carto Wong, *The p-Laplacian on the Sierpinski gasket*, Nonlinearity **17** (2004), no. 2, 595–616. MR 2039061
121. Terence Tao, *Nonlinear dispersive equations*, CBMS Regional Conference Series in Mathematics, vol. 106, Published for the Conference Board of the Mathematical Sciences, Washington, DC; by the American Mathematical Society, Providence, RI, 2006, Local and global analysis. MR 2233925
122. Michael E. Taylor, *Partial differential equations. II*, Applied Mathematical Sciences, vol. 116, Springer-Verlag, New York, 1996, Qualitative studies of linear equations. MR 1395149
123. Peter A. Tomas, *A restriction theorem for the Fourier transform*, Bull. Amer. Math. Soc. **81** (1975), 477–478. MR 358216

Part II
Functional and Harmonic Analysis on Euclidean Spaces

Chapter 2
A New Proof of Strichartz Estimates for the Schrödinger Equation in 2 + 1 Dimensions

Camil Muscalu and Itamar Oliveira

2.1 Introduction

Let S be a subset of $\mathbb{R}^n$ and $\mathrm{d}\mu$ a positive measure supported on S and of temperate growth at infinity. In [12], Strichartz considered the following problems that we quote *verbatim*:

Problem A For which values of p, $1 \leq p < 2$, is it true that $f \in L^p(\mathbb{R}^n)$ implies $\widehat{f}$ has a well-defined restriction to S in $L^2(\mathrm{d}\mu)$ with

$$\left(\int |\widehat{f}|^2 \mathrm{d}\mu\right)^{\frac{1}{2}} \leq c_p \|f\|_p ?$$

Problem B For which values of q, $2 < q \leq \infty$, is it true that the tempered distribution $F\mathrm{d}\mu$ for each $F \in L^2(\mathrm{d}\mu)$ has Fourier transform in $L^q(\mathbb{R}^n)$ with

$$\|\widehat{(F\mathrm{d}\mu)}\|_q \leq c_q \left(\int |F|^2 \mathrm{d}\mu\right)^{\frac{1}{2}} ?$$

As it is pointed out in [12], a duality argument shows that these problems are equivalent if $\frac{1}{p} + \frac{1}{q} = 1$. In the same paper, Strichartz gave a complete answer to the problems above when S is a quadratic surface given by

C. Muscalu
Department of Mathematics, Cornell University, Ithaca, NY, USA
e-mail: camil@math.cornell.edu

I. Oliveira (✉)
CNRS – Université de Nantes, Laboratoire de Mathématiques Jean Leray, Nantes, France
e-mail: oliveira.itamar.w@gmail.com

P. Alonso Ruiz et al. (eds.), *From Classical Analysis to Analysis on Fractals*, Applied and Numerical Harmonic Analysis, https://doi.org/10.1007/978-3-031-37800-3_2

$$S = \{x \in \mathbb{R}^n : R(x) = r\},$$

where $R(x)$ is a polynomial of degree two with real coefficients, r is a real constant, and $d\mu$ is a certain canonical measure on S associated with R.

In general, given a compact submanifold $S \subset \mathbb{R}^n$ and a function $f : \mathbb{R}^{d+1} \to \mathbb{C}$, the *Fourier restriction problem* asks for which pairs (p, q) one has[1]

$$\|\widehat{f}|_S\|_{L^q(S)} \lesssim \|f\|_{L^p(\mathbb{R}^{d+1})}, \tag{2.1}$$

where $\widehat{f}|_S$ is the restriction of the Fourier transform $\widehat{f}$ to S. This problem arises naturally in the study of certain Fourier summability methods and is known to be connected to questions in Geometric Measure Theory and in dispersive PDEs. We refer the reader to [13] for a detailed account of the problem.

Here, we specialize to the case $p = 2$, $d = 2$, and S the compact piece of the paraboloid parametrized by $\Gamma(x) = (x, |x|^2) \subset \mathbb{R}^{d+1}$, $x \in [0, 1]^d$. As we briefly explain in the end of this introduction, this special case is closely related to the study of the Schrödinger equation. From [12] and the necessary conditions for (2.1) to hold (see [13]), it follows that the restriction problem in this setting is equivalent up to the endpoint to showing the following:

Theorem 2.1 (Tomas–Stein/Strichartz Up to the Endpoint) *Let the Fourier extension operator $\mathcal{E}_2$ be defined on $C^0([0, 1]^2)$ by*

$$\mathcal{E}_2(g)(x_1, x_2, t) = \int_{[0,1]^2} g(\xi_1, \xi_2) e^{-2\pi i (x_1, x_2)\cdot(\xi_1, \xi_2)} e^{-2\pi i t(\xi_1^2 + \xi_2^2)} d\xi_1 d\xi_2.$$

Then,

$$\|\mathcal{E}_2 g\|_{4+\varepsilon} \lesssim_\varepsilon \|g\|_2. \tag{2.2}$$

The d-dimensional analog of the operator above is

$$\mathcal{E}_d(g)(x, t) = \int_{[0,1]^d} g(\xi) e^{-2\pi i x\cdot\xi} e^{-2\pi i t|\xi|^2} d\xi.$$

Strichartz proved in [12] that $\mathcal{E}_d$ maps L^2 to $L^{\frac{2(d+2)}{d}}$, so the only novelty of this manuscript is the argument in the case $d = 2$ without the endpoint.[2,3] We use the

[1] We start with $f \in \mathcal{S}(\mathbb{R}^{d+1})$, so both $\widehat{f}$ and $\widehat{f}|_S$ are well-defined pointwise. If (2.1) holds for such f, we extend it to L^p by density.

[2] We are not able to reach the endpoint case $\varepsilon = 0$ due to the way we interpolate between some key information later in the proof. Roughly speaking, the condition $\varepsilon > 0$ is necessary to make certain series converge.

[3] Tomas proved in [16] a dual version of this when the underlying manifold is the sphere.

framework of our earlier paper [8], in which we propose a new approach to the linear and multilinear Fourier extension problems for the paraboloid.

Remark 2.1 Given that Theorem 2.1 is known in much greater generality (with the endpoint, in arbitrary dimensions, for other underlying submanifolds, etc.), it is opportune to make a comparison between the classical approaches to it and ours. As explained in Chapter 11 of [9], an estimate such as

$$\|\mathcal{E}_d g\|_q \lesssim \|g\|_2$$

is equivalent to showing that

$$\|f * \widehat{\mu}\|_{L^q(\mathbb{R}^{d+1})} \lesssim \|f\|_{L^{q'}(\mathbb{R}^{d+1})}, \tag{2.3}$$

where μ is the measure in $\mathbb{R}^{d+1}$ defined by the integral

$$\int_{\mathbb{R}^{d+1}} h(\xi, \tau) \mathrm{d}\mu(\xi, \tau) = \int_{\mathbb{R}^d} h(\xi, |\xi|^2) \mathrm{d}\xi.$$

The curvature of the paraboloid makes $\widehat{\mu}$ decay, which implies good integrability properties for the convolution operator $f \mapsto f * \widehat{\mu}$. This observation plays an important role in the proof of (2.3).

Our strategy is different from the previous one; roughly speaking, instead of relying on a three-dimensional observation such as $\widehat{\mu}(\xi) = O(\langle \xi \rangle^{-\alpha})$ for a certain $\alpha > 0$, we exploit some sharp one-dimensional bilinear estimates to prove Theorem 2.1. The argument that we will present implies (2.2) directly, without reducing it to an inequality such as (2.3), in which the L^p norms on both sides are taken over the same Euclidean space. On the other hand, the proof relies strongly on the tensor structure embedded in the paraboloid. More explicitly, we take advantage of the simple identity

$$e^{2\pi i t |\xi|^2} = e^{2\pi i t \xi_1^2} \cdot e^{2\pi i t \xi_2^2}.$$

In short, the scheme of the proof is as follows:

(a) Bilinearize $\mathcal{E}_2$. In general, multilinearizing an oscillatory integral operator brings into play underlying geometric features of the problem. This is a common step in the study of the extension operator that goes back to Zygmund in [17] and Fefferman's thesis [3]. It also plays a central role in the work [14] of Tao, Vargas, and Vega and more recently in Bourgain and Guth's work [2].
(b) Decompose the previously mentioned bilinearization of $\mathcal{E}_2$ into certain pieces to gain access to key geometric considerations. This will be achieved via a classical Whitney decomposition.
(c) Discretize the pieces obtained in the previous step in order to conveniently place them in the framework of our earlier paper [8].

(d) Study the "Whitney pieces." Their study will be reduced to the lower dimensional bilinear analogue of (2.2).

The proof was motivated by our paper [8], in which we show that the linear and multilinear extension conjectures are true (up to the endpoint[4]) if one of the functions involved is a full tensor. One of the main ideas was to also take advantage of known bounds in the $d = 1$ case. In the ongoing process of understanding the extent to which the method of [8] applies, we learned that it could be used to prove Theorem 2.1 without any tensor hypothesis.

We hope that this new argument in the simple $d = 2$ case will foster further interest in trying to answer some very natural questions: *can one prove n-dimensional extension estimates from (linear and multilinear) lower dimensional ones?* Or at least *what does (linear and multilinear) lower dimensional extension theory reveal about its higher dimensional analogues?*

Remark 2.2 There are a number of different proofs of Strichartz estimates in dimensions $d = 1, 2$ and also of their analogues for other hypersurfaces. In particular, we highlight the papers [1, 4–6, 10, 11], which covers a wide range of tools and techniques in the analysis.

We close this short introduction with the connection between (2.2) and dispersive PDEs, as suggested by the title of the chapter. Consider the linear Schrödinger equation[5]

$$\begin{cases} iu_t - \dfrac{1}{2\pi}\Delta u = 0, \\ u(x,0) = u_0(x), \end{cases} \tag{2.4}$$

where $\widehat{u_0} \in C^\infty(B(0,1))$. The solution can be directly computed by solving an ODE after taking the Fourier transform in x, which gives

$$u(x,t) = \int_{\mathbb{R}^d} e^{2\pi i(x\cdot\xi + t|\xi|^2)}\widehat{u_0}(\xi)\mathrm{d}\xi = \mathcal{E}_d(\widehat{u_0})(-x,-t),$$

and hence estimates for $\mathcal{E}_d$ imply estimates for u. The Fourier extension theory, a scaling argument (to drop the compact support hypothesis), and the fact that $C^\infty(\mathbb{R}^d)$ is dense in $L^2(\mathbb{R}^d)$ allow us to conclude that

$$\|u\|_{L^{\frac{2(d+2)}{d}}(\mathbb{R}^{d+1})} \lesssim \|u_0\|_{L^2(\mathbb{R}^d)}, \tag{2.5}$$

[4] We obtain the endpoint in certain cases.

[5] The constant $-\frac{1}{2\pi}$ is chosen for cosmetic reasons related to the normalization of the Fourier transform.

which is the classical endpoint Strichartz estimate for the Schrödinger equation.[6] Mixed-norm variants of (2.5) are extremely useful in the study of local well-posedness of certain nonlinear dispersive PDEs, but we do not discuss them here.[7]

2.2 Reduction to a Localized Bilinear Estimate

Showing that $\mathcal{E}_2$ maps L^2 to $L^{4+\varepsilon}$ for all $\varepsilon > 0$ is equivalent to showing that

$$T(f, g) = \mathcal{E}_2(f) \cdot \mathcal{E}_2(g)$$

maps $L^2 \times L^2$ to $L^{2+\varepsilon}$ for all $\varepsilon > 0$, where now both f and g are supported on $[0, 1]^2$.

We start by decomposing $[0, 1]^2 \backslash \Delta$, where $\Delta = \{(x, x) : x \in [0, 1]\}$, into a union of non-overlapping dyadic cubes whose side-lengths are comparable to their distance to the diagonal Δ. This is known as a *Whitney decomposition* and there are many ways to achieve it. The exact way in which one does it is not crucial to our analysis, but the properties that we just described will play a central role in using lower dimensional phenomena to prove a higher dimensional result, which is our goal. For concreteness, we chose the classical construction from [14] represented in Fig. 2.1 as a reference, even though there are important differences between the roles played by the decomposition in [14] and ours (see Remark 2.3).

Let $\mathcal{D}^k_{[0,1]}$ denote the collection of closed dyadic intervals of the form $[l \cdot 2^{-k}, (l+1) \cdot 2^{-k}]$, $k \geq 0$, in $[0, 1]$. We say I and J in $\mathcal{D}^k_{[0,1]}$ are *close* and write $I \sim J$ if they have the same length and are disjoint, but their parents[8] I^* and J^* are not disjoint. We then have

$$[0, 1]^2 \backslash \Delta = \bigcup_k \bigcup_{\substack{I_1 \sim I_2 \\ I_1, I_2 \in \mathcal{D}^k_{[0,1]}}} I_1 \times I_2, \tag{2.6}$$

with $|I_1| = |I_2| \approx d(I_1, I_2)$.[9]

---

[6] We refer the reader to Chapter 11 of [9] for the details of the argument sketched above.

[7] See [7] for a more detailed account of these applications.

[8] We say that I^* is the *parent* of I if it is the only dyadic interval that contains I and has twice its length.

[9] $|I|$ denotes the length of the interval I, $d(I_1, I_2)$ is the distance between I_1 and I_2, and $\alpha \approx \beta$ means that there is a universal constant $C > 0$ such that

$$\frac{1}{C} \cdot |\alpha| \leq |\beta| \leq C \cdot |\alpha|.$$

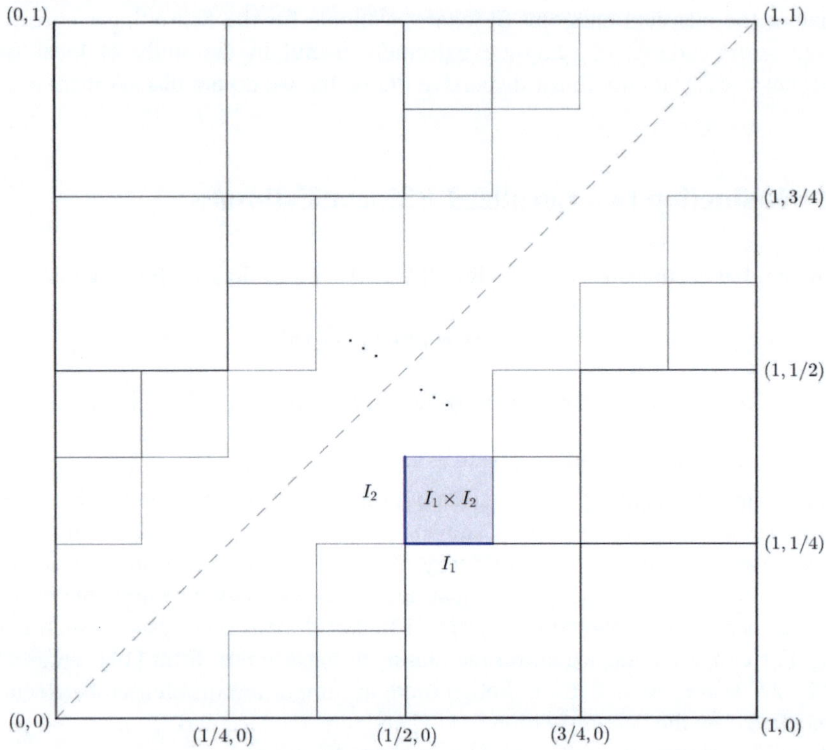

Fig. 2.1 A classical Whitney decomposition

Decomposing the supports of f and g as above gives us

$$
\begin{aligned}
&T(f, g)(x, t) \\
&= \sum_{k_1, k_2 \geq 0} \sum_{\substack{I_1 \sim I_2 \in \mathcal{D}^{k_1}_{[0,1]} \\ J_1 \sim J_2 \in \mathcal{D}^{k_2}_{[0,1]}}} \left(\int_{J_1} \int_{I_1} f(\xi_1, \xi_2) e^{-2\pi i x \cdot \xi} e^{-2\pi i t |\xi|^2} \mathrm{d}\xi_1 \mathrm{d}\xi_2 \right) \\
&\qquad \times \left(\int_{J_2} \int_{I_2} g(\eta_1, \eta_2) e^{-2\pi i x \cdot \eta} e^{-2\pi i t |\eta|^2} \mathrm{d}\eta_1 \mathrm{d}\eta_2 \right).
\end{aligned} \tag{2.7}
$$

One can further split the union in (2.6) into $O(1)$ non-overlapping unions of cubes $Q \in \mathcal{K}_\ell$

$$
\bigcup_k \bigcup_{\substack{I_1 \sim I_2 \\ I_1, I_2 \in \mathcal{D}^k_{[0,1]}}} I_1 \times I_2 = \bigcup_{O(1) \text{ indices } \ell} \left(\bigcup_{Q \in \mathcal{K}_\ell} Q \right)
$$

such that the following holds: if $Q = I_1 \times I_2 \in \mathcal{K}_\ell$, then there is no other interval $I_2' \neq I_2$ with $I_1 \times I_2' \in \mathcal{K}_\ell$ (equivalently, I_1 *determines* I_2 in $\mathcal{K}_\ell$). In the picture above, one can see that four such sub-collections $\mathcal{K}_\ell$ are enough (we just need to "separate the four diagonals" of cubes of a given scale). Because there are only $O(1)$ many of such $\mathcal{K}_\ell$, we will assume without loss of generality that the decompositions used in (2.7) also have this property. In other words, for each I_1, we will suppose that there is only one I_2 such that $I_1 \sim I_2$.

Remark 2.3 It is common in the context of the Fourier extension problem to exploit the properties of certain Whitney decompositions, notably in Tao, Vargas, and Vega's paper [14]. We highlight a fundamental difference between the employment of this tool in that work and in the present one: in [14], to prove that certain bilinear extension estimates imply linear ones, one considers a Whitney decomposition for the set

$$U \times U \backslash \widetilde{\Delta},$$

where $U \subset \mathbb{R}^d$ is a cube where both f and g are supported (similarly to the setting we had in the beginning of this section) and $\widetilde{\Delta} := \{(v, w) \in U \times U : v = w\}$. In other words, assuming $d = 2$ to place the discussion in the setting of this wok, [14] performs *a single* decomposition in a *four-dimensional* region. In contrast, we perform *two independent decompositions* in distinct *two-dimensional* regions.

Fix t and multiply $f(\xi)e^{-2\pi it|\xi|^2}$ by a positive bump $\widetilde{\varphi}_{I_1\times J_1}$ that is $\equiv 1$ on $I_1 \times J_1$ and supported on a slight enlargement of that rectangle. By expanding $f(\xi)e^{-2\pi it|\xi|^2}\widetilde{\varphi}_{I_1\times J_1}(\xi)$ in Fourier series, we obtain

$$f(\xi)e^{-2\pi it|\xi|^2}\widetilde{\varphi}_{I_1\times J_1}(\xi) = \sum_{\vec{n}\in\mathbb{Z}^2} \langle f(\cdot)e^{-2\pi it|\cdot|^2}, \varphi^{\vec{n}}_{I_1\times J_1}\rangle \varphi^{\vec{n}}_{I_1\times J_1},$$

where

$$\varphi^{\vec{n}}_{I_1\times J_1}(\xi) := \frac{1}{|I_1 \times J_1|^{\frac{1}{2}}} \cdot \varphi_{I_1\times J_1}(\xi) \cdot e^{-2\pi i \vec{n}\cdot(2^{k_1}\xi_1, 2^{k_2}\xi_2)}$$

and $\varphi_{I_1\times J_1}$ is a bump adapted to $I_1 \times J_1$.[10] Analogously,

$$g(\eta)e^{-2\pi it|\eta|^2}\widetilde{\varphi}_{I_2\times J_2}(\xi) = \sum_{\vec{n}\in\mathbb{Z}^2} \langle g(\cdot)e^{-2\pi it|\cdot|^2}, \varphi^{\vec{n}}_{I_2\times J_2}\rangle \varphi^{\vec{n}}_{I_2\times J_2}.$$

Plugging this in (2.7),

[10] Notice that the bumps $\varphi^{\vec{n}}_{I_1\times J_1}$ are L^2-*normalized*. It is *not* the case that for $\vec{n} = 0$ one has $\varphi^{0}_{I_1\times J_1} = \varphi_{I_1\times J_1}$. The latter is L^∞-normalized.

$$
\begin{aligned}
&T(f,g)(x,t)\\
&=\sum_{k_1,k_2\geq 0}\sum_{\substack{I_1\sim I_2\in\mathcal{D}^{k_1}_{[0,1]}\\ J_1\sim J_2\in\mathcal{D}^{k_2}_{[0,1]}}}\sum_{\vec{\boldsymbol{n}}_1,\vec{\boldsymbol{n}}_2\in\mathbb{Z}^2}\langle f(\cdot)e^{-2\pi it|\cdot|^2},\varphi^{\vec{\boldsymbol{n}}_1}_{I_1\times J_1}\rangle\cdot\langle g(\cdot)e^{-2\pi it|\cdot|^2},\varphi^{\vec{\boldsymbol{n}}_2}_{I_2\times J_2}\rangle\\
&\qquad\times(\varphi^{I_1\times J_1}_{\vec{\boldsymbol{n}}_1}(x)\cdot\varphi^{I_2\times J_2}_{\vec{\boldsymbol{n}}_2}(x))\\
&=\sum_{\#=0}^{\infty}\sum_{k_1,k_2\geq 0}\sum_{\substack{I_1\sim I_2\in\mathcal{D}^{k_1}_{[0,1]}\\ J_1\sim J_2\in\mathcal{D}^{k_2}_{[0,1]}}}\sum_{\|\vec{\boldsymbol{n}}_1-\vec{\boldsymbol{n}}_2\|_1=\#}\langle f(\cdot)e^{-2\pi it|\cdot|^2},\varphi^{\vec{\boldsymbol{n}}_1}_{I_1\times J_1}\rangle\\
&\qquad\cdot\langle g(\cdot)e^{-2\pi it|\cdot|^2},\varphi^{\vec{\boldsymbol{n}}_2}_{I_2\times J_2}\rangle\\
&\qquad\times(\varphi^{I_1\times J_1}_{\vec{\boldsymbol{n}}_1}(x)\cdot\varphi^{I_2\times J_2}_{\vec{\boldsymbol{n}}_2}(x)).
\end{aligned}
\tag{2.8}
$$

We then have

$$
T(f,g)(x,t)=\sum_{\#=0}^{\infty}T_{\#}(f,g)(x,t),
$$

where

$$
\begin{aligned}
&T_{\#}(f,g)(x,t)\\
&=\sum_{k_1,k_2\geq 0}\sum_{\substack{I_1\sim I_2\in\mathcal{D}^{k_1}_{[0,1]}\\ J_1\sim J_2\in\mathcal{D}^{k_2}_{[0,1]}}}\sum_{\|\vec{\boldsymbol{n}}_1-\vec{\boldsymbol{n}}_2\|_1=\#}\langle f(\cdot)e^{-2\pi it|\cdot|^2},\varphi^{\vec{\boldsymbol{n}}_1}_{I_1\times J_1}\rangle\cdot\langle g(\cdot)e^{-2\pi it|\cdot|^2},\varphi^{\vec{\boldsymbol{n}}_2}_{I_2\times J_2}\rangle\\
&\qquad\times(\varphi^{I_1\times J_1}_{\vec{\boldsymbol{n}}_1}(x)\cdot\varphi^{I_2\times J_2}_{\vec{\boldsymbol{n}}_2}(x))
\end{aligned}
$$

and[11]

$$
\varphi^{Q}_{\vec{\boldsymbol{n}}}(x):=\widehat{\varphi^{\vec{\boldsymbol{n}}}_{Q}}(x).
$$

It is important to remark that since $\varphi^{Q}_{\vec{\boldsymbol{n}}}$ is the Fourier transform of an L^2-normalized bump, it is also L^2-normalized. Observe that if $\vec{\boldsymbol{n}}_1\neq\vec{\boldsymbol{n}}_2$, the bumps $\varphi^{I_1\times J_1}_{\vec{\boldsymbol{n}}_1}$ and $\varphi^{I_2\times J_2}_{\vec{\boldsymbol{n}}_2}$ are essentially supported in different cubes, so the product

[11] We highlight the subtlety of the notation $\varphi^{Q}_{\vec{\boldsymbol{n}}}$; we swapped the indices Q and $\vec{\boldsymbol{n}}$.

$$\varphi_{\vec{n}_1}^{I_1\times J_1}(x)\cdot\varphi_{\vec{n}_2}^{I_2\times J_2}(x) \tag{2.9}$$

decays very fast as # grows. We will prove bounds for the # = 0 term, henceforth denoted by $\widetilde{T}$. The argument is the same for any other $T_{\#}$ and it comes with an operatorial bound of $O(|\#|^{-100})$ due to the fast decay of the bumps in (2.9), which is enough to sum the series in #.

Finally, we discretize the t variable. Multiply both sides of the expression of $\widetilde{T}$ by

$$1=\sum_{m\in\mathbb{Z}}\chi_m,$$

where $\chi_m(t)$ is the indicator function of the interval $[m, m+1)$. This gives

$$\begin{aligned}\widetilde{T}(f,g)(x,t) &= \sum_{k_1,k_2\geq 0}\ \sum_{\substack{I_1\sim I_2\in\mathcal{D}^{k_1}_{[0,1]}\\ J_1\sim J_2\in\mathcal{D}^{k_2}_{[0,1]}}}\ \sum_{(\vec{n},m)\in\mathbb{Z}^3}\langle f(\cdot)e^{-2\pi it|\cdot|^2},\varphi^{\vec{n}}_{I_1\times J_1}\rangle\cdot\langle g(\cdot)e^{-2\pi it|\cdot|^2},\varphi^{\vec{n}}_{I_2\times J_2}\rangle\\ &\qquad\times(\varphi_{\vec{n}}^{I_1\times J_1}(x)\cdot\varphi_{\vec{n}}^{I_2\times J_2}(x)\cdot\chi_m(t)).\end{aligned}$$

For a fixed t, one can write

$$\begin{aligned}e^{-2\pi it|\xi|^2}\cdot\varphi_{[0,1]^2}(\xi) &= e^{-2\pi im|\xi|^2}\cdot e^{-2\pi i(t-m)|\xi|^2}\varphi_{[0,1]^2}(\xi)\\ &= e^{-2\pi im|\xi|^2}\cdot\sum_{\vec{u}\in\mathbb{Z}^2}\langle e^{-2\pi i(t-m)|\cdot|^2},\varphi^{\vec{u}}_{[0,1]^2}\rangle\varphi^{\vec{u}}_{[0,1]^2}\end{aligned}$$

by expanding $e^{-2\pi i(t-m)|\xi|^2}$ as a Fourier series[12] at scale 1. Therefore, we can write $\widetilde{T}(f,g)$ as

$$\begin{aligned}&\sum_{k_1,k_2\geq 0}\ \sum_{\substack{I_1\sim I_2\in\mathcal{D}^{k_1}_{[0,1]}\\ J_1\sim J_2\in\mathcal{D}^{k_2}_{[0,1]}}}\ \sum_{(\vec{n},m)\in\mathbb{Z}^3}\ \sum_{\substack{\vec{u}\in\mathbb{Z}^2\\ \vec{v}\in\mathbb{Z}^2}}\\ &\qquad\langle f(\cdot)e^{-2\pi im|\cdot|^2},\varphi^{\vec{n}}_{I_1\times J_1}\cdot\varphi^{\vec{u}}_{[0,1]^2}\rangle\cdot\langle g(\cdot)e^{-2\pi it|\cdot|^2},\varphi^{\vec{n}}_{I_2\times J_2}\cdot\varphi^{\vec{v}}_{[0,1]^2}\rangle\\ &\qquad\times(\varphi_{\vec{n}}^{I_1\times J_1}(x)\cdot\varphi_{\vec{n}}^{I_2\times J_2}(x))\cdot C^{m,t}_{\vec{u}}\cdot C^{m,t}_{\vec{v}}\cdot\chi_m(t),\end{aligned}$$

where

[12] $\varphi_{[0,1]^2}$ denotes a bump that is $\equiv 1$ on $[0,1]^2$ and is supported in a slightly bigger box.

$$C^{m,t}_{\vec{u}} = \langle e^{-2\pi i(t-m)|\cdot|^2}, \varphi^{\vec{u}}_{[0,1]^2}\rangle,$$

$$C^{m,t}_{\vec{v}} = \langle e^{-2\pi i(t-m)|\cdot|^2}, \varphi^{\vec{v}}_{[0,1]^2}\rangle.$$

For the expression defining $\widetilde{T}$ to be nonzero, m must satisfy $|t-m| \leq 1$, and hence the Fourier coefficients $C^{m,t}_{\vec{u}}$ and $C^{m,t}_{\vec{v}}$ decay like $O(|\vec{u}|^{-100})$ and $O(|\vec{v}|^{-100})$, respectively. In addition, the extra terms

$$\varphi^{\vec{u}}_{[0,1]^2} \quad \text{and} \quad \varphi^{\vec{v}}_{[0,1]^2}$$

in the scalar products involving f and g, respectively, simply shift the integrands in frequency, and this does not affect in any way the arguments that follow.

It is then enough to treat the $\vec{u} = \vec{v} = 0$ case and the model operator (still denoted by $\widetilde{T}$ by a slight abuse of notation[13]):

$$\begin{aligned}\widetilde{T}(f,g)(x,t) &= \sum_{k_1,k_2\geq 0} \sum_{\substack{I_1\sim I_2\in\mathcal{D}^{k_1}_{[0,1]} \\ J_1\sim J_2\in\mathcal{D}^{k_2}_{[0,1]}}} \sum_{(\vec{n},m)\in\mathbb{Z}^3} \langle f(\cdot)e^{-2\pi i m|\cdot|^2}, \varphi^{\vec{n}}_{I_1\times J_1}\rangle \cdot \langle g(\cdot)e^{-2\pi i m|\cdot|^2}, \varphi^{\vec{n}}_{I_2\times J_2}\rangle \\ &\quad \times (\varphi^{I_1\times J_1}_{\vec{n}}(x)\cdot \varphi^{I_2\times J_2}_{\vec{n}}(x)\cdot \chi_m(t)).\end{aligned}$$

Fix $k_1, k_2 \geq 0$ and consider

$$\begin{aligned}\widetilde{T}_{k_1,k_2}(f,g)(x,t) &= \sum_{\substack{I_1\sim I_2\in\mathcal{D}^{k_1}_{[0,1]} \\ J_1\sim J_2\in\mathcal{D}^{k_2}_{[0,1]}}} \sum_{(\vec{n},m)\in\mathbb{Z}^3} \langle f(\cdot)e^{-2\pi i m|\cdot|^2}, \varphi^{\vec{n}}_{I_1\times J_1}\rangle \cdot \langle g(\cdot)e^{-2\pi i m|\cdot|^2}, \varphi^{\vec{n}}_{I_2\times J_2}\rangle \\ &\quad \times (\varphi^{I_1\times J_1}_{\vec{n}}(x)\cdot \varphi^{I_2\times J_2}_{\vec{n}}(x)\cdot \chi_m(t)).\end{aligned}$$

To show that $\widetilde{T}$ maps $L^p \times L^p \mapsto L^q$ (and hence the same for T), it suffices to prove that each $\widetilde{T}_{k_1,k_2}$ also maps $L^p \times L^p \mapsto L^q$ with operatorial norm $O(2^{-c_{p,q}(k_1+k_2)})$, where $c_{p,q}$ is a positive constant. This is content of Theorem 2.2 in Sect. 2.4.

[13] There is another slight abuse of notation here: observe that $\widetilde{\chi}_m(t) := C^{m,t}_{\vec{0}} \cdot \chi_m(t)$ is a smooth function supported in $[m, m+1)$, which is all that is needed in the proof. To simplify the notation, we will continue to call it $\chi_m(t)$ to lighten the notation.

2.3 The Main Building Block

The results of [8] are obtained from three fundamental building blocks, as explained in Section 4 of that paper; they allow us to acquire higher dimensional information from lower dimensional phenomena. The proof of Theorem 2.2 relies on a one-dimensional bound similar to Proposition 4.4 of [8].

Let h_1 and h_2 be functions supported on [0, 1]. To keep our notation consistent with the one from the previous section, let φ_I denote a bump adapted to the interval I and define

$$\varphi_I^n(\xi) := \frac{1}{|I|^{\frac{1}{2}}} \cdot \varphi_I(\xi) \cdot e^{-2\pi i n |I|^{-1}\xi}.$$

We have the following bound:

Proposition 2.1 *For a fixed pair* (I_1, I_2) *with* $2^{-k} = |I_1| = |I_2| \approx d(I_1, I_2)$*,*

$$\begin{aligned} S &= \sum_{n,m\in\mathbb{Z}} |\langle h_1(\cdot)e^{-2\pi i m|\cdot|^2}, \varphi_{I_1}^n\rangle|^2 \cdot |\langle h_2(\cdot)e^{-2\pi i m|\cdot|^2}, \varphi_{I_2}^n\rangle|^2 \\ &\lesssim \frac{1}{2^{-2k}} \cdot \|h_1 \cdot \varphi_{I_1}\|_2^2 \cdot \|h_2 \cdot \varphi_{I_2}\|_2^2. \end{aligned}$$

Proof Define

$$\begin{aligned} &\Phi_{I_1,I_2}^{n,m}(\xi,\eta) \\ &:= e^{-2\pi i m \xi^2} \cdot e^{2\pi i m \eta^2} \cdot \varphi_{I_1}^n(\xi) \cdot \overline{\varphi_{I_2}^n(\eta)} \\ &= e^{-2\pi i m \xi^2} \cdot e^{2\pi i m \eta^2} \cdot \left(\frac{1}{|I_1|^{\frac{1}{2}}} \cdot \varphi_{I_1}(\xi) \cdot e^{-2\pi i n 2^k \xi} \right) \cdot \left(\frac{1}{|I_2|^{\frac{1}{2}}} \cdot \varphi_{I_2}(\eta) \cdot e^{2\pi i n 2^k \eta} \right) \\ &= \frac{1}{2^{-k}} \cdot e^{-2\pi i 2^k(\xi-\eta)} \cdot e^{-2\pi i m(\xi-\eta)(\xi+\eta)} \cdot \varphi_{I_1}(\xi) \cdot \varphi_{I_2}(\eta). \end{aligned}$$

Hence, S can be rewritten as

$$S = \sum_{n,m\in\mathbb{Z}} |\langle h_1 \otimes h_2, \Phi_{I_1,I_2}^{n,m}\rangle|^2.$$

Observe that by performing the change of variables $\alpha = \xi - \eta$ and $\beta = \xi + \eta$, we obtain $|\alpha| \approx 2^{-k}$, the region $I_1 \times I_2$ gets mapped (linearly) to a compact $U \subset \mathbb{R}^2$, and

$$
\begin{aligned}
&\langle h_1 \otimes h_2, \Phi^{n,m}_{I_1,I_2} \rangle \\
&= \iint_{I_1 \times I_2} h_1(\xi) h_2(\eta) \cdot \frac{1}{2^{-k}} \cdot e^{2\pi i n 2^k(\xi-\eta)} \cdot e^{2\pi i m(\xi-\eta)(\xi+\eta)} \\
&\qquad\qquad\qquad\qquad\qquad \times \varphi_{I_1}(\xi) \cdot \varphi_{I_2}(\eta) \mathrm{d}\xi \mathrm{d}\eta \\
&\approx \iint_U h_1\left(\frac{\beta+\alpha}{2}\right) h_2\left(\frac{\beta-\alpha}{2}\right) \cdot \frac{1}{2^{-k}} \cdot e^{2\pi i n 2^k \alpha} \cdot e^{2\pi i m \alpha\beta} \\
&\qquad\qquad\qquad\qquad\qquad \times \varphi_{I_1}\left(\frac{\beta+\alpha}{2}\right) \cdot \varphi_{I_2}\left(\frac{\beta-\alpha}{2}\right) \mathrm{d}\alpha \mathrm{d}\beta \\
&\approx \frac{1}{2^{-\frac{k}{2}}} \iint_U h_1\left(\frac{\beta+\alpha}{2}\right) h_2\left(\frac{\beta-\alpha}{2}\right) \cdot \frac{1}{2^{-\frac{k}{2}}} \cdot e^{2\pi i n 2^k \alpha} \cdot \widetilde{\varphi}(\alpha) \cdot e^{2\pi i m \alpha\beta} \\
&\qquad\qquad\qquad\qquad\qquad \times \varphi_{I_1}\left(\frac{\beta+\alpha}{2}\right) \varphi_{I_2}\left(\frac{\beta-\alpha}{2}\right) \mathrm{d}\alpha \mathrm{d}\beta,
\end{aligned}
$$

where we inserted a bump $\widetilde{\varphi}$ that is 1 on the support of the integrand in α but supported on a fixed dilate of it. This is done so we can fit U inside a dilate[14] $cI_1 \times cI_2$ (with c independent of I_1 and I_2), which is possible since $|\alpha|, |\beta| \approx 2^{-k}$. The family of bumps given by

$$
\widetilde{\varphi}^n(\alpha) := \frac{1}{2^{-\frac{k}{2}}} \cdot e^{2\pi i n 2^k \alpha} \cdot \widetilde{\varphi}(\alpha)
$$

is L^2-normalized (up to a universal constant independent of k and I). Define

$$
H(\alpha, \beta) := h_1\left(\frac{\beta+\alpha}{2}\right) h_2\left(\frac{\beta-\alpha}{2}\right) \varphi_{I_1}\left(\frac{\beta+\alpha}{2}\right) \varphi_{I_2}\left(\frac{\beta-\alpha}{2}\right)
$$

and, by using Bessel twice (in n and in m, as we did in Section 7 of [8]), we conclude that

$$
\begin{aligned}
&\sum_{n,m \in \mathbb{Z}^2} |\langle h_1 \otimes h_2, \Phi^{n,m}_{I_1,I_2} \rangle|^2 \\
&\qquad \lesssim \frac{1}{2^{-k}} \cdot \sum_{m\in\mathbb{Z}} \sum_{n\in\mathbb{Z}} \left| \int_{cI_1} \left(\int_{cI_2} H(\alpha,\beta) \cdot e^{2\pi i m \alpha\beta} \mathrm{d}\beta \right) \cdot \widetilde{\varphi}^n(\alpha) \mathrm{d}\alpha \right|^2 \\
&\qquad \lesssim \frac{1}{2^{-k}} \cdot \frac{1}{2^{-k}} \cdot \|H\|_2^2 \\
&\qquad \lesssim \frac{1}{2^{-2k}} \cdot \|h_1 \cdot \varphi_{I_1}\|_2^2 \cdot \|h_2 \cdot \varphi_{I_2}\|_2^2,
\end{aligned}
$$

[14] If $c > 0$, cI denotes the interval that has the same center as I and size $c|I|$.

where this extra $\frac{1}{2^{-k}}$ factor comes from applying Bessel when summing in m since $|\alpha| \approx 2^{-k}$. □

2.4 Proof of the Main Theorem

For convenience of the reader, we start by briefly recalling and adapting the notation from our earlier work [8]. A few new objects particularly related to the problem at hand are also introduced.

- The index Q (which could be either an interval or a rectangle) in φ_Q and $\widetilde{\varphi}_Q$ indicates that these are L^∞-*normalized* bumps adapted to Q; in other words, they are $\equiv 1$ on Q, smooth, and supported on a slight enlargement of Q.
- The bumps φ_I^n with a superscript n are L^2-*normalized* and given by

$$\varphi_I^n(\xi) := \frac{1}{|I|^{\frac{1}{2}}} \cdot \varphi_I(\xi) \cdot e^{-2\pi i n |I|^{-1}\xi}.$$

Their two-dimensional analogs are

$$\varphi_{I_1\times J_1}^{\vec{n}}(\xi) := \frac{1}{|I_1 \times J_1|^{\frac{1}{2}}} \cdot \varphi_{I_1\times J_1}(\xi) \cdot e^{-2\pi i \vec{n}\cdot(2^{k_1}\xi_1, 2^{k_2}\xi_2)}.$$

- We swap Q and $\vec{n}$ to indicate the Fourier transform of $\varphi_{I_1\times J_1}^{\vec{n}}$:

$$\varphi_{\vec{n}}^{Q}(x) := \widehat{\varphi_Q^{\vec{n}}}(x).$$

- The index ξ_i in $\langle \cdot, \cdot \rangle_{\xi_i}$ indicates that the inner product is an integral in the variable ξ_i only. For instance,

$$\langle f, \varphi \rangle_{\xi_1} := \int_{\mathbb{R}} f(\xi_1, \xi_2) \cdot \overline{\varphi}(\xi_1) \mathrm{d}\xi_1 \tag{2.10}$$

is now a function of the variables ξ_2.
- The upper index in f^{ξ_i} indicates that the variable ξ_i is fixed. For instance, if f is a function of (ξ_1, ξ_2), in

$$\langle f^{\xi_1}, \varphi \rangle_{\xi_2},$$

the scalar product is an integral in ξ_2 only, and hence ξ_1 is fixed.
- The expression $\left\| \langle f, \cdot \rangle_{\xi_i} \right\|_2$ is the L^2 norm of a function in the variable ξ_l, $l \neq i$. To illustrate using (2.10),

$$\|\langle f, \varphi\rangle_{\xi_1}\|_2 = \left[\int_{\mathbb{R}} \left|\int_{\mathbb{R}} f(\xi_1, \xi_2) \cdot \overline{\varphi}(\xi_1) d\xi_1\right|^2 d\xi_2\right]^{\frac{1}{2}}.$$

- We will work on the space $\Omega = \mathbb{Z}^3 \times \mathcal{D}^{k_1}_{[0,1]} \times \mathcal{D}^{k_2}_{[0,1]} \times \mathcal{D}^{k_1}_{[0,1]} \times \mathcal{D}^{k_2}_{[0,1]}$. Because of the properties of the Whitney decomposition from Sect. 2.2, the subset of Ω that will appear in this section depends on 5 parameters, not 7; this is because a pair $(I_1, J_1) \in \mathcal{D}^{k_1}_{[0,1]} \times \mathcal{D}^{k_2}_{[0,1]}$ determines the $(I_2, J_2) \in \mathcal{D}^{k_1}_{[0,1]} \times \mathcal{D}^{k_2}_{[0,1]}$ (and vice versa) that appears in any vector $(n_1, n_2, m, I_1, J_1, I_2, J_2) \in \Omega$. In that context, if $A \subset \Omega$ and $\mathbb{1}_A$ denotes the indicator function of A, we define

$$\|\mathbb{1}_A\|_{\ell^\infty_{n_1,m,I_2}\ell^1_{n_2,J_2}} := \sup_{(n_1,m,I_2)\in\mathbb{Z}^2\times\mathcal{D}^{k_1}_{[0,1]}} \sum_{(n_2,J_2)\in\mathbb{Z}\times\mathcal{D}^{k_2}_{[0,1]}} \mathbb{1}_A(n_1, n_2, m, I_1, J_1, I_2, J_2).$$

Notice that the right-hand side does not depend on any free parameter by our previous observation. The quantity $\|\mathbb{1}_A\|_{\ell^\infty_{n_2,m,J_2}\ell^1_{n_1,J_1}}$ is defined analogously.

As explained in the end of Sect. 2.2, the following bound implies Theorem 2.1:

Theorem 2.2 *Given $\varepsilon > 0$,*

$$\|\widetilde{T}_{k_1,k_2}(f, g)\|_{2+\varepsilon'} \lesssim_\varepsilon 2^{-\frac{\varepsilon}{3}(k_1+k_2)}\|f\|_2 \cdot \|g\|_2,$$

where $\varepsilon' = \frac{8\varepsilon}{1-2\varepsilon}$.

Proof In what follows, let $E_1 \subset Q_1, E_2 \subset Q_2$, and $F \subset \mathbb{R}^{d+1}$ be measurable sets such that $|f| \le \chi_{E_1}$, $|g| \le \chi_{E_2}$, and $|H| \le \chi_F$. We define the associated trilinear form obtained by dualizing $\widetilde{T}_{k_1,k_2}$:

$$\widetilde{\Lambda}_{k_1,k_2}(f, g, H) := \sum_{\substack{I_1\sim I_2\in\mathcal{D}^{k_1}_{[0,1]}\\ J_1\sim J_2\in\mathcal{D}^{k_2}_{[0,1]}}} \sum_{(\vec{n},m)\in\mathbb{Z}^3} \langle f(\cdot)e^{-2\pi i m|\cdot|^2}, \varphi^{\vec{n}}_{I_1\times J_1}\rangle \cdot \langle g(\cdot)e^{-2\pi i m|\cdot|^2}, \varphi^{\vec{n}}_{I_2\times J_2}\rangle \cdot \langle H, \varphi^{I_1\times J_1}_{\vec{n}} \cdot \varphi^{I_2\times J_2}_{\vec{n}} \otimes \chi_m\rangle. \quad (2.11)$$

Interpolation theory shows that to prove the theorem above it suffices to show that

$$|\widetilde{\Lambda}_{k_1,k_2}(f, g, H)| \lesssim_\varepsilon 2^{-\frac{\varepsilon}{3}(k_1+k_2)}|E_1|^{\gamma_1} \cdot |E_2|^{\gamma_2} \cdot |F|^{\gamma_3} \tag{2.12}$$

for all $\varepsilon > 0$, where γ_j $(1 \le j \le 2)$ and γ_3 are in a small neighborhood of $\frac{1}{2}$ and $\frac{1}{2}+\varepsilon$, respectively.[15] To keep the notation simple, the restricted weak-type estimate that we will prove will be for the centers of such neighborhoods. In other words, we will show that

$$|\widetilde{\Lambda}_{k_1,k_2}(f, g, H)| \lesssim_\varepsilon 2^{-\frac{\varepsilon}{3}(k_1+k_2)}|E_1|^{\frac{1}{2}} \cdot |E_2|^{\frac{1}{2}} \cdot |F|^{\frac{1}{2}+\frac{2\varepsilon}{1+2\varepsilon}}$$

for all $\varepsilon > 0$, but it will be clear from the arguments that as long as we give this $\varepsilon > 0$ away, a slightly different choice of interpolation parameters yields (2.12).

We will define several level sets that encode the sizes of many quantities that will play a role in the proof. We start with the ones involving the scalar products in the trilinear form above.

$$\mathbb{A}_1^{l_1} = \left\{(\vec{\boldsymbol{n}}, m, I_1, J_1) \in \mathbb{Z}^3 \times \mathcal{D}_{[0,1]}^{k_1} \times \mathcal{D}_{[0,1]}^{k_2} : |\langle f(\cdot)e^{-2\pi i m|\cdot|^2}, \varphi_{I_1\times J_1}^{\vec{\boldsymbol{n}}}\rangle| \approx 2^{-l_1}\right\},$$

$$\mathbb{A}_2^{l_2} = \left\{(\vec{\boldsymbol{n}}, m, I_2, J_2) \in \mathbb{Z}^3 \times \mathcal{D}_{[0,1]}^{k_1} \times \mathcal{D}_{[0,1]}^{k_2} : |\langle g(\cdot)e^{-2\pi i m|\cdot|^2}, \varphi_{I_2\times J_2}^{\vec{\boldsymbol{n}}}\rangle| \approx 2^{-l_2}\right\}.$$

The two-dimensional scalar products above are not the only information that we will need to control. As we will see, some mixed-norm quantities appear naturally after using Bessel's inequality along certain directions, and we will need to capture these as well:

$$\mathbb{B}_1^{r_1} = \left\{(n_1, m, I_1) \in \mathbb{Z}^2 \times \mathcal{D}_{[0,1]}^{k_1} : \left\| \left\langle f^{\xi_2}(\cdot)e^{-2\pi i m|\cdot|^2}, \varphi_{I_1}^{n_1}\right\rangle_{\xi_1} \right\|_{L^2_{\xi_2}} \approx 2^{-r_1}\right\},$$

$$\mathbb{B}_2^{r_2} = \left\{(n_2, m, J_1) \in \mathbb{Z}^2 \times \mathcal{D}_{[0,1]}^{k_2} : \left\| \left\langle f^{\xi_1}(\cdot)e^{-2\pi i m|\cdot|^2}, \varphi_{J_1}^{n_2}\right\rangle_{\xi_2} \right\|_{L^2_{\xi_1}} \approx 2^{-r_2}\right\},$$

$$\mathbb{C}_1^{s_1} = \left\{(n_1, m, I_2) \in \mathbb{Z}^2 \times \mathcal{D}_{[0,1]}^{k_1} : \left\| \left\langle g^{\xi_2}(\cdot)e^{-2\pi i m|\cdot|^2}, \varphi_{I_2}^{n_1}\right\rangle_{\xi_1} \right\|_{L^2_{\xi_2}} \approx 2^{-s_1}\right\},$$

[15] We refer the reader to Chapter 3 of [15] for a detailed account of multilinear interpolation theory.

$$\mathbb{C}_2^{s_2} = \left\{ (n_2, m, J_2) \in \mathbb{Z}^2 \times \mathcal{D}_{[0,1]}^{k_2} : \left\| \left\langle g^{\xi_1}(\cdot)e^{-2\pi i m|\cdot|^2}, \varphi_{J_2}^{n_2} \right\rangle_{\xi_2} \right\|_{L^2_{\xi_1}} \approx 2^{-s_2} \right\}.$$

The last quantity we have to control is the one arising from the dualizing function H:[16]

$$\mathbb{D}^t = \{(\vec{n}, m, I_1, J_1, I_2, J_2) \in \Omega : |\langle H, \varphi_{\vec{n}}^{I_1 \times J_1} \cdot \varphi_{\vec{n}}^{I_2 \times J_2} \otimes \chi_m \rangle| \approx 2^{-t}\}.$$

We remark that a fixed I_1 determines I_2, as well as a fixed J_1 determines J_2, and hence the sum in (2.11) depends on five parameters (two intervals and three integers). This is because the cubes $I_1 \times I_2$ and $J_1 \times J_2$ are assumed to be in a fixed strip, as one can see from the previous Whitney decomposition.

In order to prove some crucial bounds that will play an important role later on, we will have to isolate the previous information for only one of the functions f and g. This will be done in terms of the following sets:

$$\mathbb{X}^{l_1, r_1} = \mathbb{A}_1^{l_1} \cap \{(\vec{n}, m, I_1, J_1); (n_1, m, I_1) \in \mathbb{B}_1^{r_1}\},$$

$$\mathbb{X}^{l_1, r_2} = \mathbb{A}_1^{l_1} \cap \{(\vec{n}, m, I_1, J_1); (n_2, m, J_1) \in \mathbb{B}_2^{r_2}\},$$

$$\mathbb{X}^{l_2, s_1} = \mathbb{A}_2^{l_2} \cap \{(\vec{n}, m, I_2, J_2); (n_1, m, I_2) \in \mathbb{C}_1^{s_1}\},$$

$$\mathbb{X}^{l_2, s_2} = \mathbb{A}_2^{l_2} \cap \{(\vec{n}, m, I_2, J_2); (n_2, m, J_2) \in \mathbb{C}_2^{s_2}\}.$$

In other words, the set $\mathbb{X}^{l_1, r_1}$ for instance contains all the $(\vec{n}, m, I_1, J_1)$ whose corresponding scalar product

$$\langle f(\cdot)e^{-2\pi i m|\cdot|^2}, \varphi_{I_1 \times J_1}^{\vec{n}} \rangle$$

has size about 2^{-l_1} and with (n_1, m, I_1) being such that

$$\left\| \left\langle f^{\xi_2}(\cdot)e^{-2\pi i m|\cdot|^2}, \varphi_{I_1}^{n_1} \right\rangle_{\xi_1} \right\|_{L^2_{\xi_2}}$$

has size about 2^{-r_1}.

Finally, it will also be important to encode all the previous information into one single set. This will be done with

[16] Recall that we set $\Omega = \mathbb{Z}^3 \times \mathcal{D}_{[0,1]}^{k_1} \times \mathcal{D}_{[0,1]}^{k_2} \times \mathcal{D}_{[0,1]}^{k_1} \times \mathcal{D}_{[0,1]}^{k_2}$.

$$\mathbb{X}^{\vec{l},\vec{r},\vec{s},t} := \mathbb{D}^t \cap \{(\vec{n}, m, I_1, J_1, I_2, J_2);\quad (\vec{n}, m, I_1, J_1) \in \mathbb{X}^{l_1,r_1} \cap \mathbb{X}^{l_1,r_2}, \\ (\vec{n}, m, I_2, J_2) \in \mathbb{X}^{l_2,s_1} \cap \mathbb{X}^{l_2,s_2}\},$$

where we are using the abbreviations $\vec{l} = (l_1, l_2)$, $\vec{r} = (r_1, r_2)$ and $\vec{s} = (s_1, s_2)$. Hence, we can bound the form $\widetilde{\Lambda}_{k_1,k_2}$ by

$$|\widetilde{\Lambda}_{k_1,k_2}(f, g, H)| \lesssim \sum_{\substack{\vec{l},\vec{r},\vec{s}\in\mathbb{Z}^2 \\ t\in\mathbb{Z}}} 2^{-l_1} \cdot 2^{-l_2} \cdot 2^{-t} \cdot \#\mathbb{X}^{\vec{l},\vec{r},\vec{s},t}. \tag{2.13}$$

The sums in (2.13) are not really over all integers. For instance, if $\mathbb{X}^{\vec{l},\vec{r},\vec{s},t} \neq \emptyset$, one has

$$2^{-l_1} \lesssim 2^{-\frac{(k_1+k_2)}{2}},$$

so we can assume without loss of generality that $l_1 \geq 0$ (and similarly for the other sums).

The following lemma plays a crucial role in the argument by relating the scalar and mixed-norm quantities involved in the problem. □

Lemma 2.1 *If* $\mathbb{X}^{\vec{l},\vec{r},\vec{s},t} \neq \emptyset$, *then*

$$\begin{aligned} &2^{-l_1} \lesssim 2^{-r_1}, \quad 2^{-l_1} \lesssim \frac{2^{-r_2}}{\|\mathbb{1}_{\mathbb{X}^{l_1,r_2}}\|^{\frac{1}{2}}_{\ell^\infty_{n_2,m,J_1}\ell^1_{n_1,I_1}}}, \\ &2^{-l_2} \lesssim \frac{2^{-s_1}}{\|\mathbb{1}_{\mathbb{X}^{l_2,s_1}}\|^{\frac{1}{2}}_{\ell^\infty_{n_1,m,I_2}\ell^1_{n_2,J_2}}}, \quad 2^{-l_2} \lesssim 2^{-s_2}. \end{aligned} \tag{2.14}$$

Proof By the triangle inequality and Cauchy–Schwarz,

$$\begin{aligned} 2^{-l_1} \approx |\langle f(\cdot)e^{-2\pi i m|\cdot|^2}, \varphi^{\vec{n}}_{I_1\times J_1}\rangle| &\leq \frac{1}{|J_1|^{\frac{1}{2}}} \cdot \left\| \left\langle f^{\xi_2}(\cdot)e^{-2\pi i m|\cdot|^2}, \varphi^{n_1}_{I_1} \right\rangle_{\xi_1} \right\|_{L^1_{\xi_2}(J_1)} \\ &\lesssim \frac{1}{|J_1|^{\frac{1}{2}}} \cdot \left\| \left\langle f^{\xi_2}(\cdot)e^{-2\pi i m|\cdot|^2}, \varphi^{n_1}_{I_1} \right\rangle_{\xi_1} \right\|_{L^2_{\xi_2}} \cdot |J_1|^{\frac{1}{2}} \\ &\approx 2^{-r_1}. \end{aligned}$$

The relation above between 2^{-l_1} and 2^{-r_2} is a consequence of orthogonality: for a fixed (n_2, m, J_1), define

$$\mathbb{X}^{l_1,r_2}_{(n_2,m,J_1)} := \{(n_1, I_1); \quad (\vec{n}, m, I_1, J_1) \in \mathbb{X}^{l_1,r_2}\}.$$

This way,

$$\begin{aligned}
&\#\mathbb{X}^{l_1,r_2}_{(n_2,m,J_1)} \\
&\approx 2^{2l_1} \sum_{(n_1,I_1)\in\mathbb{X}^{l_1,r_2}_{(n_2,m,J_1)}} |\langle f(\cdot)e^{-2\pi i m|\cdot|^2}, \varphi^{\vec{n}}_{I_1\times J_1}\rangle|^2 \\
&\leq 2^{2l_1} \sum_{I_1\in\mathcal{D}^{k_1}_{[0,1]}} \sum_{n_1\in\mathbb{Z}} \left| \int_0^1 \left\langle f^{\xi_1}(\cdot)e^{-2\pi i m|\cdot|^2}, \varphi^{n_2}_{J_1}\right\rangle_{\xi_2} \cdot e^{-2\pi i m \xi_1^2} \cdot \varphi^{n_1}_{I_1}(\xi_1)\cdot \mathrm{d}\xi_1 \right|^2 \\
&\leq 2^{2l_1} \sum_{I_1\in\mathcal{D}^{k_1}_{[0,1]}} \int_{\mathrm{supp}(\varphi_{I_1})} \left| \varphi_{I_1}(x_1) \left\langle f^{x_1}(\cdot)e^{-2\pi i m|\cdot|^2}, \varphi^{n_2}_{J_1}\right\rangle_{\xi_2} \right|^2 \mathrm{d}x_1 \\
&\lesssim 2^{2l_1} \int_0^1 \left| \left\langle f^{x_1}(\cdot)e^{-2\pi i m|\cdot|^2}, \varphi^{n_2}_{J_1}\right\rangle_{\xi_2} \right|^2 \mathrm{d}x_1 \\
&\approx 2^{2l_1}\cdot 2^{-2r_2},
\end{aligned}$$

where we used Bessel's inequality from the second to the third line. By taking the supremum in (n_2, m, J_1), we conclude that

$$2^{-l_1} \lesssim \frac{2^{-r_2}}{\|\mathbb{1}_{\mathbb{X}^{l_1,r_2}}\|^{\frac{1}{2}}_{\ell^\infty_{n_2,m,J_1}\ell^1_{n_1,I_1}}}.$$

The other two relations are verified analogously. □

We will need the following convex combinations of the bounds above:

$$2^{-l_1} \lesssim \frac{2^{-\frac{1}{2}r_1}\cdot 2^{-\frac{1}{2}r_2}}{\|\mathbb{1}_{\mathbb{X}^{l_1,r_2}}\|^{\frac{1}{4}}_{\ell^\infty_{n_2,m,J_1}\ell^1_{n_1,I_1}}}, \tag{2.15}$$

$$2^{-l_2} \lesssim \frac{2^{-\frac{1}{2}s_1} \cdot 2^{-\frac{1}{2}s_2}}{\|\mathbb{1}_{\mathbb{X}^{l_2,s_1}}\|^{\frac{1}{4}}_{\ell^\infty_{n_1,m,I_2}\ell^1_{n_2,J_2}}}. \tag{2.16}$$

We now concentrate on estimating the right-hand side of (2.13) by finding good bounds for $\#\mathbb{X}^{\vec{l},\vec{r},\vec{s},t}$. Observe that the Fourier transform of

$$\varphi^{I_1\times J_1}_{\vec{n}} \cdot \varphi^{I_2\times J_2}_{\vec{n}} \tag{2.17}$$

is essentially supported on $I_1 \times J_1 + I_2 \times J_2$. Since the scales of I_1, I_2, J_1, J_2 are fixed, these supports have finite overlap. Notice also that the bumps in (2.17) are L^1-normalized, and hence we conclude that the set

$$\left\{2^{\frac{(k_1+k_2)}{2}} \cdot \left(\varphi^{I_1\times J_1}_{\vec{n}} \cdot \varphi^{I_2\times J_2}_{\vec{n}}\right)\right\}_{\vec{n},I_1,J_1,I_2,J_2}$$

is an L^2-normalized almost orthogonal family.[17] This way,

$$\begin{aligned}\#\mathbb{X}^{\vec{l},\vec{r},\vec{s},t} &\lesssim 2^{2t} \cdot 2^{-(k_1+k_2)} \sum_{\substack{(\vec{n},m,I_1,J_1,I_2,J_2)\\ \in \mathbb{X}^{\vec{l},\vec{r},\vec{s},t}}} \frac{1}{2^{-(k_1+k_2)}} |\langle H, \varphi^{I_1\times J_1}_{\vec{n}} \cdot \varphi^{I_2\times J_2}_{\vec{n}} \otimes \chi_m\rangle|^2 \\ &\leq 2^{2t} \cdot 2^{-(k_1+k_2)} \cdot |F|.\end{aligned} \tag{2.18}$$

Alternatively,

[17] Recall that we are treating the term with # = 0 from (2.8). If # > 0, the family

$$\left\{\#^{50} \cdot 2^{\frac{(k_1+k_2)}{2}} \cdot \left(\varphi^{I_1\times J_1}_{\vec{n}} \cdot \varphi^{I_2\times J_2}_{\vec{n}}\right)\right\}_{\vec{n},I_1,J_1,I_2,J_2}$$

would be L^2-normalized, which would imply (2.18) with an extra factor of $\#^{-100}$. The geometric series in # is then summable.

$$
\begin{aligned}
\#\mathbb{X}^{\vec{l},\vec{r},\vec{s},t} &\lesssim 2^t \cdot \sum_{\substack{(\vec{n},m,I_1,J_1,I_2,J_2)\\ \in \mathbb{X}^{\vec{l},\vec{r},\vec{s},t}}} |\langle H, \varphi_{\vec{n}}^{I_1\times J_1}\cdot \varphi_{\vec{n}}^{I_2\times J_2}\otimes \chi_m\rangle| \\
&\leq 2^t \cdot \sum_{\substack{(I_1,J_1,I_2,J_2)\\ I_1\in\mathcal{D}^{k_1}_{[0,1]}\\ J_1\in\mathcal{D}^{k_2}_{[0,1]}}} \sum_{(\vec{n},m)\in\mathbb{Z}^3} |\langle H, \varphi_{\vec{n}}^{I_1\times J_1}\cdot \varphi_{\vec{n}}^{I_2\times J_2}\otimes \chi_m\rangle| \\
&\lesssim 2^t \cdot \sum_{\substack{(I_1,J_1,I_2,J_2)\\ I_1\in\mathcal{D}^{k_1}_{[0,1]}\\ J_1\in\mathcal{D}^{k_2}_{[0,1]}}} \frac{1}{2^{k_1+k_2}}\|H\|_1 \\
&\lesssim 2^t \cdot |F|,
\end{aligned}
\tag{2.19}
$$

where we took into account the fact that the family (2.17) is L^1-normalized and that there are $2^{k_1+k_2}$ elements in the sum over (I_1, J_1, I_2, J_2).[18]

From the definition of $\mathbb{X}^{\vec{l},\vec{r},\vec{s},t}$, it follows that

$$
\#\mathbb{X}^{\vec{l},\vec{r},\vec{s},t} \leq \|\mathbb{1}_{\mathbb{X}^{l_2,s_1}}\|_{\ell^\infty_{n_1,m,I_2}\ell^1_{n_2,J_2}} \cdot \|\mathbb{1}_{\mathbb{B}_1^{r_1}\cap\mathbb{C}_1^{s_1}}\|_{\ell^1_{n_1,m,I_1}}, \tag{2.20}
$$

$$
\#\mathbb{X}^{\vec{l},\vec{r},\vec{s},t} \leq \|\mathbb{1}_{\mathbb{X}^{l_1,r_2}}\|_{\ell^\infty_{n_2,m,J_1}\ell^1_{n_1,I_1}} \cdot \|\mathbb{1}_{\mathbb{B}_2^{r_2}\cap\mathbb{C}_2^{s_2}}\|_{\ell^1_{n_2,m,J_2}}. \tag{2.21}
$$

The next step will make it clear what is the gain obtained from our previous considerations: by squaring the extension operator $\mathcal{E}_2$ and performing appropriate Whitney decompositions, one obtains objects analogous to the bilinear extension operators discussed in [8]. For such operators, one can take advantage of *transversality* to prove better bounds than the ones satisfied by $\mathcal{E}_2$, which will be done here through Proposition 2.1. Observe that for a fixed $I_1 \in \mathcal{D}^{k_1}_{[0,1]}$ we have

$$
\begin{aligned}
&\sum_{n_1,m\in\mathbb{Z}} \left\| \left\langle f^{\xi_2}(\cdot)e^{-2\pi i m|\cdot|^2}, \varphi_{I_1}^{n_1}\right\rangle_{\xi_1}\right\|^2_{L^2_{\xi_2}} \cdot \left\| \left\langle g^{\xi_2}(\cdot)e^{-2\pi i m|\cdot|^2}, \varphi_{I_2}^{n_1}\right\rangle_{\xi_1}\right\|^2_{L^2_{\xi_2}} \\
&= \int_{[0,1]^2} \left(\sum_{n_1,m\in\mathbb{Z}} \left| \left\langle f^{\xi_2}(\cdot)e^{-2\pi i m|\cdot|^2}, \varphi_{I_1}^{n_1}\right\rangle_{\xi_1}\right|^2 \cdot \left| \left\langle g^{\widetilde{\xi_2}}(\cdot)e^{-2\pi i m|\cdot|^2}, \varphi_{I_2}^{n_1}\right\rangle_{\xi_1}\right|^2 \right) \mathrm{d}\xi_2 \mathrm{d}\widetilde{\xi_2} \\
&\lesssim \int_{[0,1]^2} \frac{1}{2^{-2k_1}} \cdot \|f^{\xi_2}\|^2_{L^2(I_1)} \cdot \|g^{\widetilde{\xi_2}}\|^2_{L^2(I_2)} \mathrm{d}\xi_2 \mathrm{d}\widetilde{\xi_2},
\end{aligned}
$$

[18] We remark that there is no condition on I_2 and J_2 in the sum over (I_1, J_1, I_2, J_2) because, again, they are determined by I_1 and J_1, respectively.

by Fubini and Proposition 2.1. This way,

$$\begin{aligned}
&\#\mathbb{B}_1^{r_1} \cap \mathbb{C}_1^{s_1} \\
&\lesssim 2^{2r_1+2s_1} \\
&\quad \times \sum_{I_1 \in \mathcal{D}_{[0,1]}^{k_1}} \sum_{n_1, m \in \mathbb{Z}} \left\| \left\langle f^{\xi_2}(\cdot) e^{-2\pi i m |\cdot|^2}, \varphi_{I_1}^{n_1} \right\rangle_{\xi_1} \right\|_{L^2_{\xi_2}}^2 \cdot \left\| \left\langle g^{\xi_2}(\cdot) e^{-2\pi i m |\cdot|^2}, \varphi_{I_2}^{n_1} \right\rangle_{\xi_1} \right\|_{L^2_{\xi_2}}^2 \\
&\lesssim 2^{2r_1+2s_1} \sum_{I_1 \in \mathcal{D}_{[0,1]}^{k_1}} \int_{[0,1]^2} \frac{1}{2^{-2k_1}} \cdot \| f^{\xi_2} \|_{L^2(I_1)}^2 \cdot \| g^{\widetilde{\xi_2}} \|_{L^2(I_2)}^2 \mathrm{d}\xi_2 \mathrm{d}\widetilde{\xi_2} \\
&\leq 2^{2r_1+2s_1} \cdot 2^{2k_1} \sum_{I_1 \in \mathcal{D}_{[0,1]}^{k_1}} \int_{I_1 \times [0,1] \times I_2 \times [0,1]} \mathbb{1}_{E_1} \otimes \mathbb{1}_{E_2} \\
&\leq 2^{2r_1+2s_1} \cdot 2^{2k_1} \cdot |E_1| \cdot |E_2|.
\end{aligned} \tag{2.22}$$

Analogously,

$$\#\mathbb{B}_2^{r_2} \cap \mathbb{C}_2^{s_2} \lesssim 2^{2r_2+2s_2} \cdot 2^{2k_2} \cdot |E_1| \cdot |E_2|. \tag{2.23}$$

Back to the multilinear form, using (2.15) and (2.16), interpolating between (2.20), (2.21), (2.18), and (2.19),

$$\begin{aligned}
&|\widetilde{\Lambda}_{k_1,k_2}(f, g, H)| \lesssim \\
&\sum_{\substack{\vec{l}, \vec{r} \\ \vec{s}, t \geq 0}} 2^{-\varepsilon l_1} \cdot \left(\frac{2^{-\frac{1}{2} r_1} \cdot 2^{-\frac{1}{2} r_2}}{\| \mathbb{1}_{\mathbb{X}^{l_1, r_2}} \|_{\ell^\infty_{n_2,m,J_1} \ell^1_{n_1,I_1}}^{\frac{1}{4}}} \right)^{1-\varepsilon} \cdot 2^{-\varepsilon l_2} \cdot \left(\frac{2^{-\frac{1}{2} s_1} \cdot 2^{-\frac{1}{2} s_2}}{\| \mathbb{1}_{\mathbb{X}^{l_2, s_1}} \|_{\ell^\infty_{n_1,m,I_2} \ell^1_{n_2,J_2}}^{\frac{1}{4}}} \right)^{1-\varepsilon} \cdot 2^{-t} \\
&\qquad \cdot \left(\| \mathbb{1}_{\mathbb{X}^{l_2,s_1}} \|_{\ell^\infty_{n_1,m,I_2} \ell^1_{n_2,J_2}} \cdot \| \mathbb{1}_{\mathbb{B}_1^{r_1} \cap \mathbb{C}_1^{s_1}} \|_{\ell^1_{n_1,m,I_1}} \right)^{\theta_1} \\
&\qquad \cdot \left(\| \mathbb{1}_{\mathbb{X}^{l_1,r_2}} \|_{\ell^\infty_{n_2,m,J_1} \ell^1_{n_1,I_1}} \cdot \| \mathbb{1}_{\mathbb{B}_2^{r_2} \cap \mathbb{C}_2^{s_2}} \|_{\ell^1_{n_2,m,J_2}} \right)^{\theta_2} \\
&\qquad \cdot \left(2^{2t} \cdot 2^{-(k_1+k_2)} \cdot |F| \right)^{\theta_3} \\
&\qquad \cdot \left(2^t \cdot |F| \right)^{\theta_4}.
\end{aligned}$$

Choosing $\theta_4 = 3\varepsilon$, $\theta_3 = \frac{1}{2} - 2\varepsilon$, $\theta_2 = \frac{1}{4} - \frac{\varepsilon}{2}$, $\theta_1 = \frac{1}{4} - \frac{\varepsilon}{2}$ and plugging in (2.22) and (2.4),

$$
\begin{aligned}
&|\widetilde{\Lambda}_{k_1,k_2}(f,g,H)| \\
&\lesssim \sum_{\substack{\vec{l},\vec{r}\\ \vec{s},t\geq 0}} 2^{-\varepsilon l_1}\cdot\left(\frac{2^{-\frac{1}{2}r_1}\cdot 2^{-\frac{1}{2}r_2}}{\|\mathbb{1}_{\mathbb{X}^{l_1,r_2}}\|^{\frac{1}{4}}_{\ell^\infty_{n_2,m,J_1}\ell^1_{n_1,I_1}}}\right)^{1-\varepsilon} \\
&\qquad\cdot 2^{-\varepsilon l_2}\cdot\left(\frac{2^{-\frac{1}{2}s_1}\cdot 2^{-\frac{1}{2}s_2}}{\|\mathbb{1}_{\mathbb{X}^{l_2,s_1}}\|^{\frac{1}{4}}_{\ell^\infty_{n_1,m,I_2}\ell^1_{n_2,J_2}}}\right)^{1-\varepsilon}\cdot 2^{-t} \\
&\qquad\cdot\left(\|\mathbb{1}_{\mathbb{X}^{l_2,s_1}}\|_{\ell^\infty_{n_1,m,I_2}\ell^1_{n_2,J_2}}\cdot 2^{2r_1+2s_1}\cdot 2^{2k_1}\cdot|E_1|\cdot|E_2|\right)^{\frac{1}{4}-\frac{\varepsilon}{2}} \\
&\qquad\cdot\left(\|\mathbb{1}_{\mathbb{X}^{l_1,r_2}}\|_{\ell^\infty_{n_2,m,J_1}\ell^1_{n_1,I_1}}\cdot 2^{2r_2+2s_2}\cdot 2^{2k_2}\cdot|E_1|\cdot|E_2|\right)^{\frac{1}{4}-\frac{\varepsilon}{2}} \\
&\qquad\cdot\left(2^{2t}\cdot 2^{-(k_1+k_2)}|F|\right)^{\frac{1}{2}-2\varepsilon} \\
&\qquad\cdot\left(2^{t}\cdot|F|\right)^{3\varepsilon} \\
&\lesssim \sum_{\substack{\vec{l},\vec{r}\\ \vec{s},t\geq 0}} 2^{-\varepsilon l_1}\cdot 2^{-\varepsilon l_2}\cdot 2^{-\frac{\varepsilon}{2}r_1}\cdot 2^{-\frac{\varepsilon}{2}r_2}\cdot 2^{-\frac{\varepsilon}{2}s_1}\cdot 2^{-\frac{\varepsilon}{2}s_2}\cdot 2^{-\varepsilon t}\cdot 2^{\varepsilon(k_1+k_2)} \\
&\qquad\cdot\left(\|\mathbb{1}_{\mathbb{X}^{l_1,r_2}}\|^{\frac{1}{4}}_{\ell^\infty_{n_2,m,J_1}\ell^1_{n_1,I_1}}\cdot\|\mathbb{1}_{\mathbb{X}^{l_2,s_1}}\|^{\frac{1}{4}}_{\ell^\infty_{n_1,m,I_2}\ell^1_{n_2,J_2}}\right)^{-\frac{\varepsilon}{4}} \\
&\qquad\cdot|E_1|^{\frac{1}{2}-\varepsilon}\cdot|E_2|^{\frac{1}{2}-\varepsilon}\cdot|F|^{\frac{1}{2}+\varepsilon}.
\end{aligned}
$$

Recall that the sums run over indices $l_1, l_2, r_1, r_2, s_1, s_2$ such that

$$
2^{-l_1} \lesssim 2^{-\frac{(k_1+k_2)}{2}},
$$

$$
2^{-l_2} \lesssim 2^{-\frac{(k_1+k_2)}{2}},
$$

by the triangle inequality and

$$
2^{-r_1} \lesssim 2^{-\frac{k_1}{2}}, \quad 2^{-r_2} \lesssim 2^{-\frac{k_2}{2}}, \quad 2^{-s_1} \lesssim 2^{-\frac{k_1}{2}}, \quad 2^{-s_2} \lesssim 2^{-\frac{k_2}{2}},
$$

since

$$
\begin{aligned}
2^{-2r_1} &\lesssim \int_0^1 \left| \int_0^1 \frac{1}{|I_1|^{\frac{1}{2}}} \cdot |f(\xi_1, \xi_2)| \cdot \varphi_{I_1}(\xi_1) \mathrm{d}\xi_1 \right|^2 \mathrm{d}\xi_2 \\
&\lesssim \int_0^1 \int_{\mathrm{supp}(\varphi_{I_1})} |\mathbb{1}_{E_1}(\xi_1, \xi_2)| \mathrm{d}\xi_1 \mathrm{d}\xi_2 \\
&\lesssim |E_1 \cap (I_1 \times [0, 1])| \\
&\leq 2^{-k_1},
\end{aligned}
$$

and analogously for r_2, s_1, s_2. Therefore, the sum above is bounded by

$$
|\widetilde{\Lambda}_{k_1,k_2}(f, g, H)| \lesssim_\varepsilon 2^{-\frac{\varepsilon}{2}(k_1+k_2)} \cdot |E_1|^{\frac{1}{2}-\frac{\varepsilon}{2}} \cdot |E_2|^{\frac{1}{2}-\frac{\varepsilon}{2}} \cdot |F|^{\frac{1}{2}+\varepsilon}. \tag{2.24}
$$

We also have the trivial estimate

$$
|\widetilde{\Lambda}_{k_1,k_2}(f, g, H)| \lesssim |E_1| \cdot |E_2| \cdot |F|. \tag{2.25}
$$

Interpolating between (2.24) and (2.25) with weights $\gamma_1 = \frac{1}{1+2\varepsilon}$ and $\gamma_2 = \frac{2\varepsilon}{1+2\varepsilon}$, respectively, gives

$$
\begin{aligned}
|\widetilde{\Lambda}_{k_1,k_2}(f, g, H)| &\lesssim_\varepsilon 2^{-\frac{\varepsilon}{2(1+2\varepsilon)}(k_1+k_2)} \cdot |E_1|^{\frac{1}{2}} \cdot |E_2|^{\frac{1}{2}} \cdot |F|^{\frac{1}{2}+\frac{2\varepsilon}{1+2\varepsilon}} \\
&\lesssim_\varepsilon 2^{-\frac{\varepsilon}{3}(k_1+k_2)} \cdot |E_1|^{\frac{1}{2}} \cdot |E_2|^{\frac{1}{2}} \cdot |F|^{\frac{1}{2}+\frac{2\varepsilon}{1+2\varepsilon}}.
\end{aligned} \tag{2.26}
$$

Restricted weak-type interpolation shows that $\widetilde{T}_{k_1,k_2}$ maps $L^2 \times L^2$ to $L^{2+\frac{8\varepsilon}{1-2\varepsilon}}$ with operatorial bound $O(2^{-\frac{\varepsilon}{3}(k_1+k_2)})$, as desired. □

Acknowledgments We thank Diogo Oliveira e Silva, Mateus Sousa, and the two anonymous referees for bringing to our attention many of the cited references and for the feedback provided, which greatly improved the original version of this chapter. C.M. was partially supported by a grant from the Simons Foundation. I.O. is supported by ERC project FAnFArE no. 637510.

References

1. J. Bennett, N. Bez, A. Carbery, D. Hundertmark, *Heat-flow monotonicity of Strichartz norms*, Analysis & PDE, Anal. PDE 2(2), 147–158, (2009)
2. J. Bourgain, L. Guth, *Bounds on oscillatory integral operators based on multilinear estimates*, Geom. Funct. Anal. 21 (2011), no. 6, 1239–1295.
3. C. Fefferman, *Inequalities for strongly singular convolution operators*. Acta Math., 124, 1970.
4. D. Foschi, *Maximizers for the Strichartz inequality*, J. Eur. Math. Soc. (JEMS) 9:4 (2007), 739–774. MR 2008k:35389 Zbl 05255314
5. F. Gonçalves, *Orthogonal polynomials and sharp estimates for the Schrödinger equation*, International Mathematics Research Notices (2019), https://doi.org/10.1093/imrn/rnx200 .

6. D. Hundertmark, V. Zharnitsky, *On sharp Strichartz inequalities in low dimensions*, Int. Math. Res. Not. (2006), id 34080. MR 2007b:35277 Zbl 1131.35308
7. F. Linares, G. Ponce, *Introduction to nonlinear dispersive equations*. Universitext. Springer, New York, 2009.
8. C. Muscalu, I. Oliveira, *A new approach to the Fourier extension problem for the paraboloid*, Accepted to Analysis & PDE, arXiv:2110.12482.
9. C. Muscalu, W. Schlag, *Classical and Multilinear Harmonic Analysis I*. Cambridge University Press, New York NY 2013.
10. T. Ozawa, Y. Tsutsumi, *Space-time estimates for null gauge forms and nonlinear Schrödinger equations*, Differential Integral Equations 11 (1998), 201–222.
11. F. Planchon, L. Vega, *Bilinear virial identities and applications*, Ann. Sci. É c. Norm. Supér. (4) 42 (2009), 261–290.
12. R. Strichartz, *Restriction of the Fourier transform to quadratic surfaces and decay of solutions of wave equations*. Duke Math. J., 70, 1977.
13. T. Tao, *Recent progress on the Restriction conjecture*. Park City proceedings, 2003.
14. T. Tao, A. Vargas, L. Vega, *A bilinear approach to the Restriction and Kakeya conjectures*. Journal of the American Mathematical Society, 11(4), 1998.
15. C. Thiele, *Wave packet analysis*. CBMS Regional Conference Series in Mathematics, vol. 105, Published for the Conference Board of the Mathematical Sciences, Washington, DC; by the American Mathematical Society, Providence, RI, 2006.
16. P. Tomas, *A Restriction theorem for the Fourier transform*. Bull. Amer. Math. Soc., (81), 1975.
17. A. Zygmund, *On Fourier coefficients and transforms of functions of two variables.*. Studia Math., 50, 1974.

Chapter 3
Modulational Instability of Classical Water Waves

Huy Q. Nguyen and Walter A. Strauss

[WS] I remember well discussing wave equations with Bob at MIT, especially his introduction in 1970 of general L^p estimates for the linear wave equations. The only previous general estimates had been based on L^2, that is, energy. In 1976, Irving Segal also noticed that one could get mixed space–time estimates for the case of one space dimension. But his proof was based on the work of Carleson and Sjölin that did not permit generalization to higher dimensions. In his 1977 paper, Bob Strichartz quickly recognized that the restriction theorem for Fourier transforms could lead to mixed space–time estimates in any dimension. These are the celebrated Strichartz estimates. Once they were in hand, I found them to be the perfect tool to study the scattering of nonlinear wave equations. That was only the first of an enormous number of applications of Bob's estimates.

3.1 Water Wave Instability

This chapter, which is based on our joint work [12], concerns the most classical kind of water wave, known as a Stokes wave, which is a periodic traveling water wave propagating in a fixed direction.

We ask whether or not a small perturbation at time $t = 0$ will remain a small perturbation for all $t > 0$. If it remains small, the Stokes wave is called stable.

H. Q. Nguyen
Department of Mathematics, University of Maryland, College Park, MD, USA
e-mail: hnguye90@umd.edu

W. A. Strauss (✉)
Department of Mathematics & Lefschetz Center for Dynamical Systems, Brown University, Providence, RI, USA
e-mail: walter_strauss@brown.edu

P. Alonso Ruiz et al. (eds.), *From Classical Analysis to Analysis on Fractals*, Applied and Numerical Harmonic Analysis, https://doi.org/10.1007/978-3-031-37800-3_3

Otherwise, it is called unstable. One often hears of stable waves; these occur only under special conditions on the perturbations or for different situations than ours, for instance, for the KdV equation. In fact, there are many kinds of instabilities in hydrodynamics. For instance, Rayleigh–Taylor instability occurs when a heavy fluid lies on top of a lighter one.

Here, we are considering a single fluid with a free surface, close to equilibrium, which is perturbed by a wave of a slightly different spatial frequency like $\delta \sin((1+\mu)x)$. This leads to the modulational (or Benjamin–Feir) instability. We take the most ideal model, the Euler equations, of a fluid which we assume is incompressible, two-dimensional, irrotational, inviscid, and under the influence of gravity. Quiescent air lies above the water. There is a free boundary S between the water and the air on which all the important interaction occurs. There are two boundary conditions on S. (i) On the surface S, the pressure is a constant (= the air pressure). (ii) The particles on S remain on S.

Precisely, we consider the fluid domain $\Omega(t) = \{(x, y) : x \in \mathbb{R}, \ -\infty < y < \eta(x,t)\}$ and its free surface $S(t) = \{(x, y) : x \in \mathbb{R}, \ y = \eta(x,t)\}$. The Euler equations together with incompressibility and irrotationality imply that there is a velocity potential ϕ, with $\mathbf{v} = \nabla\phi$ being the velocity vector, which is a harmonic function. Then, ϕ and η satisfy the water wave system (see, for instance, [11]):

$$\begin{cases} \Delta\phi = 0 \qquad \text{in } \Omega, \\ \partial_t\phi + \frac{1}{2}|\nabla\phi|^2 = -g\eta + P \qquad \text{on } S, \\ \partial_t\eta = \partial_y\phi - \partial_x\phi\,\partial_x\eta \qquad \text{on } S, \\ \nabla\phi \to 0 \text{ as } y \to -\infty. \end{cases} \tag{3.1}$$

P is atmospheric pressure, and $g > 0$ is gravity. The second equation, due to Bernoulli, expresses the fact that the pressure across S is continuous. The third equation expresses the fact that particles on S remain on S. The fourth equation says that the fluid is quiescent at great depth. There are many famous approximate models for special cases, such as the Korteweg–de Vries equation, the nonlinear Schrödinger equation, and the Whitham equation, but we do not consider them here.

A Stokes wave is a solution that is traveling; that is, it depends only on $x - ct$ and y, where c is the constant speed of propagation and is periodic in x. In 1847, Stokes expanded such waves of amplitude a formally as a power series in a. The convergence of the series for small a, and therefore a rigorous proof of existence, was not accomplished until the early 1920s independently by Nekrasov and Levi-Civita.

Is such a wave stable? It is remarkable that, beyond experimental and numerical evidence, no one explained the instability before Benjamin and Feir [1] did in 1967. They used heuristic arguments, based on showing that plane waves with slightly different wave numbers ("sidebands") can interact to get a kind of resonance that induces exponential growth in time. (Analogous research with sidebands was going on simultaneously in electromagnetic theory by Ostrovsky and others.) In fact, this modulational instability is one of the most ubiquitous phenomena in nonlinear science. There have been many numerical studies, but a rigorous proof of the instability has been elusive.

For the case of finite depth, in a *tour de force* in 1995, Bridges and Mielke [4] finally gave a rigorous proof of the linearized instability, by means of a spatial dynamical reduction to a four-dimensional center-manifold within the Hamiltonian framework. The first proof that works as well for infinite depth [12] is based instead on perturbation of spectra. For simplicity, we only discuss the case of infinite depth.

Let us take a Stokes wave of period 2π, say, and denote its amplitude by ϵ. Let μ denote a perturbation of the wave number. By this, we mean that 2π-periodic functions are multiplied by $\exp(i\mu x)$, where μ is small. Let $\mathcal{L}_{\mu,\epsilon}$ denote the linearized operator (to be precisely defined later). Our result is as follows.

Theorem 3.1 ([12]) *If ϵ is sufficiently small, then for all sufficiently small μ depending on ϵ, the operator $\mathcal{L}_{\mu,\epsilon}$ has an eigenvalue λ with positive real part. Indeed, it has the asymptotic expansion*

$$\lambda = \frac{1}{2} i\mu + \frac{1}{2\sqrt{2}}|\mu\epsilon| + O(\mu^2) + O(\mu\epsilon^2). \tag{3.2}$$

The second term is real and positive, which means that the time-dependent wave has a factor $e^{\lambda t}$ that grows at an exponential rate. Some other references that include either numerical computations or considerably greater detail in the asymptotics are [2, 3, 6, 7, 9]. It turns out that the curve λ as a function of μ is approximately a lemniscate in the complex plane. More recently the nonlinear modulational instability has been proven in [5]. A summary of recent related works can be found in [8].

The rest of this chapter is devoted to an outline of the proof of Theorem 3.1.

3.2 Linearization and Transformation

If we introduce ϕ restricted to S by defining $\psi(t, x) = \phi(t, x, \eta(t, x))$ in the moving frame with speed c, the water wave system (3.1) becomes

$$\begin{cases} \partial_t \eta = c\partial_x \eta + G(\eta)\psi, \\ \partial_t \psi = c\partial_x \psi - \frac{1}{2}|\partial_x \psi|^2 + \frac{1}{2}\frac{(G(\eta)\psi + \partial_x \psi \partial_x \eta)^2}{1+|\partial_x \eta|^2} - g\eta + P. \end{cases} \tag{3.3}$$

Here, the Dirichlet–Neumann operator $G(\eta)$ associated with Ω is defined as

$$[G(\eta) f](x) = \partial_y \theta(x, \eta(x)) - \partial_x \theta(x, \eta(x))\partial_x \eta(x),$$

where $\theta(x, y)$ solves the elliptic problem

$$\Delta\theta = 0 \text{ in } \Omega, \quad \theta|_{y=\eta(x)} = f(x), \quad \nabla\theta \in L^2(\Omega).$$

This is the Zakharov–Craig–Sulem formulation. For simplicity, we take the period to be 2π and gravity $g = 1$.

Back in the 1920s, Nekrasov and Levi-Civita proved that for a given speed $c > 0$ and air pressure $P > 0$, *there exists a curve of Stokes waves* (smooth traveling solutions) (η, ψ, c, P) parametrized by the amplitude ϵ and the air pressure $P \in \mathbb{R}$ such that both η and ψ are 2π-periodic, while η is even and ψ is odd. Many years earlier Stokes had derived the expansion

$$\begin{aligned}
\eta &= P + \epsilon \cos x + \tfrac{1}{2}\epsilon^2 \cos(2x) + O(\epsilon^3), \\
\psi &= \epsilon \sin x + \tfrac{1}{2}\epsilon^2 \sin(2x) + O(\epsilon^3), \\
c &= 1 + \tfrac{1}{2}\epsilon^2 + O(\epsilon^3)
\end{aligned} \tag{3.4}$$

for small ϵ. Of course, being in the mid-nineteenth century, he did not prove the convergence of the series. We now also know that this local curve can be extended to large ϵ all the way up to Stokes' extreme wave. Other than the trivial solutions $\eta \equiv const$, the curve is unique.

We *linearize* around a given Stokes wave (η^*, ψ^*, c^*) with amplitude ϵ. Using $\overline{\eta}, \overline{\psi}$ to denote the linearized variables of some period L and denoting $v_1 = \overline{\eta}$ and $v_2 = \overline{\psi} - B^*\overline{\eta}$ (known as Alinhac's good unknown), the linearized system becomes

$$\partial_t v_1 = \partial_x((c^* - V^*)v_1) + G(\eta^*)v_2, \tag{3.5}$$

$$\partial_t v_2 = (-1 + (c^* - V^*)\partial_x B^*)v_1 + (c^* - V^*)\partial_x v_2, \tag{3.6}$$

where

$$B^* = \frac{G(\eta^*)\psi^* + \partial_x\psi^*\partial_x\eta^*}{1 + |\partial_x\eta^*|^2}, \qquad V^* = \partial_x\psi^* - B^*\partial_x\eta^*.$$

Next, we *flatten* the free surface S by means of a conformal mapping $(x, y) \to z = z_1 + iz_2$ from the lower half plane onto Ω so that

$$z_1(x + 2\pi, y) = 2\pi + z_1(x, y) \text{ and } z_2(x + 2\pi, y) = z_2(x, y)$$

and z_1 is odd in x and z_2 is even in x. Then, we define the "Riemann stretch" $\zeta(x) = z_1(x, 0)$ on the surface and the pullback $\zeta_\sharp f = f \circ \zeta$ and we change variables

$$w_1 = \zeta'\zeta_\sharp v_1, \quad w_2 = \zeta_\sharp v_2$$

to get the reduced linear system

$$\partial_t w_1 = \partial_x(p^*(x)w_1) + |D|w_2, \tag{3.7}$$

$$\partial_t w_2 = -m^*(x)w_1 + p^*(x)\partial_x w_2. \tag{3.8}$$

This system involves the conformally flattened surface and the velocity potential on the surface, while $G(\eta)$ has become the pseudo-differential operator $|D|$. Here, the variable coefficients p^* and m^* are explicitly defined in terms of the Stokes wave (η^*, ψ^*) and the Riemann stretch ζ. [For finite depth h, we would have had to replace $|D|$ by $D\tanh(hD)$.] For small ϵ, they are very close to 1:

$$p^*(x) = 1 - 2\epsilon\cos x + O(\epsilon^2), \quad m^*(x) = 1 - 2\epsilon\cos x + O(\epsilon^2).$$

Now, we make a *long-wave perturbation*. We seek solutions of the reduced system of the form

$$w_j(x,t) = e^{\lambda t}e^{i\mu x}u_j(x) \quad \text{where } u_j(x+2\pi) = u_j(x),$$

and the Floquet parameter $\mu = \frac{m_0}{n_0}$ is a small rational number. (Rationality is assumed only for simplicity so as to retain periodicity.) Thus, $w_j(\cdot, t)$ is L-periodic with $L = n_0 2\pi \gg 2\pi$. In terms of $U = (u_1, u_2)^T$, we arrive at the eigenvalue problem

$$\lambda U = \mathcal{L}_{\mu,\epsilon}U := \begin{bmatrix} p^*(x)(i\mu + \partial_x) + \partial_x p^*(x) & |D+\mu| \\ -m^*(x) & p^*(x)(i\mu+\partial_x) \end{bmatrix} U. \tag{3.9}$$

The linear operator $\mathcal{L}_{\mu,\epsilon}$ is bounded from $(H^1(\mathbb{T}))^2$ to $(L^2(\mathbb{T}))^2$. It acts on functions $U(x) = [u_1(x), u_2(x)]$ of period 2π. Modulational instability would mean that $\mathcal{L}_{\mu,\epsilon}$ has an eigenvalue λ with positive real part.

Recall that the amplitude ϵ parametrizes the Stokes waves near the trivial wave, while the Floquet parameter μ refers to the long-wave perturbation of the 2π periodicity.

3.3 Spectrum of $\mathcal{L}_{\mu,\epsilon}$

Now, we get down to the serious analysis. First, we consider the case when $\epsilon = 0$ because then we can Fourier transform

$$\mathcal{L}_{\mu,0} = \begin{bmatrix} i\mu + \partial_x & |D+\mu| \\ -1 & i\mu + \partial_x \end{bmatrix}.$$

This matrix operator has discrete imaginary spectrum

$$\lambda = i[(\mu + k) \pm \sqrt{|\mu + k|}] =: i\omega^{\pm}_{k,\mu},$$

where $k \in \mathbb{Z}$. For small μ, the spectrum has exactly four imaginary eigenvalues near zero:

$$\{i\omega_{0,\mu}^{+}, i\omega_{0,\mu}^{-}, i\omega_{1,\mu}^{-}, i\omega_{-1,\mu}^{+}\}.$$

They vanish for $\mu = \epsilon = 0$, so that 0 is an eigenvalue of $\mathcal{L}_{0,0}$ with algebraic multiplicity 4. As μ varies, it becomes the four simple imaginary eigenvalues above.

Second, we let ϵ vary instead of μ. By the standard perturbation theory as in Kato's celebrated book [10], the spectrum of $\mathcal{L}_{0,\epsilon}$ must have an isolated part near the origin with multiplicity 4. By differentiating our equations with respect to x, ϵ, P, we can deduce that the isolated part is entirely at the origin itself, as stated in the following lemma.

Lemma 3.1 *For sufficiently small ϵ, zero is an eigenvalue of $\mathcal{L}_{0,\epsilon}$ with algebraic multiplicity four and geometric multiplicity two. The eigenvectors in the null space are*

$$U_1 = (0, 1)^T, \quad \widetilde{U}_2 = \left(\zeta' \zeta_\sharp \partial_x \eta^*, \ \zeta_\sharp(\partial_x \psi^* - B^* \partial_x \eta^*)\right)^T.$$

The generalized eigenvectors are

$$\begin{aligned} U_3 &= \left(\zeta' \zeta_\sharp \partial_\epsilon \eta, \ \zeta_\sharp(\partial_\epsilon \psi - B^* \partial_\epsilon \eta)\right)^T \\ U_4 &= \left(\zeta' \zeta_\sharp \partial_P \eta, \ \zeta_\sharp(\partial_P \psi - B^* \partial_P \eta)\right)^T. \end{aligned} \tag{3.10}$$

In fact, U_3 and U_4 satisfy

$$\mathcal{L}_{0,\epsilon} U_3 = -\partial_\epsilon c \, \widetilde{U}_2, \qquad \mathcal{L}_{0,\epsilon} U_4 = -U_1. \tag{3.11}$$

It is important to normalize the second eigenvector as

$$U_2 := \frac{1}{\epsilon}\left(\widetilde{U}_2 - U_1 \int_0^{2\pi} \widetilde{U}_2^{(2)} \frac{dx}{2\pi}\right) \tag{3.12}$$

so that both components of U_2 have mean zero. The expansions of the eigenvectors are

$$U_2 = \begin{bmatrix} \text{odd} \\ \text{even} \end{bmatrix} = \begin{bmatrix} -\sin x \\ \cos x \end{bmatrix} + \epsilon \begin{bmatrix} -2\sin(2x) \\ \cos(2x) \end{bmatrix} + O(\epsilon^2), \tag{3.13}$$

$$U_3 = \begin{bmatrix} \text{even} \\ \text{odd} \end{bmatrix} = \begin{bmatrix} \cos x \\ \sin x \end{bmatrix} + \epsilon \begin{bmatrix} 2\cos(2x) \\ \sin(2x) \end{bmatrix} + O(\epsilon^2), \tag{3.14}$$

$$U_4 = \begin{bmatrix} \text{even} \\ \text{odd} \end{bmatrix} = \begin{bmatrix} 1 \\ 0 \end{bmatrix} + \epsilon \begin{bmatrix} \cos x \\ -\sin x \end{bmatrix} + O(\epsilon^2). \tag{3.15}$$

By writing $U_2 = \begin{bmatrix}\text{odd}\\ \text{even}\end{bmatrix}$ here, we mean that the first component of U_2 is an odd function and the second component is an even function.

Finally, we fix a small amplitude $\epsilon \neq 0$ and perturb in μ. The Lyapunov–Schmidt method splits the problem into two parts, as follows. Let $\mathcal{U}$ be the four-dimensional generalized kernel of $\mathcal{L}_{0,\epsilon}$ and Π the L^2-orthogonal projection onto $\mathcal{U}^\perp$. Then, we have to solve

$$\Pi(\mathcal{L}_{\mu,\epsilon} - \lambda \mathrm{Id})U = 0, \tag{3.16}$$

$$(\mathrm{Id} - \Pi)(\mathcal{L}_{\mu,\epsilon} - \lambda \mathrm{Id})U = 0. \tag{3.17}$$

We seek $U = \sum_{j=1}^4 \alpha_j U_j + W$ with $W \in \mathcal{U}^\perp$. Then, $W = \sum_{j=1}^4 \alpha_j W_j$, where each *sideband function* W_j solves

$$\Pi(\mathcal{L}_{\mu,\epsilon} - \lambda \mathrm{Id})(U_j + W_j) = 0. \tag{3.18}$$

We recall that

$$\mathcal{L}_{\mu,\epsilon} = \begin{bmatrix} p^*(x)(i\mu + \partial_x) + \partial_x p^*(x) & |D+\mu| \\ -m^*(x) & p^*(x)(i\mu + \partial_x) \end{bmatrix}, \tag{3.19}$$

which decomposes as $\mathcal{L}_{\mu,\epsilon} = \mathcal{L}_{0,\epsilon} + \mu L_\epsilon^\dagger$, where $L_\epsilon^\dagger$ has order zero and is independent of μ. We would like to solve (3.18) for the W_j and then insert them into (3.17) to obtain a four-dimensional system. The infinite-dimensional part (3.18) has the explicit form

$$\Pi\mathcal{L}_{0,\epsilon} W_j + \Pi(\mu L_\epsilon^\dagger - \lambda \mathrm{Id})W_j = -\Pi\mathcal{L}_{0,\epsilon} U_j - \Pi(\mu L_\epsilon^\dagger - \lambda \mathrm{Id})U_j.$$

Hence, we want to invert $\Pi\mathcal{L}_{0,\epsilon}$. We can show that for any sufficiently small ϵ, $\Pi\mathcal{L}_{0,\epsilon} : H^1 \to L^2$ is a Fredholm operator with kernel $\mathcal{U}$ and range $\mathcal{U}^\perp$. (The main part of the proof is to show that the range of $\Pi\mathcal{L}_{0,\epsilon}$ is closed, which follows from the stability of Fredholmness under small perturbations.) Now, let Ξ_ϵ denote the inverse of $\Pi\mathcal{L}_{0,\epsilon}$ mapping $\mathcal{U}^\perp \to \mathcal{U}^\perp$. We can formally solve for W_j by using the smallness of μ and λ:

$$W_j = -\mu \sum_{m=0}^{\infty} (-1)^m \Big[\Xi_\epsilon \Pi(\mu L_\epsilon^\dagger - \lambda \mathrm{Id})\Big]^m \Xi_\epsilon \Pi L_\epsilon^\dagger U_j. \tag{3.20}$$

Somewhat surprisingly, the sideband functions W_j will contribute crucially to the instability. Inserting $U = \sum_{j=1}^4 \alpha_j (U_j + W_j)$ into (3.17) gives the equation

$$\sum_{j=1}^{4} \alpha_j (\mathrm{Id} - \Pi)(\mathcal{L}_{\mu,\epsilon} - \lambda \mathrm{Id})(U_j + W_j) = 0.$$

It has a nontrivial solution $(\alpha_1, \ldots, \alpha_4)$ if and only if

$$\mathcal{P}(\lambda; \mu, \epsilon) := \det(\mathbf{A}_{\mu,\epsilon} - \lambda \mathbf{I}_\epsilon + \mathbf{B}_{\mu,\epsilon}) = 0, \tag{3.21}$$

where

$$\mathbf{A}_{\mu,\epsilon} = \left(\frac{(\mathcal{L}_{\mu,\epsilon} U_j, U_k)}{(U_k, U_k)} \right)_{j,k=\overline{1,4}}, \quad \mathbf{I}_\epsilon = \left(\frac{(U_j, U_k)}{(U_k, U_k)} \right)_{j,k=\overline{1,4}} \tag{3.22}$$

and the *sideband matrix*

$$\mathbf{B}_{\mu,\epsilon} = \left(\frac{(\mathcal{L}_{\mu,\epsilon} W_j, U_k)}{(U_k, U_k)} \right)_{j,k=\overline{1,4}}. \tag{3.23}$$

To summarize, the Stokes wave (η^*, ψ^*, c^*) is unstable if there exists a small number μ such that the quartic polynomial

$$\boxed{\mathcal{P}(\lambda; \mu, \epsilon) := \det(\mathbf{A}_{\mu,\epsilon} - \lambda \mathbf{I}_\epsilon + \mathbf{B}_{\mu,\epsilon}) = 0} \tag{3.24}$$

has a small root λ with a positive real part.

3.4 Expansions

The next step is to use the preceding expansions for the coefficients of $\mathcal{L}_{\mu,\epsilon}$ and for the (generalized) eigenvectors U_j. After very long calculations, we obtain the following expansions:

$$\mathbf{A}_{\mu,\epsilon} =$$

$$\begin{bmatrix} i\mu + \mu O_\epsilon(\epsilon^2) & -i\mu\epsilon + \mu O_\epsilon(\epsilon^2) & \mu O_\epsilon(\epsilon^2) & \mu + \mu O_\epsilon(\epsilon^2) \\ -i\mu\epsilon + \mu O_\epsilon(\epsilon^2) & \frac{1}{2} i\mu + \mu O_\epsilon(\epsilon^2) & 0 & 0 \\ O_\epsilon(\epsilon^4) & -\epsilon^2 & \frac{1}{2} i\mu + \mu O_\epsilon(\epsilon^2) & -\frac{3}{2} i\mu\epsilon + \mu O(\epsilon^2) \\ -1 & 0 & -\frac{1}{2} i\mu\epsilon + \mu O_\epsilon(\epsilon^2) & i\mu + \mu O_\epsilon(\epsilon^2) \end{bmatrix} \tag{3.25}$$

(where the notation O_ϵ indicates that the bound depends only on ϵ), whence we obtain its determinant. Neglecting $\mathbf{B}_{\mu,\epsilon}$ at first, we have the following lemma:

Lemma 3.2

$$\begin{aligned}\det(\mathbf{A}_{\mu,\epsilon} - \lambda \mathbf{I}_\epsilon) = {} & (\tfrac{1}{2} i\mu - \lambda)^2 \mu \\ & + \mu(\tfrac{1}{2} i\mu - \lambda)^2 O_\epsilon(\epsilon^2) + \mu^2 (\tfrac{1}{2} i\mu - \lambda)\, O_\epsilon(\epsilon^2) \\ & + O_\epsilon(\epsilon^3)\, C(\mu, \lambda) + O(\mu^4 + |\lambda|^4),\end{aligned} \tag{3.26}$$

where $C(\mu, \lambda)$ is a cubic polynomial without a λ^3 term.

Unfortunately, this does not give us a positive real part. Therefore, we do a similar expansion for $\mathbf{B}_{\mu,\epsilon}$, which leads to the following result for the difference of the determinants.

Lemma 3.3

$$\begin{aligned}& \det(\mathbf{A}_{\mu,\epsilon} - \lambda \mathbf{I}_\epsilon + \mathbf{B}_{\mu,\epsilon}) - \det(\mathbf{A}_{\mu,\epsilon} - \lambda \mathbf{I}_\epsilon) \\ & = -\tfrac{1}{8}\mu^3 \epsilon^2 + \mu(\tfrac{1}{2} i\mu - \lambda)^2 O_\epsilon(\epsilon^2) + \mu^2 (\tfrac{1}{2} i\mu - \lambda) O_\epsilon(\epsilon^2) \\ & \quad + O_\epsilon(\epsilon^3)\, C(\mu, \lambda) + O(\mu^4 + |\lambda|^4).\end{aligned} \tag{3.27}$$

Combining (3.26) and (3.27), we obtain

$$\boxed{\begin{aligned}\mathcal{P}(\lambda; \mu, \epsilon) = {} & (\tfrac{1}{2} i\mu - \lambda)^2 \mu \\ & + \epsilon^2 \Big\{ -\tfrac{1}{8}\mu^3 + r_1 \mu (\tfrac{1}{2} i\mu - \lambda)^2 + r_2 \mu^2 (\tfrac{1}{2} i\mu - \lambda) \Big\} \\ & + O_\epsilon(\epsilon^3)\, C(\mu, \lambda) + O(\mu^4 + |\lambda|^4).\end{aligned}} \tag{3.28}$$

To leading order, $\mathcal{P}(\lambda; \mu, \epsilon)$, is cubic in the pair (λ, μ). So, if we define $\gamma = \lambda/\mu$, then $\mathcal{P}(\lambda; \mu, \epsilon) = \mu^3 \widetilde{\mathcal{P}}(\gamma; \mu, \epsilon)$, where

$$\begin{aligned}\widetilde{\mathcal{P}}(\gamma; \mu, \epsilon) = {} & (\tfrac{1}{2} i - \gamma)^2 + \epsilon^2 \Big\{ -\tfrac{1}{8} + r_1 (\tfrac{1}{2} i - \gamma)^2 + r_2 (\tfrac{1}{2} i - \gamma) \Big\} \\ & + O_\epsilon(\epsilon^3) \theta_1(\gamma) + \mu \theta_2(\gamma; \mu, \epsilon) \\ =: {} & \widetilde{\mathcal{Q}}(\gamma; \epsilon) + \mu \theta_2(\gamma; \mu, \epsilon),\end{aligned} \tag{3.29}$$

where θ_1 is quadratic, θ_2 is smooth, and r_1 and r_2 are constants. Thus, $\widetilde{\mathcal{Q}}(\gamma; \epsilon)$ is independent of μ.

The roots of $\widetilde{\mathcal{Q}}$ can be found by noting that

$$\epsilon^{-2} \widetilde{\mathcal{Q}}(\tfrac{1}{2} i + \kappa\epsilon; \epsilon) = \kappa^2 - \tfrac{1}{8} + \epsilon^2 r_1 \kappa^2 - \epsilon r_2 \kappa + O_\epsilon(\epsilon)\, \theta_1(\tfrac{1}{2} i + \kappa\epsilon). \tag{3.30}$$

At $\epsilon = 0$, we have roots $\kappa = \pm\frac{1}{\sqrt{8}}$. So, there are two local curves of roots $\kappa_\pm(\epsilon)$ near $\pm\frac{1}{\sqrt{8}}$ of $\widetilde{Q}$. Thus, there exists a small $\epsilon_0 > 0$ such that for all $\epsilon \in (-\epsilon_0, \epsilon_0) \setminus \{0\}$, the quadratic polynomial $\widetilde{Q}(\gamma; \epsilon)$ has two simple roots

$$\gamma_\pm(\epsilon) = \tfrac{1}{2}i + \epsilon\kappa_\pm(\epsilon), \tag{3.31}$$

where $\kappa_\pm : (-\epsilon_0, \epsilon_0) \to \mathbb{R}$ is smooth and $\kappa_\pm(0) = \pm\frac{1}{\sqrt{8}}$.

Now, perturbing in μ, we have $\widetilde{\mathcal{P}}(\gamma; \mu, \epsilon) = \widetilde{Q}(\gamma; \epsilon) + \mu\theta_2(\gamma; \mu, \epsilon)$. By the implicit function theorem, $\widetilde{\mathcal{P}}(\gamma; \mu, \epsilon)$ has at least two simple roots $\gamma_\pm(\mu, \epsilon)$ for all $\mu \in (0, \mu_0(\epsilon))$. Therefore, we have reached the following statement. For all $\epsilon \in (-\epsilon_0, \epsilon_0) \setminus \{0\}$ and $\mu \in (0, \mu_0(\epsilon))$, $\mathcal{P}(\lambda; \mu, \epsilon)$ has at least two simple roots of the form

$$\lambda_\pm(\mu, \epsilon) = \frac{1}{2}i\mu \pm \frac{1}{2\sqrt{2}}\mu\epsilon + \mu\epsilon^2 g_1(\epsilon) + \mu^2 g_2(\mu, \epsilon) \tag{3.32}$$

with smooth functions g_1 and g_2. This completes the proof of the modulational instability for Stokes waves of small amplitude in deep water.

Acknowledgments The work of HQN was partially supported by NSF grant DMS-2205734.

References

1. T. B. Benjamin, J. E. Feir. The disintegration of wave trains on deep water. Part 1. Theory. *J. Fluid Mech.* **27** (1967), no. 3, 417–430.
2. M. Berti, A. Maspero, P. Ventura. Full description of Benjamin-Feir instability of Stokes waves in deep water. arXiv:2109.11852, 2021.
3. M. Berti, A. Maspero, P. Ventura. Benjamin-Feir instability of Stokes waves in finite depth. arXiv:2204.00809, 2022.
4. T. J. Bridges, A. Mielke. A proof of the Benjamin-Feir instability. *Arch. Rational Mech. Anal.* **133** (1995), no. 2, 145–198.
5. Chen, G; Su, Q. Nonlinear modulational instability of the Stokes waves in 2D full water waves. arXiv:2012.15071, 2020.
6. R. Creedon, B. Deconinck. A high-order asymptotic analysis of the Benjamin-Feir instability spectrum in arbitrary depth. arXiv:2206.01817, 2022.
7. B. Deconinck, K. Oliveras. The instability of periodic surface gravity waves. *J. Fluid Mech.* **675** (2011), 141–167.
8. S. Haziot et al. Traveling water waves — the ebb and flow of two centuries. *Quart. Appl. Math.* 80 (2022), 317-401.
9. Hur, V. M.; Yang, Z. Unstable Stokes waves. arXiv:2010.10766, 2020.
10. T. Kato. *Perturbation theory for linear operators*. Die Grundlehren der mathematischen Wissenschaften, Band 132, Springer-Verlag New York, Inc., New York, 1966.
11. R.S. Johnson. *A modern introduction to the mathematical theory of water waves*. Cambridge Texts in Applied Mathematics. Cambridge University Press, Cambridge, 1997.
12. H Q. Nguyen, W. A. Strauss. Proof of modulational instability of Stokes waves in deep water. *Comm. Pure Appl. Math.* **76**(5), 1035–1084 (2022).

Chapter 4
Convergence Analysis of the Deep Galerkin Method for Weak Solutions

Yuling Jiao, Yanming Lai, Yang Wang, Haizhao Yang, and Yunfei Yang

4.1 Introduction

Deep learning [1] has achieved many breakthroughs in high-dimensional data analysis, e.g., in computer vision and natural language processing [2, 3]. Its outstanding performance has also motivated its application to solve high-dimensional PDEs, which is a challenging task for classical numerical methods, e.g., finite element methods [4] and finite difference methods [5]. The application of neural networks to solve PDEs dates back to the 1990s [6] for low-dimensional problems. In recent years, neural network-based PDE solvers were revisited for high-dimensional PDEs with tremendous successes and new development [7–11]. The key idea of these methods is to approximate the solutions of PDEs by neural networks and construct loss functions based on equations and their boundary conditions. References [8, 9] use the squared residuals on the domain as the loss function and treat boundary conditions as penalty terms, which are called physics-informed neural networks (PINNs). Inspired by the Ritz method, [10] proposes the deep Ritz method (DRM) and uses variational forms of PDEs as loss functions. The idea of the Galerkin method has also been used in [11], where they propose a minimax training procedure

Y. Jiao
School of Mathematics and Statistics, and Hubei Key Laboratory of Computational Science, Wuhan University, Wuhan, P.R. China
e-mail: yulingjiaomath@whu.edu.cn

Y. Lai · Y. Wang (✉) · Y. Yang
Department of Mathematics, The Hong Kong University of Science and Technology, Kowloon, Hong Kong
e-mail: ylaiam@connect.ust.hk; yangwang@ust.hk; yyangdc@connect.ust.hk

H. Yang
Department of Mathematics, University of Maryland, College Park, MD, USA
e-mail: hzyang@umd.edu

P. Alonso Ruiz et al. (eds.), *From Classical Analysis to Analysis on Fractals*, Applied and Numerical Harmonic Analysis, https://doi.org/10.1007/978-3-031-37800-3_4

via reformulating the problem of finding the weak solution of PDEs into minimizing an operator norm defined through a maximization problem induced by the weak formulation. Here, we call the scheme inspired by the Galerkin method DGMW for short (in the original paper [11], this method is called *Weak Adversarial Network* method and called *WAN* for short).

DGMW has the ability to solve equations that can be transformed into energy functional forms, making it applicable to a wider range of problems than DRM. However, DGMW involves training two neural networks, whereas DRM only requires one. Consequently, the theory of DGMW is more intricate, and its application requires more computations.

4.1.1 Related Works and Our Contributions

Although there are great empirical achievements of deep learning methods for PDEs in recent several years, a challenging and interesting question is to provide a rigorous error analysis such as the finite element method. Several recent efforts have been devoted to making processes along this line. The error analysis of DRM has been studied in [12–17, 17, 18]. Lu et al. [12] concerns a priori generalization analysis of the deep Ritz method with two-layer neural networks, under the a priori assumption that the exact solutions of the PDEs lie in spectral Barron space. See also [13] for handling general equations with solutions living in spectral Barron space via two-layer ReLU^k networks. Duan et al. [15], Jiao et al. [16], Lu et al. [18] studied the error analysis of the DRM in Sobolev spaces with deep networks. [19], Mishra and Molinaro [20], Shin et al. [21], Jiao et al. [22], Lu et al. [18] considered the convergence and convergence rate of PINNs.

Since the training loss of DGMW is in a minimax form and there are two networks to train, it is much more challenging to provide a theoretical guarantee for DGMW than that of DRM and PINNs. As far as we know, there is no convergence result of DGMW despite the excellent numerical performance shown in [11]. In this chapter, we give the first convergence rate analysis of DGMW to solve second-order elliptic equations with Dirichlet, Neumann, and Robin boundary conditions, respectively, with deep neural networks in Sobolev spaces. Our results show how to set the hyper-parameters of depth and width to achieve the desired convergence rate in terms of the number of training samples. The main contributions of this chapter are summarized as follows:

- We derive novel error decomposition results for DGMW, which is of independent interest for minimax training with deep networks.
- We establish the first convergence rate of the DGMW with Dirichlet, Neumann, and Robin boundary conditions. $\forall \epsilon > 0$, we prove that if we set the number of samples as $O(\epsilon^{-d \log d})$ and the depth, width, and the bound of the weights in the two networks to be

$$\mathcal{D} \leq O(\log d), \quad \mathcal{W} \leq \epsilon^{-d}, \quad B_\theta \leq O(\epsilon^{\frac{-9d-8}{2}}),$$

then the H^1 norm error of DGMW in expectation is smaller than ϵ.

4.1.2 *Organization*

The outline of the rest of this chapter is as follows. In Sect. 4.2, the error decomposition of the DGMW is given, while the details of approximation error and statistical error are presented in Sects. 4.3 and 4.4, respectively. We devote Sect. 4.5 to the convergence rate of the DGMW. Finally, we give a conclusion and extension in Sect. 4.6.

We end up this section with some notations used throughout this chapter. Let $\mathcal{D} \in \mathbb{N}^+$. A function $\mathbf{f} : \mathbb{R}^d \to \mathbb{R}^{n_\mathcal{D}}$ implemented by a neural network is defined by

$$\begin{aligned} &\mathbf{f}_0(\mathbf{x}) = \mathbf{x}, \\ &\mathbf{f}_\ell(\mathbf{x}) = \rho\left(A_\ell \mathbf{f}_{\ell-1} + \mathbf{b}_\ell\right) \qquad \text{for } \ell = 1, \ldots, \mathcal{D}-1, \\ &\mathbf{f} := \mathbf{f}_\mathcal{D}(\mathbf{x}) = A_\mathcal{D} \mathbf{f}_{\mathcal{D}-1} + \mathbf{b}_\mathcal{D}, \end{aligned} \tag{4.1}$$

where $A_\ell = \left(a_{ij}^{(\ell)}\right) \in \mathbb{R}^{n_\ell \times n_{\ell-1}}$ and $\mathbf{b}_\ell = \left(b_i^{(\ell)}\right) \in \mathbb{R}^{n_\ell}$. ρ is called the activation function and acts componentwise. $\mathcal{D}$ is called the depth of the network and $\mathcal{W} := \max\{n_\ell : \ell = 1, \cdots, \mathcal{D}\}$ is called the width of the network. $\phi = \{A_\ell, \mathbf{b}_\ell\}_\ell$ are called the weight parameters. For convenience, we denote $\mathfrak{n}_i$, $i = 1, \cdots, \mathcal{D}$, as the number of nonzero weights on the first i layers in the representation (4.1). Clearly, $\mathfrak{n}_\mathcal{D}$ is the total number of nonzero weights. Sometimes we denote a function implemented by a neural network as $\mathbf{f}_\rho$ for short. We use the notation $\mathcal{N}_\rho\left(\mathcal{D}, \mathfrak{n}_\mathcal{D}, B_\theta\right)$ to refer to the collection of functions implemented by a ρ-neural network with depth $\mathcal{D}$, the total number of nonzero weights $\mathfrak{n}_\mathcal{D}$, and each weight being bounded by B_θ.

4.2 Error Decomposition

We consider the following second-order divergence form in the elliptic equation:

$$-\sum_{i,j=1}^{d} \partial_j(a_{ij}\partial_i u) + \sum_{i=1}^{d} b_i \partial_i u + cu = f \quad \text{in } \Omega \tag{4.2}$$

with three kinds of boundary conditions:

$$u = 0 \text{ on } \partial\Omega \tag{4.3a}$$

$$\sum_{i,j=1}^{d} a_{ij}\partial_i u n_j = g \text{ on } \partial\Omega \tag{4.3b}$$

$$\alpha u + \beta \sum_{i,j=1}^{d} a_{ij}\partial_i u n_j = g \text{ on } \partial\Omega, \quad \alpha, \beta \in \mathbb{R}, \beta \neq 0 \tag{4.3c}$$

which are called the Dirichlet, Neumann, and Robin boundary conditions, respectively. Note for Dirichlet's problem, we only consider the homogeneous boundary condition here since the inhomogeneous case can be turned into a homogeneous case by translation. We also remark that the Neumann condition (4.3b) is covered by Robin condition (4.3c). Hence, in the following, we only consider the Dirichlet problem and the Robin problem.

We make the following assumptions on the known terms in the equation:

(A1) $f \in L^2(\Omega)$, $g \in L^2(\partial\Omega)$, $a_{ij} \in C(\bar{\Omega})$, $b_i, c \in L^\infty(\Omega)$, $c > 0$.
(A2) There exists $\lambda, \Lambda > 0$ such that $\lambda|\xi|^2 \leq \sum_{i,j=1}^{d} a_{ij}\xi_i\xi_j \leq \Lambda|\xi|^2, \quad \forall x \in \Omega, \xi \in \mathbb{R}^d$.
(A3) $4\lambda c > d \max_{1\leq i\leq d} \|b_i\|^2_{L^\infty(\Omega)}$.

In the following, we abbreviate $C\Big(\|f\|_{L^2(\Omega)}, \|g\|_{L^2(\partial\Omega)}, \|a_{ij}\|_{C(\bar{\Omega})}, \|b_i\|_{L^\infty(\Omega)}, \|c\|_{L^\infty(\Omega)}, \lambda\Big)$, constants depending on the known terms in equation, as $C(coe)$ for simplicity.

Under the above assumptions, a coercivity result is easily acquired.

Lemma 4.1 *Let (A1)-(A3) hold. For any* $u \in H^1(\Omega)$,

$$\sum_{i,j=1}^{d}(a_{ij}\partial_i u, \partial_j u) + \sum_{i=1}^{d}(b_i\partial_i u, u) + (cu, u) \geq C(d, coe)\|u\|^2_{H^1(\Omega)}.$$

Proof Applying Hölder and Cauchy's inequality and choosing δ such that

$$\frac{d \max_{1\leq i\leq d} \|b_i\|^2_{L^\infty(\Omega)}}{4c} < \delta < \lambda,$$

we have

$$\begin{aligned}
&\sum_{i,j=1}^{d}(a_{ij}\partial_i u, \partial_j u) + \sum_{i=1}^{d}(b_i\partial_i u, u) + (cu, u) \\
&\geq \lambda|u|^2_{H^1(\Omega)} + c\|u\|^2_{L^2(\Omega)} - \sqrt{d} \max_{1\leq i\leq d} \|b_i\|_{L^\infty(\Omega)}\|u\|_{L^2(\Omega)}|u|_{H^1(\Omega)} \\
&\geq (\lambda-\delta)|u|^2_{H^1(\Omega)} + \left(c - \frac{d \max_{1\leq i\leq d} \|b_i\|^2_{L^\infty(\Omega)}}{4\delta}\right)\|u\|^2_{L^2(\Omega)} \geq C(coe)\|u\|^2_{H^1(\Omega)}.
\end{aligned}$$

The coercivity ensures the existence and uniqueness of the weak solution of the Dirichlet problem and the Robin problem. Specifically, for problem (4.2), (4.3a), the variational problems is as follows: find $u \in H_0^1(\Omega)$ such that

$$\sum_{i,j=1}^{d} (a_{ij}\partial_i u, \partial_j v) + \sum_{i=1}^{d} (b_i \partial_i u, v) + (cu, v) = (f, v) \quad \forall v \in H_0^1(\Omega). \tag{4.4}$$

Lemma 4.2 *Let (A1)–(A3) hold. Let u_D be the solution of the problem (4.4). Then, $u_D \in H^2(\Omega)$.*

Proof See [23]. □

For problem (4.2), (4.3c), the variational problem is as follows: find $u \in H^1(\Omega)$ such that

$$\sum_{i,j=1}^{d} (a_{ij}\partial_i u, \partial_j v) + \sum_{i=1}^{d} (b_i \partial_i u, v) + (cu, v) + \frac{\alpha}{\beta}(T_0 u, T_0 v)|_{\partial\Omega}$$
$$= (f, v) + \frac{1}{\beta}(g, T_0 v)|_{\partial\Omega}, \quad \forall v \in H^1(\Omega), \tag{4.5}$$

where T_0 is a zero-order trace operator.

Lemma 4.3 *Let (A1)-(A3) hold. Let u_R be the solution of the problem (4.5). Then, $u_R \in H^2(\Omega)$ and $\|u_R\|_{H^2(\Omega)} \le \frac{C(coe)}{\beta}$ for any $\beta > 0$.*

Proof See [24]. □

Intuitively, when $\alpha = 1$, $g = 0$, and $\beta \to 0$, we expect that the solution of the Robin problem converges to the solution of the Dirichlet problem. Hence, we only need to consider the Robin problem since the Dirichlet problem can be handled through a limiting process. The next lemma verifies this assertion.

Lemma 4.4 *Let (A1)–(A3) hold. Let $\alpha = 1$ and $g = 0$. Let u_D be the solution of problem (4.4) and u_R the solution of problem (4.5). There holds*

$$\|u_R - u_D\|_{H^1(\Omega)} \le C(d, \Omega, coe)\beta^{1/2}.$$

Proof By the definition of u_R and u_D, we have for any $v \in H^1(\Omega)$

$$\sum_{i,j=1}^{d} (a_{ij}\partial_i u_R, \partial_j v) + \sum_{i=1}^{d} (b_i \partial_i u_R, v) + (cu_R, v) + \frac{1}{\beta}(T_0 u_R, T_0 v)|_{\partial\Omega} = (f, v) \tag{4.6}$$

$$\sum_{i,j=1}^{d}(a_{ij}\partial_i u_D, \partial_j v) + \sum_{i=1}^{d}(b_i\partial_i u_D, v) + (cu_D, v)$$

$$= (f, v) + \sum_{i,j=1}^{d}\int_{\partial\Omega} a_{ij}\partial_i u_D T_0 v n_j ds, \tag{4.7}$$

where n_j is the jth component of $\mathbf{n}$, the outward pointing unit normal vector along $\partial\Omega$. Subtracting (4.7) from (4.6) and choosing $v = u_R - u_D$, we have

$$\sum_{i,j=1}^{d}(a_{ij}\partial_i(u_R - u_D), \partial_j(u_R - u_D)) + \sum_{i=1}^{d}(b_i\partial_i(u_R - u_D), (u_R - u_D))$$
$$+ (c(u_R - u_D), (u_R - u_D)) + \frac{1}{\beta}(T_0(u_R - u_D), T_0(u_R - u_D))|_{\partial\Omega}$$
$$= \sum_{i,j=1}^{d}\int_{\partial\Omega} a_{ij}\partial_i u_D T_0(u_R - u_D)n_j ds \tag{4.8}$$

where we use the fact that $T_0 u_D = 0$. For the term in the right-hand side of (4.8), by the Hölder inequality and Cauchy's inequality, we have

$$\sum_{i,j=1}^{d}\int_{\partial\Omega} a_{ij}\partial_i u_D T_0(u_R - u_D)n_j ds$$
$$\leq \max_{1\leq i,j\leq d}\|a_{ij}\|_{C(\bar{\Omega})} d^{3/2}|T_0 u_D|_{H^1(\partial\Omega)}\|T_0(u_R - u_D)\|_{L^2(\partial\Omega)}$$
$$\leq \frac{1}{4}\beta\left(\max_{1\leq i,j\leq d}\|a_{ij}\|_{C(\bar{\Omega})}\right)^2 d^3|T_0 u_D|^2_{H^1(\partial\Omega)} + \frac{1}{\beta}\|T_0(u_R - u_D)\|^2_{L^2(\partial\Omega)}$$
$$\leq \frac{1}{4}\beta\left(\max_{1\leq i,j\leq d}\|a_{ij}\|_{C(\bar{\Omega})}\right)^2 d^3 C(\Omega)\|u_D\|^2_{H^2(\Omega)} + \frac{1}{\beta}\|T_0(u_R - u_D)\|^2_{L^2(\partial\Omega)} \tag{4.9}$$

where in the final step we apply the trace theorem

$$\|T_0 v\|_{L^2(\partial\Omega)} \leq C(\Omega)\|v\|_{H^1(\Omega)}$$

See more details in [25]. Now, combining Lemma 4.1, (4.8), and (4.9) yields the result. □

Remark A result similar to Lemma 4.4 is given in [13, Lemma 5.4], where the authors study the PDE that can be turned into a corresponding energy functional form.

Define

$$\mathcal{L}(u, v) := \sum_{i,j=1}^{d} (a_{ij}\partial_i u, \partial_j v) + \sum_{i=1}^{d}(b_i\partial_i u, v) + (cu, v) + \frac{\alpha}{\beta}(T_0 u, T_0 v)|_{\partial\Omega}$$
$$- (f, v) - \frac{1}{\beta}(g, T_0 v)|_{\partial\Omega}.$$

It is clear that if u is the solution of problem (4.5), then it solves the following optimization problem:

$$\inf_{u\in H^1(\Omega)} \sup_{\substack{v\in H^1(\Omega)\\ \|v\|_{H^1(\Omega)}\le 1}} \mathcal{L}(u, v). \tag{4.10}$$

Note that $\mathcal{L}(u, v)$ can be equivalently written as

$$\mathcal{L}(u, v) = |\Omega|\mathbb{E}_{X\sim U(\Omega)}$$
$$\times \left(\sum_{i,j=1}^{d} (a_{ij}\partial_i u\partial_j v)(X) + \sum_{i=1}^{d}(b_i\partial_i uv)(X) + (cuv)(X) - (fv)(X) \right)$$
$$+ \frac{|\partial\Omega|}{\beta}\mathbb{E}_{Y\sim U(\partial\Omega)} \left(\frac{\alpha}{2}(T_0 u T_0 v)(Y) - (gT_0 v)(Y) \right),$$

where $U(\Omega)$ and $U(\partial\Omega)$ are uniform distributions on Ω and $\partial\Omega$, respectively. We then introduce a discrete version of $\mathcal{L}$ defined on $C^1(\Omega) \times C^1(\Omega)$:

$$\widehat{\mathcal{L}}(u, v) := \frac{|\Omega|}{N}\sum_{k=1}^{N}$$
$$\times \left(\sum_{i,j=1}^{d} (a_{ij}\partial_i u\partial_j v)(X_k) + \sum_{i=1}^{d}(b_i\partial_i uv)(X_k) + (cuv)(X_k) - (fv)(X_k) \right)$$
$$+ \frac{|\partial\Omega|}{\beta M}\sum_{k=1}^{M} \left(\frac{\alpha}{2}(T_0 u T_0 v)(Y_k) - (gT_0 v)(Y_k) \right),$$

where $\{X_k\}_{k=1}^N$ and $\{Y_k\}_{k=1}^M$ are i.i.d. random variables according to $U(\Omega)$ and $U(\partial\Omega)$, respectively. We now consider a minimax problem with respect to $\widehat{\mathcal{L}}$:

$$\inf_{u\in\mathcal{P}} \sup_{v\in\mathcal{P}} \widehat{\mathcal{L}}(u, v), \tag{4.11}$$

where $\mathcal{P} \subset C^1(\Omega)$ refers to the parameterized function class. Finally, we call a (random) solver $\mathcal{A}$, say SGD, to minimize $\sup_{v\in\mathcal{P}} \widehat{\mathcal{L}}(\cdot, v)$ and denote the output of $\mathcal{A}$, say $u_{\phi_{\mathcal{A}}}$, as the final solution.

In order to study the difference between the weak solution of PDE (4.2) (u_R and u_D) and the solution of empirical loss generated by a random solver ($u_{\phi_{\mathcal{A}}}$), we first define for any $u \in H^1(\Omega)$

$$\begin{aligned}
\mathcal{L}_0(u) &:= \sup_{\substack{v\in H^2(\Omega)\\ \|v\|_{H^2(\Omega)}\le 1}} \mathcal{L}(u, v)\\
\mathcal{L}_1(u) &:= \sup_{v\in\mathcal{P}} \mathcal{L}(u, v)\\
\mathcal{L}_2(u) &:= \sup_{v\in\mathcal{P}} \widehat{\mathcal{L}}(u, v).
\end{aligned}$$

The following result decomposes the total error into three parts, enabling us to apply different methods to deal with different kinds of errors.

Proposition 4.1 *Let (A1)–(A3) hold. Assume that* $\mathcal{P} \subset C^1(\Omega)\bigcap H^2(\Omega)$ *and* $\|u\|_{H^1(\Omega)} \le \mathcal{M}$ *for all* $u \in \mathcal{P}$. *Let* u_R *and* u_D *be the solution of the problem* (4.5) *and* (4.4), *respectively. Let* $u_{\phi_{\mathcal{A}}}$ *be the solution of problem* (4.11) *generated by a random solver.*

(1) There holds

$$\|u_{\phi\mathcal{A}} - u_R\|_{H^1(\Omega)} \le C(d, \Omega, coe)\left(\mathcal{E}_{app} + \mathcal{E}_{sta} + \mathcal{E}_{opt}\right)$$

with

$$\mathcal{E}_{app} := \frac{\mathcal{M}}{\beta} \sup_{\substack{v_1\in H^2(\Omega)\\ \|v_1\|_{H^2(\Omega)}\le 1}} \inf_{v_2\in\mathcal{P}} \|v_1 - v_2\|_{H^1(\Omega)} + \frac{\mathcal{M}}{\beta} \inf_{\bar{u}\in\mathcal{P}} \|\bar{u} - u_R\|_{H^1(\Omega)} \tag{4.12}$$

$$\mathcal{E}_{sta} := 2 \sup_{u\in\mathcal{P}} |\mathcal{L}_1(u) - \mathcal{L}_2(u)| \tag{4.13}$$

$$\mathcal{E}_{opt} := \mathcal{L}_2\left(u_{\phi_{\mathcal{A}}}\right) - \inf_{u\in\mathcal{P}} \mathcal{L}_2(u). \tag{4.14}$$

(2) Set $\alpha = 1$ *and* $g = 0$. *There holds*

$$\|u_{\phi\mathcal{A}} - u_R\|_{H^1(\Omega)} \le C(d, \Omega, coe)\left(\mathcal{E}_{app} + \mathcal{E}_{sta} + \mathcal{E}_{opt} + \mathcal{E}_{pen}\right),$$

where $\mathcal{E}_{app}, \mathcal{E}_{sta}, \mathcal{E}_{opt}$ *are given by* (4.12), (4.13), (4.14), *and*

$$\mathcal{E}_{pen} := \|u_R - u_D\|_{H^1(\Omega)}.$$

Proof We only prove (1) since (2) is a direct result of (1) and the triangle inequality.

Letting $\bar{u}$ be any element in $\mathcal{P}$, we have

$$\begin{aligned}
&\mathcal{L}_0\left(u_{\phi_{\mathcal{A}}}\right) - \mathcal{L}_0(u_R)\\
=&\mathcal{L}_0\left(u_{\phi_{\mathcal{A}}}\right) - \mathcal{L}_1\left(u_{\phi_{\mathcal{A}}}\right) + \mathcal{L}_1\left(u_{\phi_{\mathcal{A}}}\right) - \mathcal{L}_2\left(u_{\phi_{\mathcal{A}}}\right) + \mathcal{L}_2\left(u_{\phi_{\mathcal{A}}}\right) - \inf_{u\in\mathcal{P}} \mathcal{L}_2(u)\\
&+ \inf_{u\in\mathcal{P}} \mathcal{L}_2(u) - \mathcal{L}_2(\bar{u}) + \mathcal{L}_2(\bar{u}) - \mathcal{L}_1(\bar{u}) + \mathcal{L}_1(\bar{u}) - \mathcal{L}_1(u_R)\\
\le& [\mathcal{L}_0\left(u_{\phi_{\mathcal{A}}}\right) - \mathcal{L}_1\left(u_{\phi_{\mathcal{A}}}\right)] + [\mathcal{L}_1(\bar{u}) - \mathcal{L}_1(u_R)] + 2\sup_{u\in\mathcal{P}} |\mathcal{L}_1(u) - \mathcal{L}_2(u)|\\
&+ \left[\mathcal{L}_2\left(u_{\phi_{\mathcal{A}}}\right) - \inf_{u\in\mathcal{P}} \mathcal{L}_2(u)\right],
\end{aligned}$$

where we use the fact that $\mathcal{L}_0(u_R) = \mathcal{L}_1(u_R) = 0$. Since $\bar{u}$ can be any element in $\mathcal{P}$, we take the infimum of $\bar{u}$ on both sides of the above display

$$\begin{aligned}
\mathcal{L}_0\left(u_{\phi_{\mathcal{A}}}\right) - \mathcal{L}_0(u_R) \le& [\mathcal{L}_0\left(u_{\phi_{\mathcal{A}}}\right) - \mathcal{L}_1\left(u_{\phi_{\mathcal{A}}}\right)] + \inf_{\bar{u}\in\mathcal{P}} [\mathcal{L}_1(\bar{u}) - \mathcal{L}_1(u_R)]\\
&+ 2\sup_{u\in\mathcal{P}} |\mathcal{L}_1(u) - \mathcal{L}_2(u)| + \left[\mathcal{L}_2\left(u_{\phi_{\mathcal{A}}}\right) - \inf_{u\in\mathcal{P}} \mathcal{L}_2(u)\right].
\end{aligned} \tag{4.15}$$

Now, for the term on the left-hand side of (4.15), by Lemma 4.1, we have

$$\begin{aligned}
\mathcal{L}_0\left(u_{\phi_{\mathcal{A}}}\right) - \mathcal{L}_0(u_R) =& \sup_{\substack{v\in H^2(\Omega)\\ \|v\|_{H^2(\Omega)}\le 1}} \left[\mathcal{L}(u_{\phi_{\mathcal{A}}}, v) - \mathcal{L}(u_R, v)\right]\\
=& \sup_{\substack{v\in H^2(\Omega)\\ \|v\|_{H^2(\Omega)}\le 1}} \Bigg[\sum_{i,j=1}^{d} (a_{ij}\partial_i(u_{\phi_{\mathcal{A}}} - u_R), \partial_j v) + \sum_{i=1}^{d} (b_i\partial_i(u_{\phi_{\mathcal{A}}} - u_R), v)\\
&+ (c(u_{\phi_{\mathcal{A}}} - u_R), v) + \frac{\alpha}{\beta}(T_0(u_{\phi_{\mathcal{A}}} - u_R), T_0 v)|_{\partial\Omega}\Bigg]\\
\ge& \sum_{i,j=1}^{d} \left(a_{ij}\partial_i(u_{\phi_{\mathcal{A}}} - u_R), \frac{\partial_j(u_{\phi_{\mathcal{A}}} - u_R)}{\|u_{\phi_{\mathcal{A}}} - u_R\|_{H^1(\Omega)}}\right)
\end{aligned}$$

$$
\begin{aligned}
&+\sum_{i=1}^{d}\left(b_i\partial_i(u_{\phi_{\mathcal{A}}}-u_R),\frac{u_{\phi_{\mathcal{A}}}-u_R}{\|u_{\phi_{\mathcal{A}}}-u_R\|_{H^1(\Omega)}}\right)\\
&+\left(c(u_{\phi_{\mathcal{A}}}-u_R),\frac{u_{\phi_{\mathcal{A}}}-u_R}{\|u_{\phi_{\mathcal{A}}}-u_R\|_{H^1(\Omega)}}\right)\\
&+\frac{\alpha}{\beta}\left(T_0(u_{\phi_{\mathcal{A}}}-u_R),\frac{T_0(u_{\phi_{\mathcal{A}}}-u_R)}{\|u_{\phi_{\mathcal{A}}}-u_R\|_{H^1(\Omega)}}\right)|_{\partial\Omega}\\
&\geq C(d,coe)\|u_{\phi_{\mathcal{A}}}-u_R\|_{H^1(\Omega)},
\end{aligned}
\tag{4.16}
$$

where the first step is due to the fact that $\mathcal{L}_0(u_R) = 0$. For the first term on the right-hand side of (4.15),

$$
\begin{aligned}
\mathcal{L}_0\left(u_{\phi_{\mathcal{A}}}\right)-\mathcal{L}_1\left(u_{\phi_{\mathcal{A}}}\right) &= \sup_{\substack{v\in H^2(\Omega)\\ \|v\|_{H^2(\Omega)}\leq 1}}\mathcal{L}(u_{\phi_{\mathcal{A}}},v)-\sup_{v\in\mathcal{P}}\mathcal{L}(u_{\phi_{\mathcal{A}}},v)\\
&= \sup_{\substack{v_1\in H^2(\Omega)\\ \|v_1\|_{H^2(\Omega)}\leq 1}}\inf_{v_2\in\mathcal{P}}\mathcal{L}(u_{\phi_{\mathcal{A}}},v_1-v_2)\\
&\leq \sup_{\substack{v_1\in H^2(\Omega)\\ \|v_1\|_{H^2(\Omega)}\leq 1}}\inf_{v_2\in\mathcal{P}}\frac{1}{\beta}C(d,\Omega,coe)\|u_{\phi_{\mathcal{A}}}\|_{H^1(\Omega)}\|v_1-v_2\|_{H^1(\Omega)}\\
&+\frac{1}{\beta}C(d,\Omega,coe)\|v_1-v_2\|_{H^1(\Omega)}\\
&\leq \frac{\mathcal{M}}{\beta}C(d,\Omega,coe)\sup_{\substack{v_1\in H^2(\Omega)\\ \|v_1\|_{H^2(\Omega)}\leq 1}}\inf_{v_2\in\mathcal{P}}\|v_1-v_2\|_{H^1(\Omega)}.
\end{aligned}
\tag{4.17}
$$

For the second term on the right-hand side of (4.15),

$$
\mathcal{L}_1(\bar{u})-\mathcal{L}_1(u_R)=\sup_{v\in\mathcal{P}}\left[\mathcal{L}(\bar{u},v)-\mathcal{L}(u_R,v)\right]\leq\frac{\mathcal{M}}{\beta}C(d,\Omega,coe)\|\bar{u}-u_R\|_{H^1(\Omega)}.
\tag{4.18}
$$

Combining (4.15)–(4.18) yields the result. □

4.3 Approximation Error

In this section, we study the approximation error $\mathcal{E}_{app}$ defined in (4.12). Clearly, we first need a neural network approximation result in Sobolev spaces. In this field, [26] is a comprehensive study concerning a variety of activation functions, including ReLU, sigmoidal type functions, etc. The key idea in [26] to build the upper bound in Sobolev spaces is to construct an approximate partition of unity.

Denote $\mathcal{F}_{s,p,d} := \left\{ f \in W^{s,p}\left([0,1]^d\right) : \|f\|_{W^{s,p}([0,1]^d)} \leq 1 \right\}$.

Theorem 4.1 (Proposition 4.8, [26]) *Let $p \geq 1$, $s, k, d \in \mathbb{N}^+$, $s \geq k+1, \bar{k} \geq k$. Let ρ be $\max\{0, x\}^{\bar{k}}$(ReLU$^{\bar{k}}$), $\frac{1}{1+e^{-x}}$(logistic function), or $\frac{e^x - e^{-x}}{e^x + e^{-x}}$(tanh function). For any $\epsilon > 0$ and $f \in \mathcal{F}_{s,p,d}$, there exists a neural network f_ρ with depth $C\log(d+s)$ such that*

$$\|f - f_\rho\|_{W^{k,p}([0,1]^d)} \leq \epsilon.$$

(1) If $\rho = \max\{0, x\}^{\bar{k}}$, then the number of non-zero weights of f_ρ is bounded by $C(d, s, p, k)\epsilon^{-d/(s-k)}$. (2) If $\rho = \frac{1}{1+e^{-x}}$ or $\frac{e^x - e^{-x}}{e^x + e^{-x}}$, then the number of non-zero weights of f_ρ is bounded by $C(d, s, p, k)\epsilon^{-d/(s-k-\mu k)}$. Moreover, in case (2), the value of weights is bounded in absolute value by

$$C(d, s, p, k)\epsilon^{-2-\frac{2(d/p+d+k+\mu k)+d/p+d}{s-k-\mu k}},$$

where μ is an arbitrarily small positive number.

Remark 4.1 The bounds in the theorem can be found in the proof of [26, Proposition 4.8], except that they did not explicitly give the bound on the depth. In their proof, they partition $[0, 1]^d$ into small patches, approximate f by a sum of localized polynomial $\sum_m \phi_m p_m$, and approximately implement $\sum_m \phi_m p_m$ by a neural network, where the bump functions $\{\phi_m\}$ form an approximately partition of unity and $p_m = \sum_{|\alpha|<s} c_{f,m,\alpha} x^\alpha$ are the averaged Taylor polynomials. As shown in [26], ϕ_m can be approximated by the products of the d-dimensional output of a neural network with constant layers. And the identity map $I(x) = x$ and the product function $\times(a, b) = ab$ can also be approximated by neural networks with constant layers. In order to approximate $\phi_m x^\alpha$, we need to implement $d + s - 1$ products. Hence, the required depth can be bounded by $C\log(d + s)$.

Since the region $[0, 1]^d$ is larger than the region Ω we consider (recall we assume without loss of generality that $\Omega \subset [0, 1]^d$ at the beginning), we need the following extension result.

Lemma 4.5 *Let $k \in \mathbb{N}^+$, $1 \leq p < \infty$. There exists a linear operator E from $W^{k,p}(\Omega)$ to $W_0^{k,p}\left([0,1]^d\right)$ and $Eu = u$ in Ω.*

Proof See Theorem 7.25 in [24]. □

From Lemma 4.3 and (4.12), we know that we need to approximate functions in $H^1(\Omega)$ and our target functions lie in $H^2(\Omega)$. We immediately obtain the result we desire from Theorem 4.1 and Lemma 4.5.

Theorem 4.2 *Let* ρ *be* $\frac{1}{1+e^{-x}}$*(logistic function) or* $\frac{e^x-e^{-x}}{e^x+e^{-x}}$*(tanh function). For any sufficiently small* $\epsilon > 0$*, set the parameterized function class*

$$\mathcal{P}:=\mathcal{N}_\rho\left(C\log(d+1), C(d,coe)(\beta^2\epsilon)^{\frac{-d}{1-\mu}}, C(d,coe)(\beta^2\epsilon)^{\frac{-9d-8}{2-2\mu}}\right)\bigcap B_{H^1(\Omega)}(0,2),$$

where $B_{H^1(\Omega)}(0,2) := \{f \in H^1(\Omega) : \|f\|_{H^1(\Omega)} \leq 2\}$*, and then* $\mathcal{E}_{app} \leq \epsilon$ *with* $\mathcal{E}_{app}$ *defined by (4.12).*

Proof Set $k=1$, $s=2$, $p=2$ in Theorem 4.1 and use the fact $\|f - f_\rho\|_{H^2(\Omega)} \leq \|Ef - f_\rho\|_{H^2([0,1]^d)}$ with E being the extension operator in Lemma 4.5, and we conclude that for any $0 < \delta \leq 1$ and $f \in H^2(\Omega)$ with $\|f\|_{H^2(\Omega)} \leq 1$, there exists a neural network f_ρ with depth $C\log(d+1)$ and the number of weights $C(d)\delta^{-d/(1-\mu)}$ such that

$$\|f - f_\rho\|_{H^1(\Omega)} \leq \delta \tag{4.19}$$

and the value of weights are bounded by $C(d)\delta^{-(9d+8)/(2-2\mu)}$, where μ is an arbitrarily small positive number. Denote

$$\mathcal{P}_\delta^0 := \{f_\rho : f \in H^2(\Omega), \quad \|f\|_{H^2(\Omega)} \leq 1\}.$$

Clearly,

$$\mathcal{P}_\delta^0 \subset \mathcal{P}_\delta^1 := \mathcal{N}_\rho\left(C\log(d+1), C(d)\delta^{-d/(1-\mu)}, C(d)\delta^{-(9d+8)/(2-2\mu)}\right).$$

In addition, for any $f_\rho \in \mathcal{P}_\delta^0$,

$$\|f_\rho\|_{H^1(\Omega)} \leq \|f_\rho - f\|_{H^1(\Omega)} + \|f\|_{H^1(\Omega)} \leq \delta + 1 \leq 2.$$

Therefore,

$$\mathcal{P}_\delta^0 \subset \mathcal{P}_\delta^1 \bigcap B_{H^1(\Omega)}(0,2)$$

with $B_{H^1(\Omega)}(0,2) := \{f \in H^1(\Omega) : \|f\|_{H^1(\Omega)} \leq 2\}$.

Now, we set the parameterized function class

$$\mathcal{P} = \mathcal{P}_\delta^B := \mathcal{P}_\delta^1 \bigcap B_{H^1(\Omega)}(0,2)$$

and estimate the approximation error $\mathcal{E}_{app}$ defined by (4.12). We first normalize the second term in (4.12).

$$\begin{aligned}\inf_{\bar{u}\in\mathcal{P}_\delta^B}\|\bar{u}-u_R\|_{H^1(\Omega)} &= \|u_R\|_{H^1(\Omega)}\inf_{\bar{u}\in\mathcal{P}_\delta^B}\left\|\frac{\bar{u}}{\|u_R\|_{H^1(\Omega)}}-\frac{u_R}{\|u_R\|_{H^1(\Omega)}}\right\|_{H^1(\Omega)}\\ &= \|u_R\|_{H^1(\Omega)}\inf_{\bar{u}\in\mathcal{P}_\delta^B}\left\|\bar{u}-\frac{u_R}{\|u_R\|_{H^1(\Omega)}}\right\|_{H^1(\Omega)}\\ &\le \frac{C(coe)}{\beta}\inf_{\bar{u}\in\mathcal{P}_\delta^B}\left\|\bar{u}-\frac{u_R}{\|u_R\|_{H^1(\Omega)}}\right\|_{H^1(\Omega)}\end{aligned}$$

where in the third step we apply Lemma 4.3. Hence,

$$\begin{aligned}\mathcal{E}_{app} \le \frac{2}{\beta}&\sup_{\substack{v_1\in H^2(\Omega)\\ \|v_1\|_{H^2(\Omega)}\le 1}}\inf_{v_2\in\mathcal{P}_\delta^B}\|v_1-v_2\|_{H^1(\Omega)}\\ &+\frac{2C(coe)}{\beta^2}\inf_{\bar{u}\in\mathcal{P}_\delta^B}\left\|\bar{u}-\frac{u_R}{\|u_R\|_{H^1(\Omega)}}\right\|_{H^1(\Omega)}.\end{aligned}\tag{4.20}$$

Setting $\delta = C(coe)\beta^2\epsilon$ and combining (4.19) and (4.20) yield the result. □

4.4 Statistical Error

In this section, we study the statistical error $\mathcal{E}_{sta}$ defined by (4.13).

Lemma 4.6 *For the statistical error $\mathcal{E}_{sta}$, there holds*

$$\mathcal{E}_{sta} \le \sum_{k=1}^{6} I_k$$

with

$$I_1 := 2|\Omega|\sup_{u,v\in\mathcal{P}}\left|\mathbb{E}_{X\sim U(\Omega)}\sum_{i,j=1}^{d}(a_{ij}\partial_i u\partial_j v)(X)-\frac{1}{N}\sum_{k=1}^{N}\sum_{i,j=1}^{d}(a_{ij}\partial_i u\partial_j v)(X_k)\right|$$

$$I_2 := 2|\Omega|\sup_{u,v\in\mathcal{P}}\left|\mathbb{E}_{X\sim U(\Omega)}\sum_{i=1}^{d}(b_i\partial_i uv)(X)-\frac{1}{N}\sum_{k=1}^{N}\sum_{i=1}^{d}(b_i\partial_i uv)(X_k)\right|$$

$$I_3 := 2|\Omega| \sup_{u,v\in\mathcal{P}} \left| \mathbb{E}_{X\sim U(\Omega)}(cuv)(X) - \frac{1}{N}\sum_{k=1}^{N}(cuv)(X_k) \right|$$

$$I_4 := 2|\Omega| \sup_{u,v\in\mathcal{P}} \left| \mathbb{E}_{X\sim U(\Omega)}(fv)(X) - \frac{1}{N}\sum_{k=1}^{N}(fv)(X_k) \right|$$

$$I_5 := 2\frac{|\partial\Omega|}{\beta} \sup_{u,v\in\mathcal{P}} \left| \mathbb{E}_{Y\sim U(\partial\Omega)}\frac{\alpha}{2}(T_0uT_0v)(Y) - \frac{1}{M}\sum_{k=1}^{M}\frac{\alpha}{2}(T_0uT_0v)(Y_k) \right|$$

$$I_6 := 2\frac{|\partial\Omega|}{\beta} \sup_{u,v\in\mathcal{P}} \left| \mathbb{E}_{Y\sim U(\partial\Omega)}(gT_0v)(Y) - \frac{1}{M}\sum_{k=1}^{M}(gT_0v)(Y_k) \right|.$$

Proof We have

$$\begin{aligned}
\mathcal{E}_{sta} &= 2\sup_{u\in\mathcal{P}} |\mathcal{L}_1(u) - \mathcal{L}_2(u)| = 2\sup_{u\in\mathcal{P}} \left| \sup_{v\in\mathcal{P}} \mathcal{L}(u,v) - \sup_{v\in\mathcal{P}} \widehat{\mathcal{L}}(u,v) \right| \\
&\le 2\sup_{u\in\mathcal{P}}\sup_{v\in\mathcal{P}} \left| \mathcal{L}(u,v) - \widehat{\mathcal{L}}(u,v) \right| \le \sum_{k=1}^{6} I_k,
\end{aligned}$$

where the third step is due to the fact that

$$\sup_{v\in\mathcal{P}} \mathcal{L}(u,v) - \sup_{v\in\mathcal{P}} \widehat{\mathcal{L}}(u,v) \le \sup_{v\in\mathcal{P}} \left[\mathcal{L}(u,v) - \widehat{\mathcal{L}}(u,v)\right] \le \sup_{v\in\mathcal{P}} \left|\mathcal{L}(u,v) - \widehat{\mathcal{L}}(u,v)\right|$$

$$\sup_{v\in\mathcal{P}} \widehat{\mathcal{L}}(u,v) - \sup_{v\in\mathcal{P}} \mathcal{L}(u,v) \le \sup_{v\in\mathcal{P}} \left[\widehat{\mathcal{L}}(u,v) - \mathcal{L}(u,v)\right] \le \sup_{v\in\mathcal{P}} \left|\mathcal{L}(u,v) - \widehat{\mathcal{L}}(u,v)\right|.$$

By the technique of symmetrization, we can bound the difference between continuous loss and empirical loss (i.e., $I_1, \cdots, I_6$) by the Rademacher complexity. We first introduce the Rademacher complexity.

Definition 4.1 The Rademacher complexity of a set $A \subseteq \mathbb{R}^N$ is defined as

$$\mathfrak{R}_N(A) = \mathbb{E}_{\{\sigma_i\}_{k=1}^N}\left[\sup_{a\in A}\frac{1}{N}\sum_{k=1}^{N}\sigma_k a_k\right],$$

where $\{\sigma_k\}_{k=1}^N$ are N i.i.d. Rademacher variables with $\mathbb{P}(\sigma_k = 1) = \mathbb{P}(\sigma_k = -1) = \frac{1}{2}$. The Rademacher complexity of a function class $\mathcal{F}$ associated with a random sample $\{X_k\}_{k=1}^N$ is defined as

$$\mathfrak{R}_N(\mathcal{F}) = \mathbb{E}_{\{X_k,\sigma_k\}_{k=1}^N}\left[\sup_{u\in\mathcal{F}} \frac{1}{N}\sum_{k=1}^N \sigma_k u(X_k)\right].$$

Lemma 4.7 *There holds*

$$\mathbb{E}_{\{X_k\}_{k=1}^N} I_i \le 4|\Omega|\mathfrak{R}_N(\mathcal{F}_i), \quad i = 1, \cdots, 4$$

$$\mathbb{E}_{\{Y_k\}_{k=1}^M} I_i \le \frac{4|\partial\Omega|}{\beta}\mathfrak{R}_M(\mathcal{F}_i), \quad i = 5, 6$$

with

$$\mathcal{F}_1 := \left\{\sum_{i,j=1}^d a_{ij}\partial_i u \partial_j v : u, v \in \mathcal{P}\right\}, \qquad \mathcal{F}_2 := \left\{\sum_{i=1}^d b_i \partial_i u v : u, v \in \mathcal{P}\right\}$$

$$\mathcal{F}_3 := \{cuv : u, v \in \mathcal{P}\}, \qquad \mathcal{F}_4 := \{fv : u, v \in \mathcal{P}\}$$

$$\mathcal{F}_5 := \left\{\frac{\alpha}{2} T_0 u T_0 v : u, v \in \mathcal{P}\right\}, \qquad \mathcal{F}_6 := \{g T_0 v : u, v \in \mathcal{P}\}.$$

Proof We only present the proof with respect to I_3 since other inequalities can be shown similarly. We take $\{\widetilde{X_k}\}_{k=1}^N$ as an independent copy of $\{X_k\}_{k=1}^N$, and then

$$\begin{aligned}
I_3 &= 2|\Omega| \sup_{u,v\in\mathcal{P}} \left|\mathbb{E}_{X\sim U(\Omega)}(cuv)(X) - \frac{1}{N}\sum_{k=1}^N (cuv)(X_k)\right| \\
&\le \frac{2|\Omega|}{N} \sup_{u,v\in\mathcal{P}} \left|\mathbb{E}_{\{\widetilde{X_k}\}_{k=1}^N} \sum_{k=1}^N \left[(cuv)(\widetilde{X_k}) - (cuv)(X_k)\right]\right| \\
&\le \frac{2|\Omega|}{N}\mathbb{E}_{\{\widetilde{X_k}\}_{k=1}^N} \sup_{u,v\in\mathcal{P}} \left|\sum_{k=1}^N \left[(cuv)(\widetilde{X_k}) - (cuv)(X_k)\right]\right|.
\end{aligned}$$

Hence,

$$\begin{aligned}
\mathbb{E}_{\{X_k\}_{k=1}^N} I_3 &\le \frac{2|\Omega|}{N}\mathbb{E}_{\{X_k,\widetilde{X_k}\}_{k=1}^N} \sup_{u,v\in\mathcal{P}} \left|\sum_{k=1}^N \left[(cuv)(\widetilde{X_k}) - (cuv)(X_k)\right]\right| \\
&= \frac{2|\Omega|}{N}\mathbb{E}_{\{X_k,\widetilde{X_k},\sigma_k\}_{k=1}^N} \sup_{u,v\in\mathcal{P}} \left|\sum_{k=1}^N \sigma_k \left[(cuv)(\widetilde{X_k}) - (cuv)(X_k)\right]\right| \\
&= \frac{2|\Omega|}{N}\mathbb{E}_{\{X_k,\widetilde{X_k},\sigma_k\}_{k=1}^N} \sup_{u,v\in\mathcal{P}}
\end{aligned}$$

$$\max\left\{\sum_{k=1}^{N}\sigma_k\left[(cuv)(\widetilde{X_k})-(cuv)(X_k)\right],\right.$$
$$\left.\sum_{k=1}^{N}\sigma_k\left[(cuv)(X_k)-(cuv)(\widetilde{X_k})\right]\right\},$$

where the second step is due to the fact that the insertion of Rademacher variables does not change the distribution. We note that

$$\begin{aligned}
&\mathbb{E}_{\{X_k,\widetilde{X_k},\sigma_k\}_{k=1}^N}\sup_{u,v\in\mathcal{P}}\sum_{k=1}^{N}\sigma_k\left[(cuv)(\widetilde{X_k})-(cuv)(X_k)\right]\\
&\le \mathbb{E}_{\{X_k,\widetilde{X_k},\sigma_k\}_{k=1}^N}\sup_{u,v\in\mathcal{P}}\sum_{k=1}^{N}\sigma_k(cuv)(\widetilde{X_k})+\mathbb{E}_{\{X_k,\widetilde{X_k},\sigma_k\}_{k=1}^N}\sup_{u,v\in\mathcal{P}}\sum_{k=1}^{N}-\sigma_k(cuv)(X_k)\\
&=2\mathbb{E}_{\{X_k,\sigma_k\}_{k=1}^N}\sup_{u,v\in\mathcal{P}}\sum_{k=1}^{N}\sigma_k(cuv)(X_k).
\end{aligned}$$

Similarly,

$$\begin{aligned}
&\mathbb{E}_{\{X_k,\widetilde{X_k},\sigma_k\}_{k=1}^N}\sup_{u,v\in\mathcal{P}}\sum_{k=1}^{N}\sigma_k\left[(cuv)(X_k)-(cuv)(\widetilde{X_k})\right]\\
&\le 2\mathbb{E}_{\{X_k,\sigma_k\}_{k=1}^N}\sup_{u,v\in\mathcal{P}}\sum_{k=1}^{N}\sigma_k(cuv)(X_k).
\end{aligned}$$

Therefore,

$$\mathbb{E}_{\{X_k\}_{k=1}^N}I_3\le 4|\Omega|\mathfrak{R}_N(\mathcal{F}_3).$$

In order to bound Rademacher complexities, we need the concept of covering numbers.

Definition 4.2 An ϵ-cover of a set T in a metric space (S,τ) is a subset $T_c\subset S$ such that for each $t\in T$, there exists a $t_c\in T_c$ such that $\tau(t,t_c)\le\epsilon$. The ϵ-covering number of T, denoted as $C(\epsilon,T,\tau)$, is defined to be the minimum cardinality among all ϵ-cover of T with respect to the metric τ.

In Euclidean space, we can establish an upper bound of the covering number for a bounded set easily.

Lemma 4.8 *Suppose that $T\subset\mathbb{R}^d$ and $\|t\|_2\le B$ for $t\in T$, then*

$$C(\epsilon, T, \|\cdot\|_2) \leq \left(\frac{2B\sqrt{d}}{\epsilon}\right)^d.$$

Proof Let $m = \left\lfloor \frac{2B\sqrt{d}}{\epsilon} \right\rfloor$ and define

$$T_c = \left\{-B + \frac{\epsilon}{\sqrt{d}}, -B + \frac{2\epsilon}{\sqrt{d}}, \cdots, -B + \frac{m\epsilon}{\sqrt{d}}\right\}^d,$$

then for $t \in T$, there exists $t_c \in T_c$ such that

$$\|t - t_c\|_2 \leq \sqrt{\sum_{i=1}^{d} \left(\frac{\epsilon}{\sqrt{d}}\right)^2} = \epsilon.$$

Hence,

$$C(\epsilon, T, \|\cdot\|_2) \leq |T_c| = m^d \leq \left(\frac{2B\sqrt{d}}{\epsilon}\right)^d.$$

A Lipschitz parameterization allows us to translate a cover of the function space into a cover of the parameter space. Such a property plays an essential role in our analysis of statistical error.

Lemma 4.9 *Let $\mathcal{F}$ be a parameterized class of functions: $\mathcal{F} = \{f(x;\theta) : \theta \in \Theta\}$. Let $\|\cdot\|_\Theta$ be a norm on Θ and let $\|\cdot\|_{\mathcal{F}}$ be a norm on $\mathcal{F}$. Suppose that the mapping $\theta \mapsto f(x;\theta)$ is L-Lipschitz, that is,*

$$\left\|f(x;\theta) - f\left(x;\widetilde{\theta}\right)\right\|_{\mathcal{F}} \leq L\left\|\theta - \widetilde{\theta}\right\|_\Theta,$$

then for any $\epsilon > 0$, $C\left(\epsilon, \mathcal{F}, \|\cdot\|_{\mathcal{F}}\right) \leq C\left(\epsilon/L, \Theta, \|\cdot\|_\Theta\right)$.

Proof Suppose that $C\left(\epsilon/L, \Theta, \|\cdot\|_\Theta\right) = n$ and $\{\theta_i\}_{i=1}^n$ is an ϵ/L-cover of Θ. Then, for any $\theta \in \Theta$, there exists $1 \leq i \leq n$ such that

$$\|f(x;\theta) - f(x;\theta_i)\|_{\mathcal{F}} \leq L\|\theta - \theta_i\|_\Theta \leq \epsilon.$$

Hence, $\{f(x;\theta_i)\}_{i=1}^n$ is an ϵ-cover of $\mathcal{F}$, implying that $C\left(\epsilon, \mathcal{F}, \|\cdot\|_{\mathcal{F}}\right) \leq n$. □

To find the relation between the Rademacher complexity and covering number, we first need Massart's finite class lemma stated below.

Lemma 4.10 *For any finite set $A \subset \mathbb{R}^N$ with diameter $D = \sup_{a\in A}\|a\|_2$,*

$$\mathfrak{R}_N(A) \leq \frac{D}{N}\sqrt{2\log|A|}.$$

Proof See, for example, [27, Lemma 26.8]. □

Lemma 4.11 *Let $\mathcal{F}$ be a function class and $\|f\|_\infty \leq B$ for any $f \in \mathcal{F}$, and we have*

$$\mathfrak{R}_N(\mathcal{F}) \leq \inf_{0<\delta<B/2} \left(4\delta + \frac{12}{\sqrt{N}} \int_\delta^{B/2} \sqrt{\log C(\epsilon, \mathcal{F}, \|\cdot\|_\infty)} d\epsilon \right).$$

This lemma is a classical result in statistical machine learning. The basic idea behind it is chaining. The readers can find more details in [28]. We provide the proof below for completeness.

Proof We apply the chaining method. Set $\epsilon_k = 2^{-k+1}B$. Let $\mathcal{F}_k$ be an ϵ_k-cover of $\mathcal{F}$ and $|\mathcal{F}_k| = C(\epsilon_k, \mathcal{F}, \|\cdot\|_\infty)$. Hence, for any $u \in \mathcal{F}$, there exists $u_k \in \mathcal{F}_k$ such that $\|u - u_k\|_\infty \leq \epsilon_k$. Let K be a positive integer determined later. We have

$$\begin{aligned}
\mathfrak{R}_N(\mathcal{F}) &= \mathbb{E}_{\{\sigma_i, X_i\}_{i=1}^N} \left[\sup_{u \in \mathcal{F}} \frac{1}{N} \sum_{i=1}^N \sigma_i u(X_i) \right] \\
&= \mathbb{E}_{\{\sigma_i, X_i\}_{i=1}^N} \left[\frac{1}{N} \sup_{u \in \mathcal{F}} \sum_{i=1}^N \sigma_i \left(u(X_i) - u_K(X_i) \right) \right. \\
&\quad \left. + \sum_{j=1}^{K-1} \sum_{i=1}^N \sigma_i \left(u_{j+1}(X_i) - u_j(X_i) \right) + \sum_{i=1}^N \sigma_i u_1(X_i) \right] \\
&\leq \mathbb{E}_{\{\sigma_i, X_i\}_{i=1}^N} \left[\sup_{u \in \mathcal{F}} \frac{1}{N} \sum_{i=1}^N \sigma_i \left(u(X_i) - u_K(X_i) \right) \right] \\
&\quad + \sum_{j=1}^{K-1} \mathbb{E}_{\{\sigma_i, X_i\}_{i=1}^N} \left[\sup_{u \in \mathcal{F}} \frac{1}{N} \sum_{i=1}^N \sigma_i \left(u_{j+1}(X_i) - u_j(X_i) \right) \right] \\
&\quad + \mathbb{E}_{\{\sigma_i, X_i\}_{i=1}^N} \left[\frac{1}{N} \sup_{u \in \mathcal{F}_1} \frac{1}{N} \sum_{i=1}^N \sigma_i u(X_i) \right].
\end{aligned}$$

We can choose $\mathcal{F}_1 = \{0\}$ to eliminate the third term. For the first term,

$$\begin{aligned}
&\mathbb{E}_{\{\sigma_i, X_i\}_{i=1}^N} \sup_{u \in \mathcal{F}} \frac{1}{N} \left[\sum_{i=1}^N \sigma_i \left(u(X_i) - u_K(X_i) \right) \right] \\
&\leq \mathbb{E}_{\{\sigma_i, X_i\}_{i=1}^N} \sup_{u \in \mathcal{F}} \frac{1}{N} \sum_{i=1}^N |\sigma_i| \, \|u - u_K\|_\infty \leq \epsilon_K.
\end{aligned}$$

For the second term, for any fixed samples $\{X_i\}_{i=1}^N$, we define

$$V_j := \{(u_{j+1}(X_1) - u_j(X_1), \ldots, u_{j+1}(X_N) - u_j(X_N)) \in \mathbb{R}^N : u \in \mathcal{F}\}.$$

Then, for any $v^j \in V_j$,

$$\begin{aligned}\|v^j\|_2 &= \left(\sum_{i=1}^n \left|u_{j+1}(X_i) - u_j(X_i)\right|^2\right)^{1/2} \leq \sqrt{n}\left\|u_{j+1} - u_j\right\|_\infty \\ &\leq \sqrt{n}\left\|u_{j+1} - u\right\|_\infty + \sqrt{n}\left\|u_j - u\right\|_\infty = \sqrt{n}\epsilon_{j+1} + \sqrt{n}\epsilon_j = 3\sqrt{n}\epsilon_{j+1}.\end{aligned}$$

Applying Lemma 4.10, we have

$$\begin{aligned}&\sum_{j=1}^{K-1} \mathbb{E}_{\{\sigma\}_{i=1}^N}\left[\sup_{u\in\mathcal{F}} \frac{1}{N}\sum_{i=1}^N \sigma_i\left(u_{j+1}(X_i) - u_j(X_i)\right)\right] \\ &= \sum_{j=1}^{K-1} \mathbb{E}_{\{\sigma_i\}_{i=1}^N}\left[\sup_{v^j\in V_j} \frac{1}{N}\sum_{i=1}^N \sigma_i v_i^j\right] \leq \sum_{j=1}^{K-1} \frac{3\epsilon_{j+1}}{\sqrt{N}}\sqrt{2\log|V_j|}.\end{aligned}$$

By the definition of V_j, we know that $\left|V_j\right| \leq \left|\mathcal{F}_j\right|\left|\mathcal{F}_{j+1}\right| \leq \left|\mathcal{F}_{j+1}\right|^2$. Hence,

$$\sum_{j=1}^{K-1} \mathbb{E}_{\{\sigma_i, X_i\}_{i=1}^N}\left[\sup_{u\in\mathcal{F}} \frac{1}{N}\sum_{i=1}^N \sigma_i\left(u_{j+1}(X_i) - u_j(X_i)\right)\right] \leq \sum_{j=1}^{K-1} \frac{6\epsilon_{j+1}}{\sqrt{N}}\sqrt{\log\left|\mathcal{F}_{j+1}\right|}.$$

Now, we obtain

$$\begin{aligned}\mathfrak{R}_N(\mathcal{F}) &\leq \epsilon_K + \sum_{j=1}^{K-1} \frac{6\epsilon_{j+1}}{\sqrt{N}}\sqrt{\log\left|\mathcal{F}_{j+1}\right|} \\ &= \epsilon_K + \frac{12}{\sqrt{N}}\sum_{j=1}^{K-1}(\epsilon_{j+1} - \epsilon_{j+2})\sqrt{\log C(\epsilon_{j+1}, \mathcal{F}, \|\cdot\|_\infty)} \\ &\leq \epsilon_K + \frac{12}{\sqrt{N}}\int_{\epsilon_{K+1}}^{B/2} \sqrt{\log C(\epsilon, \mathcal{F}, \|\cdot\|_\infty)}d\epsilon.\end{aligned}$$

We conclude the lemma by choosing K such that $\epsilon_{K+2} < \delta \leq \epsilon_{K+1}$ for any $0 < \delta < B/2$. □

From Lemma 4.9, we know that the key step to bound $C(\epsilon, \mathcal{F}_i, \|\cdot\|_\infty)$ with $\mathcal{F}_i$ defined in Lemma 4.7 is to compute the upper bound of Lipschitz constant of class $\mathcal{F}_i$, which is done in Lemmas 4.12–4.15.

Lemma 4.12 *Let $\mathcal{D}, \mathfrak{n}_{\mathcal{D}}, n_i \in \mathbb{N}^+$, $n_{\mathcal{D}} = 1$, $B_\theta \geq 1$, and ρ be a bounded Lipschitz continuous function with $B_\rho, L_\rho \leq 1$. Assume that the parameterized function class $\mathcal{P} \subset \mathcal{N}_\rho\left(\mathcal{D}, \mathfrak{n}_{\mathcal{D}}, B_\theta\right)$. For any $f(x;\theta) \in \mathcal{P}$, $f(x;\theta)$ is $\sqrt{\mathfrak{n}_{\mathcal{D}}} B_\theta^{\mathcal{D}-1}\left(\prod_{i=1}^{\mathcal{D}-1} n_i\right)$-Lipschitz continuous with respect to variable θ, i.e.,*

$$\left|f(x;\theta) - f(x;\widetilde{\theta})\right| \leq \sqrt{\mathfrak{n}_{\mathcal{D}}} B_\theta^{\mathcal{D}-1}\left(\prod_{i=1}^{\mathcal{D}-1} n_i\right) \left\|\theta - \widetilde{\theta}\right\|_2, \quad \forall x \in \Omega.$$

Proof For $\ell = 2, \cdots, \mathcal{D}$ (the argument for the case of $\ell = \mathcal{D}$ is slightly different),

$$\begin{aligned}
\left|f_q^{(\ell)} - \widetilde{f}_q^{(\ell)}\right| &= \left|\rho\left(\sum_{j=1}^{n_{\ell-1}} a_{qj}^{(\ell)} f_j^{(\ell-1)} + b_q^{(\ell)}\right) - \rho\left(\sum_{j=1}^{n_{\ell-1}} \widetilde{a}_{qj}^{(\ell)} \widetilde{f}_j^{(\ell-1)} + \widetilde{b}_q^{(\ell)}\right)\right| \\
&\leq L_\rho \left|\sum_{j=1}^{n_{\ell-1}} a_{qj}^{(\ell)} f_j^{(\ell-1)} - \sum_{j=1}^{n_{\ell-1}} \widetilde{a}_{qj}^{(\ell)} \widetilde{f}_j^{(\ell-1)} + b_q^{(\ell)} - \widetilde{b}_q^{(\ell)}\right| \\
&\leq L_\rho \sum_{j=1}^{n_{\ell-1}} \left|a_{qj}^{(\ell)}\right| \left|f_j^{(\ell-1)} - \widetilde{f}_j^{(\ell-1)}\right| + L_\rho \sum_{j=1}^{n_{\ell-1}} \left|a_{qj}^{(\ell)} - \widetilde{a}_{qj}^{(\ell)}\right| \left|\widetilde{f}_j^{(\ell-1)}\right| \\
&\quad + L_\rho \left|b_q^{(\ell)} - \widetilde{b}_q^{(\ell)}\right| \\
&\leq B_\theta L_\rho \sum_{j=1}^{n_{\ell-1}} \left|f_j^{(\ell-1)} - \widetilde{f}_j^{(\ell-1)}\right| + B_\rho L_\rho \sum_{j=1}^{n_{\ell-1}} \left|a_{qj}^{(\ell)} - \widetilde{a}_{qj}^{(\ell)}\right| \\
&\quad + L_\rho \left|b_q^{(\ell)} - \widetilde{b}_q^{(\ell)}\right| \\
&\leq B_\theta \sum_{j=1}^{n_{\ell-1}} \left|f_j^{(\ell-1)} - \widetilde{f}_j^{(\ell-1)}\right| + \sum_{j=1}^{n_{\ell-1}} \left|a_{qj}^{(\ell)} - \widetilde{a}_{qj}^{(\ell)}\right| + \left|b_q^{(\ell)} - \widetilde{b}_q^{(\ell)}\right|.
\end{aligned}$$

For $\ell = 1$,

$$\begin{aligned}
\left|f_q^{(1)} - \widetilde{f}_q^{(1)}\right| &= \left|\rho\left(\sum_{j=1}^{n_0} a_{qj}^{(1)} x_j + b_q^{(1)}\right) - \rho\left(\sum_{j=1}^{n_0} \widetilde{a}_{qj}^{(1)} x_j + \widetilde{b}_q^{(1)}\right)\right| \\
&\leq \sum_{j=1}^{n_0} \left|a_{qj}^{(1)} - \widetilde{a}_{qj}^{(1)}\right| + \left|b_q^{(1)} - \widetilde{b}_q^{(1)}\right| = \sum_{j=1}^{\mathfrak{n}_1} \left|\theta_j - \widetilde{\theta}_j\right|.
\end{aligned}$$

For $\ell = 2$,

$$\begin{aligned}\left|f_q^{(2)}-\widetilde{f}_q^{(2)}\right| &\le B_\theta \sum_{j=1}^{n_1}\left|f_j^{(1)}-\widetilde{f}_j^{(1)}\right| + \sum_{j=1}^{n_1}\left|a_{qj}^{(2)}-\widetilde{a}_{qj}^{(2)}\right| + \left|b_q^{(2)}-\widetilde{b}_q^{(2)}\right| \\ &\le B_\theta \sum_{j=1}^{n_1}\sum_{k=1}^{\mathfrak{n}_1}\left|\theta_k-\widetilde{\theta}_k\right| + \sum_{j=1}^{n_1}\left|a_{qj}^{(2)}-\widetilde{a}_{qj}^{(2)}\right| + \left|b_q^{(2)}-\widetilde{b}_q^{(2)}\right| \\ &\le n_1 B_\theta \sum_{j=1}^{\mathfrak{n}_2}\left|\theta_j-\widetilde{\theta}_j\right|.\end{aligned}$$

Assuming that for $\ell \ge 2$,

$$\left|f_q^{(\ell)}-\widetilde{f}_q^{(\ell)}\right| \le \left(\prod_{i=1}^{\ell-1} n_i\right) B_\theta^{\ell-1}\sum_{j=1}^{\mathfrak{n}_\ell}\left|\theta_j-\widetilde{\theta}_j\right|,$$

we have

$$\begin{aligned}\left|f_q^{(\ell+1)}-\widetilde{f}_q^{(\ell+1)}\right| &\le B_\theta \sum_{j=1}^{n_\ell}\left|f_j^{(\ell)}-\widetilde{f}_j^{(\ell)}\right| + \sum_{j=1}^{n_\ell}\left|a_{qj}^{(\ell+1)}-\widetilde{a}_{qj}^{(\ell+1)}\right| \\ &\quad + \left|b_q^{(\ell+1)}-\widetilde{b}_q^{(\ell+1)}\right| \\ &\le B_\theta \sum_{j=1}^{n_\ell}\left(\prod_{i=1}^{\ell-1} n_i\right) B_\theta^{\ell-1}\sum_{k=1}^{\mathfrak{n}_1}\left|\theta_k-\widetilde{\theta}_k\right| + \sum_{j=1}^{n_\ell}\left|a_{qj}^{(\ell+1)}-\widetilde{a}_{qj}^{(\ell+1)}\right| \\ &\quad + \left|b_q^{(\ell+1)}-\widetilde{b}_q^{(\ell+1)}\right| \\ &\le \left(\prod_{i=1}^{\ell} n_i\right) B_\theta^{\ell}\sum_{j=1}^{\mathfrak{n}_{\ell+1}}\left|\theta_j-\widetilde{\theta}_j\right|.\end{aligned}$$

Hence, by induction and the Hölder inequality, we conclude that

$$|f-\widetilde{f}| \le \left(\prod_{i=1}^{\mathcal{D}-1} n_i\right) B_\theta^{\mathcal{D}-1}\sum_{j=1}^{\mathfrak{n}_\mathcal{D}}\left|\theta_j-\widetilde{\theta}_j\right| \le \sqrt{\mathfrak{n}_\mathcal{D}}B_\theta^{\mathcal{D}-1}\left(\prod_{i=1}^{\mathcal{D}-1} n_i\right)\left\|\theta-\widetilde{\theta}\right\|_2.$$

Lemma 4.13 *Let* $\mathcal{D}, \mathfrak{n}_\mathcal{D}, n_i \in \mathbb{N}^+$, $n_\mathcal{D} = 1$, $B_\theta \ge 1$, *and* ρ *be a function such that* ρ' *is bounded by* $B_{\rho'}$. *Assume that the parameterized function class* $\mathcal{P} \subset \mathcal{N}_\rho\left(\mathcal{D}, \mathfrak{n}_\mathcal{D}, B_\theta\right)$. *Let* $p = 1, \cdots, d$. *We have*

$$\left|\partial_{x_p} f_q^{(\ell)}\right| \le \left(\prod_{i=1}^{\ell-1} n_i\right) \left(B_\theta B_{\rho'}\right)^\ell, \quad \ell = 1, 2, \cdots, \mathcal{D}-1,$$

$$\left|\partial_{x_p} f\right| \le \left(\prod_{i=1}^{\mathcal{D}-1} n_i\right) B_\theta^{\mathcal{D}} B_{\rho'}^{\mathcal{D}-1}.$$

Proof For $\ell = 1, 2, \cdots, \mathcal{D}-1$,

$$\begin{aligned}
\left|\partial_{x_p} f_q^{(\ell)}\right| &= \left|\sum_{j=1}^{n_{\ell-1}} a_{qj}^{(\ell)} \partial_{x_p} f_j^{(\ell-1)} \rho'\left(\sum_{j=1}^{n_{\ell-1}} a_{qj}^{(\ell)} f_j^{(\ell-1)} + b_q^{(\ell)}\right)\right| \\
&\le B_\theta B_{\rho'} \sum_{j=1}^{n_{\ell-1}} \left|\partial_{x_p} f_j^{(\ell-1)}\right| \\
&\le \left(B_\theta B_{\rho'}\right)^2 \sum_{k=1}^{n_{\ell-1}} \sum_{j=1}^{n_{\ell-2}} \left|\partial_{x_p} f_j^{(\ell-2)}\right| = n_{\ell-1} \left(B_\theta B_{\rho'}\right)^2 \sum_{j=1}^{n_{\ell-2}} \left|\partial_{x_p} f_j^{(\ell-2)}\right| \\
&\le \cdots \le \left(\prod_{i=2}^{\ell-1} n_i\right) \left(B_\theta B_{\rho'}\right)^{\ell-1} \sum_{j=1}^{n_1} \left|\partial_{x_p} f_j^{(1)}\right| \\
&\le \left(\prod_{i=2}^{\ell-1} n_i\right) \left(B_\theta B_{\rho'}\right)^{\ell-1} \sum_{j=1}^{n_1} B_\theta B_{\rho'} = \left(\prod_{i=1}^{\ell-1} n_i\right) \left(B_\theta B_{\rho'}\right)^\ell.
\end{aligned}$$

The bound for $\left|\partial_{x_p} f\right|$ can be derived similarly. □

Lemma 4.14 *Let $\mathcal{D}, \mathfrak{n}_{\mathcal{D}}, n_i \in \mathbb{N}^+$, $n_{\mathcal{D}} = 1$, $B_\theta \ge 1$, and ρ be a function such that ρ, ρ' are bounded by $B_\rho, B_{\rho'} \le 1$ and have Lipschitz constants $L_\rho, L_{\rho'} \le 1$, respectively. Assume that the parameterized function class $\mathcal{P} \subset \mathcal{N}_\rho\left(\mathcal{D}, \mathfrak{n}_{\mathcal{D}}, B_\theta\right)$. Then, for any $f(x;\theta) \in \mathcal{P}$, $p = 1, \cdots, d$, $\partial_{x_p} f(x;\theta)$ is $\sqrt{\mathfrak{n}_{\mathcal{D}}}(\mathcal{D}+1) B_\theta^{2\mathcal{D}} \left(\prod_{i=1}^{\mathcal{D}-1} n_i\right)^2$-Lipschitz continuous with respect to variable θ, i.e.,*

$$\left|\partial_{x_p} f(x;\theta) - \partial_{x_p} f(x;\widetilde{\theta})\right| \le \sqrt{\mathfrak{n}_{\mathcal{D}}}(\mathcal{D}+1) B_\theta^{2\mathcal{D}} \left(\prod_{i=1}^{\mathcal{D}-1} n_i\right)^2 \left\|\theta - \widetilde{\theta}\right\|_2, \quad \forall x \in \Omega.$$

Proof For $\ell = 1$,

$$
\begin{aligned}
&\left|\partial_{x_p} f_q^{(1)} - \partial_{x_p} \widetilde{f}_q^{(1)}\right| \\
=&\left|a_{qp}^{(1)} \rho'\left(\sum_{j=1}^{n_0} a_{qj}^{(1)} x_j + b_q^{(1)}\right) - \widetilde{a}_{qp}^{(1)} \rho'\left(\sum_{j=1}^{n_0} \widetilde{a}_{qj}^{(1)} x_j + \widetilde{b}_q^{(1)}\right)\right| \\
\leq&\left|a_{qp}^{(1)} - \widetilde{a}_{qp}^{(1)}\right| \left|\rho'\left(\sum_{j=1}^{n_0} a_{qj}^{(1)} x_j + b_q^{(1)}\right)\right| + \left|\widetilde{a}_{qp}^{(1)}\right| \left|\rho'\left(\sum_{j=1}^{n_0} a_{qj}^{(1)} x_j + b_q^{(1)}\right)\right. \\
&\left. - \rho'\left(\sum_{j=1}^{n_0} \widetilde{a}_{qj}^{(1)} x_j + \widetilde{b}_q^{(1)}\right)\right| \\
\leq& B_{\rho'}\left|a_{qp}^{(1)} - \widetilde{a}_{qp}^{(1)}\right| + B_\theta L_{\rho'} \sum_{j=1}^{n_0}\left|a_{qj}^{(1)} - \widetilde{a}_{qj}^{(1)}\right| + B_\theta L_{\rho'}\left|b_q^{(1)} - \widetilde{b}_q^{(1)}\right| \\
\leq&\ 2B_\theta \sum_{k=1}^{\mathfrak{n}_1}\left|\theta_k - \widetilde{\theta}_k\right|.
\end{aligned}
$$

For $\ell \geq 2$, we establish the recurrence relation:

$$
\begin{aligned}
&\left|\partial_{x_p} f_q^{(\ell)} - \partial_{x_p} \widetilde{f}_q^{(\ell)}\right| \\
&\leq \sum_{j=1}^{n_{\ell-1}}\left|a_{qj}^{(\ell)}\right| \\
&\left|\partial_{x_p} f_j^{(\ell-1)}\right| \left|\rho'\left(\sum_{j=1}^{n_{\ell-1}} a_{qj}^{(\ell)} f_j^{(\ell-1)} + b_q^{(\ell)}\right) - \rho'\left(\sum_{j=1}^{n_{\ell-1}} \widetilde{a}_{qj}^{(\ell)} \widetilde{f}_j^{(\ell-1)} + \widetilde{b}_q^{(\ell)}\right)\right| \\
&\quad + \sum_{j=1}^{n_{\ell-1}}\left|a_{qj}^{(\ell)} \partial_{x_p} f_j^{(\ell-1)} - \widetilde{a}_{qj}^{(\ell)} \partial_{x_p} \widetilde{f}_j^{(\ell-1)}\right| \left|\rho'\left(\sum_{j=1}^{n_{\ell-1}} \widetilde{a}_{qj}^{(\ell)} \widetilde{f}_j^{(\ell-1)} + \widetilde{b}_q^{(\ell)}\right)\right| \\
&\leq B_\theta L_{\rho'} \sum_{j=1}^{n_{\ell-1}}\left|\partial_{x_p} f_j^{(\ell-1)}\right| \\
&\left(\sum_{j=1}^{n_{\ell-1}}\left|a_{qj}^{(\ell)} f_j^{(\ell-1)} - \widetilde{a}_{qj}^{(\ell)} \widetilde{f}_j^{(\ell-1)}\right| + \left|b_q^{(\ell)} - \widetilde{b}_q^{(\ell)}\right|\right)
\end{aligned}
$$

$$
\begin{aligned}
&+ B_{\rho'} \sum_{j=1}^{n_{\ell-1}} \left| a_{qj}^{(\ell)} \partial_{x_p} f_j^{(\ell-1)} - \widetilde{a}_{qj}^{(\ell)} \partial_{x_p} \widetilde{f}_j^{(\ell-1)} \right| \\
&\le B_\theta L_{\rho'} \sum_{j=1}^{n_{\ell-1}} \left| \partial_{x_p} f_j^{(\ell-1)} \right| \\
&\left(B_\rho \sum_{j=1}^{n_{\ell-1}} \left| a_{qj}^{(\ell)} - \widetilde{a}_{qj}^{(\ell)} \right| + B_\theta \sum_{j=1}^{n_{\ell-1}} \left| f_j^{(\ell-1)} - \widetilde{f}_j^{(\ell-1)} \right| + \left| b_q^{(\ell)} - \widetilde{b}_q^{(\ell)} \right| \right) \\
&+ B_{\rho'} B_\theta \sum_{j=1}^{n_{\ell-1}} \left| \partial_{x_p} f_j^{(\ell-1)} - \partial_{x_p} \widetilde{f}_j^{(\ell-1)} \right| + B_{\rho'} \sum_{j=1}^{n_{\ell-1}} \left| a_{qj}^{(\ell)} - \widetilde{a}_{qj}^{(\ell)} \right| \left| \partial_{x_p} \widetilde{f}_j^{(\ell-1)} \right| \\
&\le B_\theta \sum_{j=1}^{n_{\ell-1}} \left| \partial_{x_p} f_j^{(\ell-1)} \right| \left(\sum_{j=1}^{n_{\ell-1}} \left| a_{qj}^{(\ell)} - \widetilde{a}_{qj}^{(\ell)} \right| + B_\theta \sum_{j=1}^{n_{\ell-1}} \left| f_j^{(\ell-1)} - \widetilde{f}_j^{(\ell-1)} \right| + \left| b_q^{(\ell)} - \widetilde{b}_q^{(\ell)} \right| \right) \\
&+ B_\theta \sum_{j=1}^{n_{\ell-1}} \left| \partial_{x_p} f_j^{(\ell-1)} - \partial_{x_p} \widetilde{f}_j^{(\ell-1)} \right| + \sum_{j=1}^{n_{\ell-1}} \left| a_{qj}^{(\ell)} - \widetilde{a}_{qj}^{(\ell)} \right| \left| \partial_{x_p} \widetilde{f}_j^{(\ell-1)} \right| \\
&\le B_\theta \left(\prod_{i=1}^{\ell-1} n_i \right) B_\theta^\ell \\
&\left(\sum_{j=1}^{n_{\ell-1}} \left| a_{qj}^{(\ell)} - \widetilde{a}_{qj}^{(\ell)} \right| + B_\theta \sum_{j=1}^{n_{\ell-1}} \left(\prod_{i=1}^{\ell-2} n_i \right) B_\theta^{\ell-2} \sum_{k=1}^{n_{\ell-1}} \left| \theta_k - \widetilde{\theta}_k \right| + \left| b_q^{(\ell)} - \widetilde{b}_q^{(\ell)} \right| \right) \\
&+ B_\theta \sum_{j=1}^{n_{\ell-1}} \left| \partial_{x_p} f_j^{(\ell-1)} - \partial_{x_p} \widetilde{f}_j^{(\ell-1)} \right| + \sum_{j=1}^{n_{\ell-1}} \left| a_{qj}^{(\ell)} - \widetilde{a}_{qj}^{(\ell)} \right| \left(\prod_{i=1}^{\ell-2} n_i \right) B_\theta^{\ell-1} \\
&\le B_\theta \sum_{j=1}^{n_{\ell-1}} \left| \partial_{x_p} f_j^{(\ell-1)} - \partial_{x_p} \widetilde{f}_j^{(\ell-1)} \right| + B_\theta^{2\ell} \left(\prod_{i=1}^{\ell-1} n_i \right)^2 \sum_{k=1}^{n_\ell} \left| \theta_k - \widetilde{\theta}_k \right|.
\end{aligned}
$$

For $\ell = 2$,

$$
\begin{aligned}
\left| \partial_{x_p} f_q^{(2)} - \partial_{x_p} \widetilde{f}_q^{(2)} \right| &\le B_\theta \sum_{j=1}^{n_1} \left| \partial_{x_p} f_j^{(1)} - \partial_{x_p} \widetilde{f}_j^{(1)} \right| + B_\theta^4 n_1^2 \sum_{k=1}^{n_2} \left| \theta_k - \widetilde{\theta}_k \right| \\
&\le 2 B_\theta^2 n_1 \sum_{k=1}^{n_1} \left| \theta_k - \widetilde{\theta}_k \right| + B_\theta^4 n_1^2 \sum_{k=1}^{n_2} \left| \theta_k - \widetilde{\theta}_k \right| \\
&\le 3 B_\theta^4 n_1^2 \sum_{k=1}^{n_2} \left| \theta_k - \widetilde{\theta}_k \right|.
\end{aligned}
$$

Assuming that for $\ell \geq 2$,

$$\left|\partial_{x_p} f_q^{(\ell)} - \partial_{x_p} \widetilde{f}_q^{(\ell)}\right| \leq (\ell+1) B_\theta^{2\ell} \left(\prod_{i=1}^{\ell-1} n_i\right)^2 \sum_{k=1}^{\mathfrak{n}_\ell} \left|\theta_k - \widetilde{\theta}_k\right|,$$

we have

$$\begin{aligned}
&\left|\partial_{x_p} f_q^{(\ell+1)} - \partial_{x_p} \widetilde{f}_q^{(\ell+1)}\right| \\
\leq & B_\theta \sum_{j=1}^{n_\ell} \left|\partial_{x_p} f_j^{(\ell)} - \partial_{x_p} \widetilde{f}_j^{(\ell)}\right| + B_\theta^{2\ell+2} \left(\prod_{i=1}^{\ell} n_i\right)^2 \sum_{k=1}^{\mathfrak{n}_{\ell+1}} \left|\theta_k - \widetilde{\theta}_k\right| \\
\leq & B_\theta \sum_{j=1}^{n_\ell} (\ell+1) B_\theta^{2\ell} \left(\prod_{i=1}^{\ell-1} n_i\right)^2 \sum_{k=1}^{\mathfrak{n}_\ell} \left|\theta_k - \widetilde{\theta}_k\right| + B_\theta^{2\ell+2} \left(\prod_{i=1}^{\ell} n_i\right)^2 \sum_{k=1}^{\mathfrak{n}_{\ell+1}} \left|\theta_k - \widetilde{\theta}_k\right| \\
\leq & (\ell+2) B_\theta^{2\ell+2} \left(\prod_{i=1}^{\ell} n_i\right)^2 \sum_{k=1}^{\mathfrak{n}_{\ell+1}} \left|\theta_k - \widetilde{\theta}_k\right|.
\end{aligned}$$

Hence, by induction and the Hölder inequality, we conclude that

$$\begin{aligned}
\left|\partial_{x_p} f - \partial_{x_p} \widetilde{f}\right| &\leq (\mathcal{D}+1) B_\theta^{2\mathcal{D}} \left(\prod_{i=1}^{\mathcal{D}-1} n_i\right)^2 \sum_{k=1}^{\mathfrak{n}_{\mathcal{D}}} \left|\theta_k - \widetilde{\theta}_k\right| \\
&\leq \sqrt{\mathfrak{n}_{\mathcal{D}}} (\mathcal{D}+1) B_\theta^{2\mathcal{D}} \left(\prod_{i=1}^{\mathcal{D}-1} n_i\right)^2 \left\|\theta - \widetilde{\theta}\right\|_2.
\end{aligned}$$

Lemma 4.15 *Let $\mathcal{D}, \mathfrak{n}_{\mathcal{D}}, n_i \in \mathbb{N}^+$, $n_{\mathcal{D}} = 1$, $B_\theta \geq 1$, and ρ be a function such that ρ, ρ' are bounded by $B_\rho, B_{\rho'} \leq 1$ and have Lipschitz constants $L_\rho, L_{\rho'} \leq 1$, respectively. Assume that the parameterized function class $\mathcal{P} \subset N_\rho\left(\mathcal{D}, \mathfrak{n}_{\mathcal{D}}, B_\theta\right)$. Then, $\mathcal{F}_1, \mathcal{F}_2, \mathcal{F}_3, \mathcal{F}_5$ are parameterized function classes with respect to parameter set $\Theta \times \Theta$ and $\mathcal{F}_4, \mathcal{F}_6$ are parameterized function classes with respect to parameter set Θ with $\Theta := \{\theta \in \mathbb{R}^{\mathfrak{n}_{\mathcal{D}}} : \|\theta\|_2 \leq B_\theta\}$. In addition, for any $f_i(x;\theta), f_i(x;\widetilde{\theta}) \in \mathcal{F}_i$, $i = 1, \cdots, 6$, we have*

$$\begin{aligned}
|f_i(x;\theta)| &\leq B_i, \quad \forall x \in \Omega, \\
|f_i(x;\theta) - f_i(x;\widetilde{\theta})| &\leq L_i \|\theta - \widetilde{\theta}\|_2, \quad \forall x \in \Omega,
\end{aligned}$$

with

$$B_1 = C(coe)d^2 B_\theta^{2\mathcal{D}} \left(\prod_{i=1}^{\mathcal{D}-1} n_i\right)^2, \quad B_2 = C(coe)d(n_{\mathcal{D}-1}+1)B_\theta^{\mathcal{D}+1}\left(\prod_{i=1}^{\mathcal{D}-1} n_i\right),$$

$$B_3 = C(coe)(n_{\mathcal{D}-1}+1)^2 B_\theta^2, \qquad B_4 = C(coe)(n_{\mathcal{D}-1}+1)B_\theta,$$

$$B_5 = \frac{\alpha}{2}(n_{\mathcal{D}-1}+1)^2 B_\theta^2, \qquad B_6 = C(coe)(n_{\mathcal{D}-1}+1)B_\theta$$

and

$$L_1 = C(coe)d^2\sqrt{2\mathfrak{n}_{\mathcal{D}}}(\mathcal{D}+1)B_\theta^{3\mathcal{D}}\left(\prod_{i=1}^{\mathcal{D}-1} n_i\right)^3,$$

$$L_2 = C(coe)d\sqrt{2\mathfrak{n}_{\mathcal{D}}}(\mathcal{D}+1)(n_{\mathcal{D}-1}+1)B_\theta^{2\mathcal{D}+1}\left(\prod_{i=1}^{\mathcal{D}-1} n_i\right)^2,$$

$$L_3 = C(coe)\sqrt{2\mathfrak{n}_{\mathcal{D}}}(n_{\mathcal{D}-1}+1)B_\theta^{\mathcal{D}}\left(\prod_{i=1}^{\mathcal{D}-1} n_i\right),$$

$$L_4 = C(coe)\sqrt{\mathfrak{n}_{\mathcal{D}}}B_\theta^{\mathcal{D}-1}\left(\prod_{i=1}^{\mathcal{D}-1} n_i\right),$$

$$L_5 = \frac{\alpha}{2}\sqrt{2\mathfrak{n}_{\mathcal{D}}}(n_{\mathcal{D}-1}+1)B_\theta^{\mathcal{D}}\left(\prod_{i=1}^{\mathcal{D}-1} n_i\right),$$

$$L_6 = C(coe)\sqrt{\mathfrak{n}_{\mathcal{D}}}B_\theta^{\mathcal{D}-1}\left(\prod_{i=1}^{\mathcal{D}-1} n_i\right).$$

Proof The result is a direct consequence of Lemmas 4.12, 4.13, and 4.14 and standard calculation. □

Now, we state our main result with respect to statistical error $\mathcal{E}_{sta}$.

Theorem 4.3 *Let $\mathcal{D}, \mathfrak{n}_{\mathcal{D}}, n_i \in \mathbb{N}^+$, $n_{\mathcal{D}} = 1$, $B_\theta \geq 1$, and ρ be a function such that ρ, ρ' are bounded by $B_\rho, B_{\rho'} \leq 1$ and have Lipschitz constants $L_\rho, L_{\rho'} \leq 1$, respectively. Assume that the parameterized function class $\mathcal{P} \subset \mathcal{N}_\rho\left(\mathcal{D}, \mathfrak{n}_{\mathcal{D}}, B_\theta\right)$. Then, if $N = M$, we have for $\mathcal{E}_{sta}$ defined in (4.13)*

$$\mathbb{E}_{\{X_i\}_{i=1}^N, \{Y_j\}_{j=1}^M} \mathcal{E}_{sta} \leq \frac{C(\Omega, coe, \alpha)}{\beta} \frac{d^3 \mathcal{D}^{\frac{1}{2}} \mathfrak{n}_{\mathcal{D}}^{\frac{7}{2}\mathcal{D}-\frac{3}{2}} B_\theta^{\frac{7}{2}\mathcal{D}+\frac{1}{2}}}{N^{\frac{1}{4}}}.$$

Proof From Lemmas 4.8, 4.9, and 4.11, we have

$$\begin{aligned}\mathfrak{R}_N(\mathcal{F}_i) &\le \inf_{0<\delta<B_i/2}\left(4\delta + \frac{12}{\sqrt{N}}\int_\delta^{B_i/2}\sqrt{\log C(\epsilon,\mathcal{F}_i,\|\cdot\|_\infty)}d\epsilon\right)\\
&\le \inf_{0<\delta<B_i/2}\left(4\delta + \frac{12}{\sqrt{N}}\int_\delta^{B_i/2}\sqrt{\mathfrak{n}_{\mathcal{D}}\log\left(\frac{2L_iB_\theta\sqrt{\mathfrak{n}_{\mathcal{D}}}}{\epsilon}\right)}d\epsilon\right)\\
&\le \inf_{0<\delta<B_i/2}\left(4\delta + \frac{6\sqrt{\mathfrak{n}_{\mathcal{D}}}B_i}{\sqrt{N}}\sqrt{\log\left(\frac{2L_iB_\theta\sqrt{\mathfrak{n}_{\mathcal{D}}}}{\delta}\right)}\right).\end{aligned}$$

Choosing $\delta = 1/\sqrt{N} < B_i/2$ and applying Lemma 4.15, we have for $i = 1, \cdots, 4$

$$\begin{aligned}\mathfrak{R}_N(\mathcal{F}_i) &\le \frac{4}{\sqrt{N}} + \frac{6\sqrt{\mathfrak{n}_{\mathcal{D}}}B_i}{\sqrt{N}}\sqrt{\log\left(2L_iB_\theta\sqrt{\mathfrak{n}_{\mathcal{D}}}\sqrt{N}\right)}\\
&\le \frac{C(coe,\alpha)}{\sqrt{N}}\cdot d^2\sqrt{\mathfrak{n}_{\mathcal{D}}}\left(\prod_{i=1}^{\mathcal{D}-1}n_i\right)^2 B_\theta^{2\mathcal{D}}\\
&\quad\sqrt{\log\left(d^2\mathfrak{n}_{\mathcal{D}}(\mathcal{D}+1)B_\theta^{3\mathcal{D}+1}\left(\prod_{i=1}^{\mathcal{D}-1}n_i\right)^3\sqrt{N}\right)}\\
&\le \frac{C(coe,\alpha)}{\sqrt{N}}\cdot d^2\mathfrak{n}_{\mathcal{D}}^{2\mathcal{D}-\frac{1}{2}}B_\theta^{2\mathcal{D}}\sqrt{\log\left(d^2\mathcal{D}\mathfrak{n}_{\mathcal{D}}^{3\mathcal{D}-2}B_\theta^{3\mathcal{D}+1}\sqrt{N}\right)}\\
&\le \frac{C(coe,\alpha)d^3\mathcal{D}^{\frac{1}{2}}\mathfrak{n}_{\mathcal{D}}^{\frac{7}{2}\mathcal{D}-\frac{3}{2}}B_\theta^{\frac{7}{2}\mathcal{D}+\frac{1}{2}}}{N^{\frac{1}{4}}}.\end{aligned}\tag{4.21}$$

Similarly, for $i = 5, 6$,

$$\mathfrak{R}_M(\mathcal{F}_i) \le \frac{C(coe,\alpha)d^3\mathcal{D}^{\frac{1}{2}}\mathfrak{n}_{\mathcal{D}}^{\frac{7}{2}\mathcal{D}-\frac{3}{2}}B_\theta^{\frac{7}{2}\mathcal{D}+\frac{1}{2}}}{M^{\frac{1}{4}}}.\tag{4.22}$$

Combining Lemmas 4.6 and 4.7, (4.21), and (4.22), we obtain, if $N = M$,

$$\mathbb{E}_{\{X_i\}_{i=1}^N,\{Y_j\}_{j=1}^M}\mathcal{E}_{sta} \le \frac{C(\Omega,coe,\alpha)}{\beta}\frac{d^3\mathcal{D}^{\frac{1}{2}}\mathfrak{n}_{\mathcal{D}}^{\frac{7}{2}\mathcal{D}-\frac{3}{2}}B_\theta^{\frac{7}{2}\mathcal{D}+\frac{1}{2}}}{N^{\frac{1}{4}}}.$$

4.5 Convergence Rate for the Galerkin Method

Now, we state our main result.

Theorem 4.4 *Let (A1)–(A3) hold. Assume that* $\mathcal{E}_{opt} = 0$. *Let* ρ *be logistic function* $\frac{1}{1+e^{-x}}$ *or tanh function* $\frac{e^x-e^{-x}}{e^x+e^{-x}}$. *Let* $u_{\phi_{\mathcal{A}}}$ *be the solution of problem* (4.11) *generated by a random solver* $\mathcal{A}$.

(1) Let u_R *be the weak solution of the Robin problem* (4.2), (4.3c). *Assume that* $\epsilon > 0$ *is sufficiently small. Set the parameterized function class*

$$\mathcal{P}:=\mathcal{N}_\rho\left(C\log(d+1), C(d, coe, \beta)\epsilon^{\frac{-d}{1-\mu}}, C(d, coe, \beta)\epsilon^{\frac{-9d-8}{2-2\mu}}\right)\bigcap B_{H^1(\Omega)}(0,2),$$

where $\mu > 0$ *can be any arbitrarily small number and* $B_{H^1(\Omega)}(0,2) := \{f \in H^1(\Omega) : \|f\|_{H^1(\Omega)} \le 2\}$. *Set the number of samples*

$$N = M = C(d, \Omega, coe, \alpha, \beta)\epsilon^{-Cd\log d},$$

then

$$\mathbb{E}_{\{X_i\}_{i=1}^N,\{Y_j\}_{j=1}^M}\|u_{\phi_{\mathcal{A}}} - u_R\|_{H^1(\Omega)} \le \epsilon.$$

(2) Let u_D *be the weak solution of the Dirichlet problem* (4.2)(4.3a). *Set* $\alpha = 1, g = 0$. *Assume that* $\epsilon > 0$ *is sufficiently small. Set the penalty parameter* $\beta = C(d, \Omega, coe)\epsilon^2$. *Set the parameterized function class*

$$\mathcal{P} := \mathcal{N}_\rho\left(C\log(d+1), C(d, coe)\epsilon^{\frac{-d}{1-\mu}}, C(d, coe)\epsilon^{\frac{-9d-8}{2-2\mu}}\right)\bigcap B_{H^1(\Omega)}(0,2)$$

where $\mu > 0$ *can be any arbitrarily small number and* $B_{H^1(\Omega)}(0,2) := \{f \in H^1(\Omega) : \|f\|_{H^1(\Omega)} \le 2\}$. *Set the number of samples*

$$N = M = C(d, \Omega, coe)\epsilon^{-Cd\log d},$$

then

$$\mathbb{E}_{\{X_i\}_{i=1}^N,\{Y_j\}_{j=1}^M}\|u_{\phi_{\mathcal{A}}} - u_D\|_{H^1(\Omega)} \le \epsilon.$$

Proof We set the approximation error $\mathcal{E}_{app}$ as $\frac{\epsilon}{2}$ in Theorem 4.2 and the statistical error $\mathcal{E}_{sta}$ as $\frac{\epsilon}{2}$ in Theorem 4.3. Combining Proposition 4.1, Theorem 4.2, and Theorem 4.3 yields (1).

We set the approximation error $\mathcal{E}_{app}$ as $\frac{\epsilon}{3}$ in Theorem 4.2, the statistical error $\mathcal{E}_{sta}$ as $\frac{\epsilon}{3}$ in Theorem 4.3, and the penalty error $\mathcal{E}_{pen}$ as $\frac{\epsilon}{3}$ in Lemma 4.4. Combining Proposition 4.1, Theorem 4.2, Theorem 4.3, and Lemma 4.4 yields (2). □

4.6 Conclusions and Extensions

This chapter analyzes the convergence rate of the deep Galerkin method (DGMW) for second-order elliptic equations in $\mathbb{R}^d$ with Dirichlet, Neumann, and Robin boundary conditions, respectively. We provide the first $O(n^{-1/d})$ convergence rate of DGMW by properly choosing the depth and width of the two networks in terms of the number of training samples n. We will extend the current analysis to the Friedrichs learning method [29] in our future work.

Acknowledgments The work of Y. Jiao is supported in part by the National Science Foundation of China under Grant 11871474 and by the research fund of KLATASDSMOE. The work of Y. Wang is supported in part by the Hong Kong Research Grant Council grants 16308518 and 16317416 and HK Innovation Technology Fund ITS/044/18FX, as well as Guangdong-Hong Kong-Macao Joint Laboratory for Data-Driven Fluid Mechanics and Engineering Applications. The work of H. Yang was partially supported by the US National Science Foundation under Award DMS-2244988, DMS-2206333, and the Office of Naval Research Award N00014-23-1-2007.

References

1. Ian Goodfellow, Yoshua Bengio, Aaron Courville, and Yoshua Bengio. *Deep learning*. MIT press, Cambridge, 2016.
2. Alex Krizhevsky, Ilya Sutskever, and Geoffrey E. Hinton. ImageNet classification with deep convolutional neural networks. *Communications of the ACM*, 60(6):84–90, 2017.
3. Ilya Sutskever, Oriol Vinyals, and Quoc V. Le. Sequence to sequence learning with neural networks. In Z. Ghahramani, M. Welling, C. Cortes, N. Lawrence, and K. Q. Weinberger, editors, *Advances in Neural Information Processing Systems*, volume 27. Curran Associates, Inc., 2014.
4. T. J. Hughes. *The Finite Element Method: Linear Static and Dynamic Finite Element Analysis*. Courier Corporation, 2012.
5. J.W. Thomas. *Numerical Partial Differential Equations: Finite Difference Methods*, volume 22. Springer Science & Business Media, 2013.
6. Isaac E Lagaris, Aristidis Likas, and Dimitrios I Fotiadis. Artificial neural networks for solving ordinary and partial differential equations. *IEEE transactions on neural networks*, 9(5):987–1000, 1998.
7. Weinan E, Jiequn Han, and Arnulf Jentzen. Deep learning-based numerical methods for high-dimensional parabolic partial differential equations and backward stochastic differential equations. *Communications in Mathematics and Statistics*, 5:349–380, 2017.
8. Maziar Raissi, Paris Perdikaris, and George Em Karniadakis. Physics-informed neural networks: A deep learning framework for solving forward and inverse problems involving nonlinear partial differential equations. *Journal of Computational Physics*, 378:686–707, 2019.
9. Justin A. Sirignano and K. Spiliopoulos. DGM: A deep learning algorithm for solving partial differential equations. *Journal of Computational Physics*, 375:1339–1364, 2018.

10. E. Weinan and Bing Yu. The deep ritz method: A deep learning-based numerical algorithm for solving variational problems. *Communications in Mathematics and Statistics*, 6(1):1–12, 2017.
11. Yaohua Zang, Gang Bao, Xiaojing Ye, and Haomin Zhou. Weak adversarial networks for high-dimensional partial differential equations. *Journal of Computational Physics*, 411:109409, 2020.
12. Jianfeng Lu, Yulong Lu, and Min Wang. A priori generalization analysis of the deep ritz method for solving high dimensional elliptic equations. *arXiv preprint arXiv:2101.01708*, 2021.
13. Jinchao Xu. The finite neuron method and convergence analysis. *arXiv preprint arXiv:2010.01458*, 2020.
14. Jonathan W. Siegel and Jinchao Xu. Approximation rates for neural networks with general activation functions. *Neural Networks*, 128:313–321, 2020.
15. Chenguang Duan, Yuling Jiao, Yanming Lai, Dingwei Li, Xiliang Lu, and Zhijian Yang. Convergence rate analysis for deep ritz method. *arXiv preprint arXiv:2103.13330*, 2021.
16. Yuling Jiao, Yanming Lai, Yisu Lo, Yang Wang, and Yunfei Yang. Error analysis of deep ritz methods for elliptic equations. *arXiv preprint arXiv:2107.14478*, 2021.
17. Patrick Dondl, Johannes Müller, and Marius Zeinhofer. Uniform convergence guarantees for the deep ritz method for nonlinear problems. *arXiv preprint arXiv:2111.05637*, 2021.
18. Yiping Lu, Haoxuan Chen, Jianfeng Lu, Lexing Ying, and Jose Blanchet. Machine learning for elliptic PDEs: Fast rate generalization bound, neural scaling law and minimax optimality. *arXiv preprint arXiv:2110.06897*, 2021.
19. Yeonjong Shin, Jerome Darbon, and George Em Karniadakis. On the convergence and generalization of physics informed neural networks. *arXiv preprint arXiv:2004.01806*, 2020.
20. Siddhartha Mishra and Roberto Molinaro. Estimates on the generalization error of physics informed neural networks (PINNs) for approximating PDEs. *arXiv preprint arXiv:2006.16144*, 2020.
21. Yeonjong Shin, Zhongqiang Zhang, and George Em Karniadakis. Error estimates of residual minimization using neural networks for linear PDEs. *arXiv preprint arXiv:2010.08019*, 2020.
22. Yuling Jiao, Yanming Lai, Dingwei Li, Xiliang Lu, Fengru Wang, Jerry Zhijian Yang, et al. A rate of convergence of physics informed neural networks for the linear second order elliptic PDEs. *Communications in Computational Physics*, 31(4):1272–1295, 2022.
23. Lawrence C Evans. Partial differential equations. *Graduate studies in mathematics*, 19(2), 1998.
24. David Gilbarg and Neil S Trudinger. *Elliptic partial differential equations of second order*. springer, 2015.
25. Robert A Adams and John JF Fournier. *Sobolev spaces*. Elsevier, 2003.
26. Ingo Gühring and Mones Raslan. Approximation rates for neural networks with encodable weights in smoothness spaces. *Neural Networks*, 134:107–130, 2021.
27. Shai Shalev-Shwartz and Shai Ben-David. *Understanding machine learning: From theory to algorithms*. Cambridge university press, 2014.
28. Aad van der Vaart, and Jon Wellner. *Weak Convergence and Empirical Processes: With Applications to Statistics*. Springer Science & Business Media, 1996.
29. Fan Chen, Jianguo Huang, Chunmei Wang, and Haizhao Yang. Friedrichs learning: Weak solutions of partial differential equations via deep learning. *arXiv preprint arXiv:2012.08023*, 2020.

Chapter 5
The 4-Player Gambler's Ruin Problem

Kathryn O'Connor and Laurent Saloff-Coste

5.1 Introduction

Consider the following multiplayer version of the classical gambler's ruin problem. There are k players and a fixed total amount of chips, N, distributed between them with player i holding x_i chips. At each turn, a pair $ij = \{i, j\}$, $1 \le i < j \le k$, is picked uniformly at random and the chosen two players play a fair Heads-or-Tails game and exchange one chip as a result. We consider this game until the time τ when one of the k players is left with no chips. Let $\mathbf{s} = (s_1, \ldots, s_k)$ be the distribution of chips at the start of the game. Let $\mathbf{X}^n = (X_1^n, \ldots, X_k^n)$ be the distribution of the chips after n games, $n \le \tau$.

For a review of this problem and the associated literature, the reader is referred to [5] and [7]. Here, we only recall that a (continuous) version of this problem is discussed in [1]. The aim of this chapter is to complement [5, 7] (see also the recent work [4], motivated by Diaconis and Ethier [5]) with a careful discussion of the application of the results of [7] to the 4-player version in a spirit similar to the treatment of the 3-player version in [5]. The general goal is to provide an informative analytic description of the large N behavior of

$$\mathbf{P}(\mathbf{X}_\tau = \mathbf{z} | \mathbf{X}_0 = \mathbf{s}), \ \mathbf{s} \in \mathfrak{S}_N, \ \mathbf{z} \in \overline{\mathfrak{S}}_N \setminus \mathfrak{S}_N,$$

uniformly over all $\mathbf{s}$, $\mathbf{z}$, where

$$\overline{\mathfrak{S}}_N = \left\{ (x_i)_1^k : 0 \le x_i, \sum_1^k x_i = N \right\}, \quad \mathfrak{S}_N = \left\{ (x_i)_1^k : 0 < x_i, \sum_1^k x_i = N \right\}.$$

K. O'Connor · L. Saloff-Coste (✉)
Department of Mathematics, Cornell University, Ithaca, NY, USA
e-mail: kfo23@cornell.edu; lps2@cornell.edu

P. Alonso Ruiz et al. (eds.), *From Classical Analysis to Analysis on Fractals*, Applied and Numerical Harmonic Analysis, https://doi.org/10.1007/978-3-031-37800-3_5

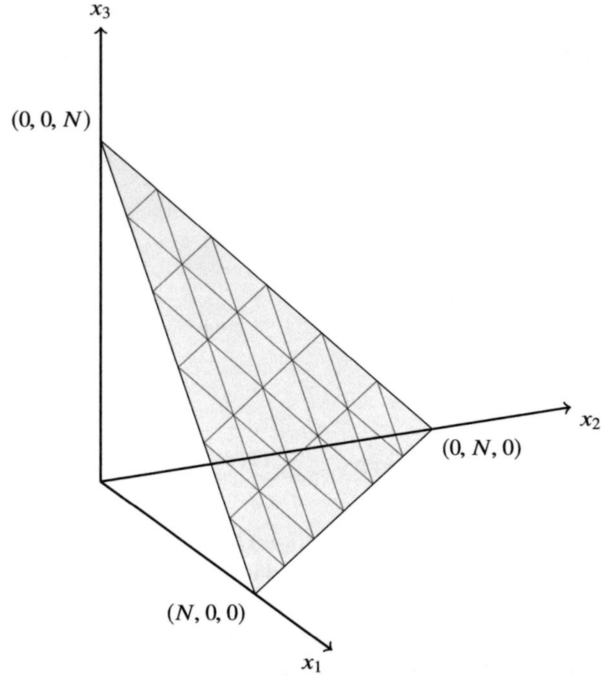

Fig. 5.1 The triangle for the 3-player game in the plane $\sum_1^3 x_i = N$. Here, $N = 6$

Here, all points have integer coordinates. See Figs. 5.1 and 5.2. Note that the point of $\overline{\mathfrak{S}_N} \setminus \mathfrak{S}_N$ which has more than one 0 coordinates cannot actually be reached by $(X_n)_{0 \le n \le \tau}$. For this reason, it is convenient to introduce

$$\partial \mathfrak{S}_N = \left\{ \mathbf{z} = (z_1, \ldots, z_k) \in \overline{\mathfrak{S}_N} \setminus \mathfrak{S}_N : \text{ exactly one } z_i = 0, 1 \le i \le k \right\}$$

and

$$\mathfrak{T}_{N,i} = \partial \mathfrak{S}_N \cap \{x_i = 0\}.$$

Among the particular questions one can ask, one test question identified in [5] concerns the behavior of

$$\mathbf{P}\left(\mathbf{X}_\tau \in \mathfrak{T}_{N,k} | \mathbf{X}_0 = \mathbf{s}\right) \text{ and } \mathbf{P}\left(\mathbf{X}_\tau = \mathbf{z} | \mathbf{X}_0 = \mathbf{s} \text{ and } \mathbf{X}_\tau \in \mathfrak{T}_{N,k}\right)$$

when $\mathbf{s} = (s_i)_1^k$ satisfies $s_k = N - k + 1$. In words, what is the probability that a very dominant player at time 0 (i.e., a player with $N - k + 1$ chips) be the first to lose all their chips? Given that the very dominant player at time 0 is the first to lose, what is the probability distribution describing the allocation of chips among the remaining players at the time the former dominant player loses? When f and g are two positive functions defined on the same domain, we will use the notation

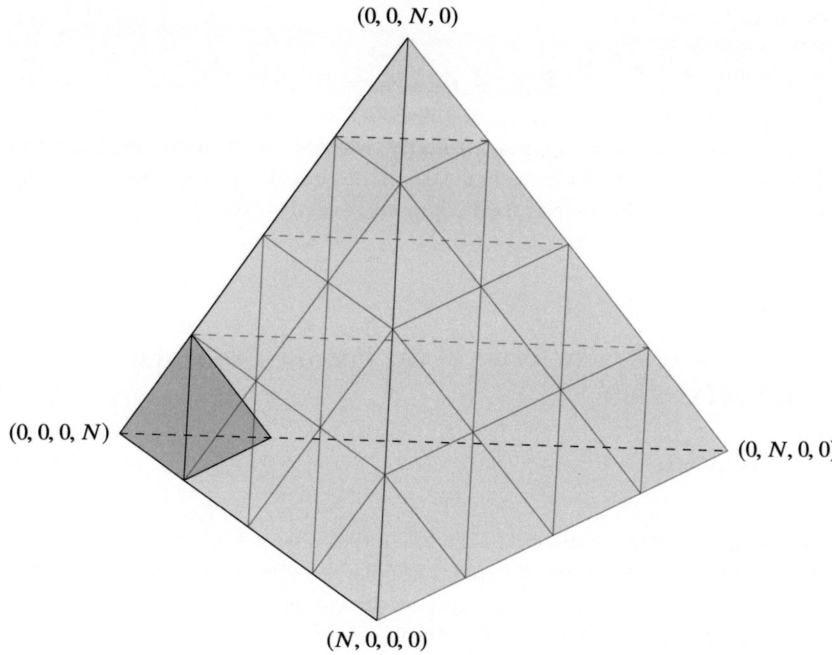

Fig. 5.2 The three-dimensional 4-player tetrahedron ($N = 4$). The picture shows only the vertices that are on the two visible front faces and dashed horizontal lines on the back face. In 4-space, the left-most corner hangs from the x_4-axis and the right-most face lies in the $\{x_4 = 0\} \cap \{\sum_1^4 x_i = N\}$ plane. The entire tetrahedron lies in the $\{\sum_1^4 x_i = N\}$ 3-space. Compare to the 3-player game shown in Fig. 5.1

$f \asymp g$ to signify that there are constants $0 < c \le C < +\infty$ such that $cf \le g \le Cg$ on the common domain of f and g.

Theorem 5.1 *Referring to the 4-player game with player* 4 *being the very dominant player at the start of the game* ($s_4 = N - 3, s_1, s_2, s_3 = 1, \mathbf{s}_N^* = (1, 1, 1, N - 3)$)*, there exists* $\alpha > 0$ *(the approximate value of* α *is* $\alpha \approx 5.68$*) such that*

$$\mathbf{P}(\mathbf{X}_\tau \in \mathfrak{T}_{N,4} | \mathbf{X}_0 = \mathbf{s}_N^*) \asymp N^{-\alpha}$$

and for $\mathbf{z} = (z_1, z_2, z_3, 0) \in \mathfrak{T}_{N,4}$. *Moreover, with* $\beta = \pi/\arccos(1/3) \approx 2.55$,

$$\mathbf{P}(\mathbf{X}_\tau = \mathbf{z} | \mathbf{X}_0 = \mathbf{s} \text{ and } \mathbf{X}_\tau \in \mathfrak{T}_{N,4}) \asymp \frac{z_1 z_2 z_3 [(z_1 + z_2)(z_1 + z_3)(z_2 + z_3)]^{\beta - 2}}{N^{3\beta - 1}}.$$

For simplicity, Theorem 5.1 addresses only a very extreme situation and it is desirable to treat other cases as well. This is done in Sect. 5.5. The given value of α is from a numerical computation which will be briefly discussed at the end of the chapter. Note that there is no simple heuristic allowing one to guess correctly the behavior of

$$\mathbf{P}(\mathbf{X}_\tau \in \mathfrak{T}_{N,4} | \mathbf{X}_0 = \mathbf{s}_N^*).$$

The method used here applies similarly when $k \geq 5$ with some additional complications (the number of exponents such as α and β that one needs to compute is $k-2$ and only one of them can be computed exactly). We focus here on the case $k = 4$.

5.2 Key Estimates in Terms of the Perron–Frobenius Eigenfunction

5.2.1 Notation

Throughout, we always assume that the integers N and k are such that $3 \leq k \leq N$. In fact, we always assume that k is fixed and we are interested in what happens for that fixed k when N is large. We let $(e_i)_1^k$ be the canonical basis of $\mathbb{R}^k$. The process $(\mathbf{X}^n)_{n\geq 0}$ is a Markov chain taking place in the simplex

$$\overline{\mathfrak{S}}_N = \left\{ (x_i)_1^k \in \mathbb{Z}^k : 0 \leq x_i, \sum_1^k x_i = N \right\},$$

started at $\mathbf{s}$ in

$$\mathfrak{S}_N = \left\{ (x_i)_1^k \in \mathbb{Z}^k : 0 < x_i, \sum_1^k x_i = N \right\}$$

and stopped at the random time τ when it hits the boundary $\partial\mathfrak{S}_N \subset \overline{\mathfrak{S}}_N \setminus \mathfrak{S}_N$ defined earlier. This Markov process is based on the global Markov kernel

$$\widetilde{K}(\mathbf{x}, \mathbf{y}) = \begin{cases} \frac{1}{2\binom{k}{2}} & \text{if } \mathbf{y} - \mathbf{x} \in \{\pm(e_i - e_j) : 1 \leq i < j \leq k\} \\ 0 & \text{otherwise,} \end{cases}$$

in the lattice

$$\mathfrak{L} = \left\{ \mathbf{x} = (x_i)_1^k \in \mathbb{Z}^k : \sum_1^n x_i = N \right\}$$

in the hyperplane $\left\{ \mathbf{x} = (x_i)_1^k \in \mathbb{R}^k : \sum_1^n x_i = N \right\}$, and its killed version, $K = \widetilde{K}_{\mathfrak{S}_N}$ defined by

$$K(\mathbf{x},\mathbf{y}) = \widetilde{K}(\mathbf{x},\mathbf{y})\mathbf{1}_{\mathfrak{S}_N}(\mathbf{x})\mathbf{1}_{\mathfrak{S}_N}(\mathbf{y}). \tag{5.1}$$

This kernel defines an operator (by abuse of notation, we call it K) acting on functions on the finite set $\mathfrak{S}_N$ by

$$K\phi(\mathbf{x}) = \sum_{\mathbf{y}\in\mathfrak{S}_N} K(\mathbf{x},\mathbf{y})\phi(\mathbf{y}).$$

We consider this operator acting on the Hilbert space $L^2(\mathfrak{S}_N)$ equipped with

$$\langle\phi,\psi\rangle = \sum_{\mathbf{x}\in\mathfrak{S}_N}\phi(\mathbf{x})\psi(\mathbf{x}), \quad \|\phi\|_2^2 = \langle\phi,\phi\rangle.$$

The sub-Markovian kernel K is irreducible and aperiodic in $\mathfrak{S}_N$ in the sense that there exists m such that $K^m(\mathbf{x},\mathbf{y}) > 0$ for all $\mathbf{x},\mathbf{y}\in\mathfrak{S}_N$. It thus follows from the Perron–Frobenius theorem that there is a unique $\phi_0 : \mathfrak{S}_N \to (0,+\infty)$ such that

$$K\phi_0 = \beta_0\phi_0 \;\; \|\phi_0\|_2 = 1, \;\; \text{and } \beta_0 = \sup\{\|K\phi\|_2 : \|\phi\|_2 = 1\}. \tag{5.2}$$

Note that, at the exit time τ, the process $(X_n)_{0\le n\le\tau}$ started at $X_0\in\mathfrak{S}_N$ is at a point in $\partial\mathfrak{S}_N$ for the first time. For any point $\mathbf{z}\in\partial\mathfrak{S}_N$, let $\mathcal{N}_{\mathbf{z}}$ be the non-empty finite set of those points in $\mathfrak{S}_N$ which are neighbors of $\mathbf{z}$ in the sense that $\widetilde{K}(\mathbf{y},\mathbf{z}) > 0$ if $\mathbf{y}\in\mathcal{N}_{\mathbf{z}}$, namely

$$\mathcal{N}_{\mathbf{z}} = \{\mathbf{y}\in\mathfrak{S}_N : \widetilde{K}(\mathbf{y},\mathbf{z}) > 0\}, \quad \mathbf{z}\in\partial\mathfrak{S}_N.$$

We will often use the notation $\mathbf{y}_{\mathbf{z}}$ to denote an arbitrary point in $\mathcal{N}_{\mathbf{z}}$.

5.2.2 *Estimates in Terms of* ϕ_0

According to [6, 7], what is needed to answer questions such as the ones considered in Theorem 5.1 is estimates describing the boundary behavior of the Perron–Frobenius function ϕ_0. Indeed, for $k \geq 3$, one can apply [7, (6.2),Theorem 6.7] and these results give the following very general estimates (by "very general," we mean that these estimates hold in much greater generality than the present setting of the k-player game. See [6, 7].) In all these estimates, the dimension k is fixed and all implied constants may depend on k. The key point is that these estimates are uniform in N. The first lemma gives the basic size estimate for ϕ_0 in the middle of $\mathfrak{S}_N$ and captures the fact that ϕ_0 behaves in a very tame fashion in the middle of $\mathfrak{S}_N$. This is the result of a basic Harnack-type inequality that applies to ϕ_0. In these statements, we can choose (with some small adjustments) to use the Euclidean distance to compute the distance between points or the graph distance in the lattice supporting $\mathfrak{S}_N$ in the hyperplane $\{\sum x_i = N\}$ (we use the latter). The various implied constants will then have to be adjusted depending of the choice, and these adjustments depend on the dimension k.

Lemma 5.1 *The normalized Perron–Frobenius eigenvalue β_0 and eigenfunction ϕ_0 satisfy*

$$1 - \beta_0 \asymp N^{-2}$$

and, with $\mathbf{o}_N$ denoting any one of the points in $\mathfrak{S}_N$ closest to the middle point $(N/k, \cdots, N/k)$,

$$\phi_0(\mathbf{o}_N) \asymp N^{-(k-1)/2}, \quad \sum_{\mathbf{s}\in\mathfrak{S}_N} \phi_0(\mathbf{s}) \asymp N^{(k-1)/2}.$$

Moreover, for any $a \in (0, 1)$, there is a constant A for which

$$A^{-1}\phi(\mathbf{o}_N) \le \phi_0(\mathbf{x}) \le A\phi_0(\mathbf{o}_N)$$

for all $\mathbf{x} \in \mathfrak{S}_N$ such that $d(\mathbf{x}, \partial\mathfrak{S}_N) \ge aN$.

The estimate of ϕ_0 given in this lemma is for points that are away from the boundary $\partial\mathfrak{S}_N$. In that region, ϕ_0 behaves like a constant. Even though this lemma does not capture this fact, it is also true that $\phi_0(\mathbf{o}_N) \asymp \|\phi_0\|_\infty$. See, e.g., [7, (5.7) and Theorem 6.5] and [6, Theorem 8.9]

Theorem 5.2 ([7, (6.2)]) *Fix $k \ge 3$. Referring to the k-player game, consider arbitrary points $\mathbf{s} \in \mathfrak{S}_N$ and $\mathbf{z} \in \partial\mathfrak{S}_N$. We have*

$$\mathbf{P}(\mathbf{X}_\tau = \mathbf{z}|\mathbf{X}_0 = \mathbf{s}) \asymp \phi_0(\mathbf{s})\phi_0(\mathbf{y}_z)\left(N^2 + \frac{1}{\phi_0(\mathbf{s}_d)^2 d^{k-3}}\right),$$

where $\mathbf{y}_\mathbf{z} \in \mathcal{N}_\mathbf{z}$, $d = d(\mathbf{s}, \mathbf{z})$, and $\mathbf{s}_d$ is any point in $\mathfrak{S}_N$ such that

$$d(\mathbf{s}, \mathbf{s}_d) \le C_k d \text{ and } d(\mathbf{s}_d, \partial\mathfrak{S}_N) \ge c_k d$$

for some appropriately chosen constants $0 < c_k \le C_k < +\infty$.

Given Lemma 5.1, Theorem 5.1 gives

$$\mathbf{P}(\mathbf{X}_\tau = \mathbf{z}|\mathbf{X}_0 = \mathbf{s}) \asymp_{\epsilon,k} N^2\phi_0(\mathbf{s})\phi_0(\mathbf{y}_z) \tag{5.3}$$

for each fixed $\epsilon \in (0, 1)$ and all points $\mathbf{s} \in \mathfrak{S}_N$ and $\mathbf{z} \in \partial\mathfrak{S}_N$ with $d(\mathbf{s}, \mathbf{z}) > \epsilon N$. This is because $d = d(\mathbf{s}, \mathbf{z}) \asymp N$ and thus $\phi_0(\mathbf{s}_d)^2 d^{k-3} \asymp N^{-2}$.

Estimate (5.3) is what is needed to obtain Theorem 5.1 because, in that theorem, the starting point $\mathbf{s}_N = (1, 1, 1, N-3)$ is at a distance of order N of the face $\mathfrak{T}_{N,4} = \partial\mathfrak{S}_N \cap \{z_4 = 0\}$. See, e.g., Figs. 5.2 and 5.3.

In all other cases, that is, whenever $d(\mathbf{s}, \mathbf{z}) \leq \epsilon N$, we have

$$\mathbf{P}(\mathbf{X}_\tau = \mathbf{z} | \mathbf{X}_0 = \mathbf{s}) \asymp_{\epsilon,k} \frac{\phi_0(\mathbf{s})\phi_0(\mathbf{y}_z)}{\phi_0(\mathbf{s}_d)^2 d^{k-3}}, \quad d = d(\mathbf{s}, \mathbf{z}). \tag{5.4}$$

Note that, if desired, the results in [7] also provide detailed information in the same spirit (as a function of $t > 0$) for hitting probabilities under the additional requirement that $\tau < t$.

5.3 Estimating ϕ_0

Theorem 5.2 makes it clear that having a detailed understanding of ϕ_0 is what is needed to obtain good hitting probability estimates. We will give explicit estimates for the case $k = 3, 4$, and 5. In the case $k = 3$, [7] gives an explicit exact formula for ϕ_0 as well as the two-sided estimate ([7, (5.10)])

$$\phi_0((s_1, s_2, s_3)) \asymp N^{-7}(s_1 + s_2)(s_1 + s_3)(s_2 + s_3)s_1 s_2 s_3.$$

Unfortunately, in the case $k > 3$, there are no reasons to expect exact formulas expressible in simple terms and we need to rely on a much more sophisticated analysis. One reason for the existence of an explicit formula in dimension 2 is that equilateral triangles pave the plane (with great symmetries), while regular simplexes never pave $\mathbb{R}^d$ in dimension greater than 2. In dimension 2, this allows for the extension of ϕ_0 to the entire relevant lattice and its computation using Fourier expansion (sine and cosine functions). See, e.g., [7], but this is of course well known.

5.3.1 *From the Simplex to the Lattice Cone*

The simplex $\mathfrak{S}_N$ lies in the lattice $\mathfrak{L}$ generated by the vectors $(e_i - e_j)$, $1 \leq i \neq j \leq k$, in the hyperplane $\{\mathbf{x} = (x_1, \ldots, x_k) : \sum_1^k x_i = N\}$. In that hyperplane, we can consider the continuous open half-cone (here $\mathbf{x} \in \mathbb{R}^k$)

$$C = \left\{ \mathbf{x} = (t\xi_1, \ldots, t\xi_{k-1}, N(1-t)) : t > 0, \xi_i > 0, 1 \leq i \leq k-1, \sum_1^{k-1} \xi_i = N \right\}$$

and the lattice cone

$$\mathfrak{C}_{\mathfrak{L}} = C \cap \left\{ \mathbf{x} = (x_1, \ldots, x_k) \in \mathbb{Z}^k : \mathbf{x} - Ne_k = \sum_{1 \leq i < j \leq k} m_{ij}(e_i - e_j), m_{ij} \in \mathbb{Z} \right\}.$$

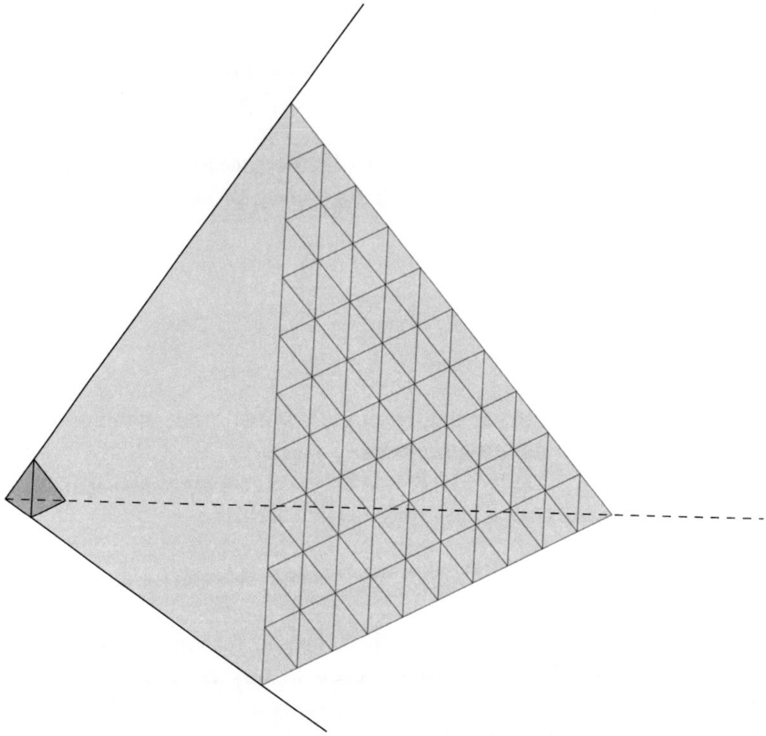

Fig. 5.3 A larger three-dimensional 4-player tetrahedron ($N = 10$), part of a lattice cone (not shown but easily imagined), and sitting in a continuous cone. The edges of the small dark gray tetrahedron indicate the basic steps of the lattice. Contrary to what happens in the two-dimensional case, the large tetrahedron is NOT paved by isometric copies of the small tetrahedron since the only space-filling regular polyhedron is the cube

Note that the tip of this cone is the point $\mathbf{t}_N$ with coordinates

$$t_1 = \cdots = t_{k-1} = 0, \ t_k = N.$$

See Fig. 5.3. Note also that this description of the cone is NOT a description in polar coordinate, that is, the point $(\xi_1, \ldots, \xi_{k-1})$ is not on a sphere centered at $\mathbf{t}_N$ in the $\{\mathbf{x} : \sum_1^N x_i = N\}$. It is in the $(k-2)$-dimensional face of the simplex that lies in $\{x_k = 0\}$.

Now, in addition to the Perron–Frobenius eigenfunction ϕ_0 defined in $\mathfrak{S}_N$, consider a function u defined on the lattice $\mathfrak{L}$ satisfying

$$\begin{cases} \sum_{\mathbf{y}} \widetilde{K}(\mathbf{x}, \mathbf{y}) u(\mathbf{y}) = u(\mathbf{x}) \text{ for } \mathbf{x} \in \mathfrak{C}_{\mathfrak{L}} \\ u(\mathbf{x}) = 0 \text{ when } \mathbf{x} \in \mathfrak{L} \setminus \mathfrak{C}_{\mathfrak{L}} \text{ and } u > 0 \text{ in } \mathfrak{C}_{\mathfrak{L}}. \end{cases} \tag{5.5}$$

In words, u is a positive $\widetilde{K}$-harmonic function in $\mathfrak{C}_{\mathfrak{L}}$ which vanishes on $\mathfrak{L} \setminus \mathfrak{C}_{\mathfrak{L}}$ (that is, in particular, on the boundary of $\mathfrak{C}_{\mathfrak{L}}$.

Definition 5.1 We call harmonic profile for the lattice cone $\mathfrak{C}_{\mathfrak{L}}$ any function u satisfying (5.5).

Such a function exists and is unique up to multiplication by a positive constant. In other words, any two harmonic profiles are constant positive multiples of each other. Existence is easily proved by using an exhaustion by finite subsets in which one can solve a properly defined Dirichlet problem (after normalization, point-wise convergence of some subsequence follows from a standard and general argument using a very local Harnack-type inequality for graphs. Uniqueness requires more work. Because graphical harmonic functions are also harmonic functions for the associated cable process along the edges, uniqueness can be deduced from the results in [8] set in the context of local Dirichlet spaces. See also Theorem 3.2 below). For different viewpoints on the function u, see also [2–4].

The following key property holds. It explains the important role played by the profile: any other local positive solution vanishing at the boundary can be controlled using the profile. Consider the following notation. Let $B = B(\mathbf{x}, r)$ be the Euclidean ball of radius r around $\mathbf{x}$ and set

$$\mathfrak{B}_{\mathfrak{L}} = B \cap \mathfrak{C}_{\mathfrak{L}}, \quad \mathfrak{B}'_{\mathfrak{L}} = B(\mathbf{x}, r/2) \cap \mathfrak{C}_{\mathfrak{L}},$$

$$\overline{\mathfrak{B}}_{\mathfrak{L}} = \{\mathbf{x} \in \mathfrak{L} : \mathbf{x} \text{ has a lattice neighbor in } \mathfrak{B}_{\mathfrak{L}}\},$$

and

$$\partial^* \mathfrak{B}_{\mathfrak{L}} = \{\mathbf{x} \in \mathfrak{L} \setminus \mathfrak{C}_{\mathfrak{L}} : \mathbf{x} \text{ has a lattice neighbor in } \mathfrak{B}_{\mathfrak{L}}\}.$$

In words, $\mathfrak{B}_{\mathfrak{L}}$ is the trace of $B(\mathbf{x}, r)$ in the lattice cone $\mathfrak{C}_{\mathfrak{L}}$, $\mathfrak{B}'_{\mathfrak{L}}$ is the trace of $B(\mathbf{x}, r/2)$ in that lattice cone, $\overline{\mathfrak{B}}_{\mathfrak{L}}$ is $\mathfrak{B}_{\mathfrak{L}}$ together with all the lattice points that have a lattice neighbor in $\mathfrak{B}_{\mathfrak{L}}$, and $\partial^* \mathfrak{B}_{\mathfrak{L}}$ is the part of $\overline{\mathfrak{B}}_{\mathfrak{L}} \setminus \mathfrak{B}_{\mathfrak{L}}$ that lies on the boundary of the cone C.

Theorem 5.3 *Fix a constant C_0. Let $\mathbf{x}$ be a point in $\mathfrak{C}_{\mathfrak{L}}$ and $r > 0$. Let v and w be two positive functions in $\mathfrak{B}_{\mathfrak{L}}$ which are defined on $\overline{\mathfrak{B}}_{\mathfrak{L}}$ and vanish along $\partial^* \mathfrak{B}_{\mathfrak{L}}$. Assume that v and w are solutions of*

$$(I - \widetilde{K})v = pv, \quad (I - \widetilde{K})w = qw \text{ in } \mathfrak{B}_{\mathfrak{L}},$$

where p, q are functions satisfying $|p|, |q| \leq C_0/r^2$. Then, there is a constant C (depending on k and C_0 but not on N, $\mathbf{x}$, r, v, w) such that

$$\sup_{\mathfrak{B}'_{\mathfrak{L}}} \left\{ \frac{v}{w} \right\} \leq C \inf_{\mathfrak{B}'_{\mathfrak{L}}} \left\{ \frac{v}{w} \right\}.$$

This is a deep and rather intricate theorem and we explain where it comes from. First, without loss of generality, we can assume that $p = 0$ as the general result is a consequence of this special case applied twice. In the context of strictly local Dirichlet spaces, analogous results are given in [9, 10]. The direct proof in the discrete case goes as follows. Assume $p = 0$. Use the function v to perform a Doob transform by setting $\widetilde{K}_v(\mathbf{x}, \mathbf{y}) = \frac{1}{v(\mathbf{x})}\widetilde{K}(\mathbf{x}, \mathbf{y})v(\mathbf{y})$, $\mathbf{x}, \mathbf{y} \in \mathfrak{B}$. This is a reversible Markov chain in $\mathfrak{B}_{\mathfrak{L}}$ with reversible measure v^2. Moreover, it is a Harnack chain (see [6, 7]). A version of the elliptic Harnack inequality for the solution w/v of $(I - \widetilde{K})(w/v) = q(w/v)$ gives the desired result.

Corollary 5.1 *Let u be a harmonic profile for $\mathfrak{C}_{\mathfrak{L}}$. There exists a constant $C_k \geq 1$ such that setting*

$$\mathfrak{B}_N = \mathfrak{S}_N \cap \{x_k \geq N/k\},$$

$$u(\mathbf{x})\frac{\phi_0(\mathbf{y})}{C_k u(\mathbf{y})} \leq \phi_0(\mathbf{x}) \leq u(\mathbf{x})\frac{C_k\phi_0(\mathbf{y})}{u(\mathbf{y})} \text{ for all } \mathbf{x}, \mathbf{y} \in \mathfrak{B}_N.$$

This corollary is a special case of [6, Theorem 8.13]. It also follows from Theorem 5.3 above. The function ϕ_0 has the same behavior in each of the corners of the simplex $\mathfrak{S}_N$, and, in each of these corners, this behavior is comparable to the behavior of the function u near the tip of the cone $\mathfrak{C}$. The region $\mathfrak{B}_N$ is pictured in Fig. 5.5.

5.3.2 From the Lattice Cone to the Continuous Cone

In general, there is no easy way to compute the behavior of the lattice cone profile u. However, consider the following continuous version of the profile.

Definition 5.2 Let $V = V_\Omega = \{x : x = r\theta, r > 0, \theta \in \Omega \subseteq \mathbb{S}^{d-1}\}$ be an open half-cone in Euclidean space $\mathbb{R}^d$ with base Ω. Call harmonic profile for (the continuous half-cone) V, any function $h_V : V \to (0, +\infty)$ which is harmonic in V (that is, C^2 is in V and satisfies $\Delta h_V = 0$ in V) and vanishes continuously at the boundary of the cone.

Without assumptions on the base Ω, the profile may not be unique up to a multiplicative constant, e.g., if the base Ω has more than one connected component. Here, we are only interested with the case when Ω is a nice connected polygonal subset of the sphere $\mathbb{S}^{d-1}$ in $\mathbb{R}^d$. See, e.g., Fig. 5.4.

Proposition 5.1 (Folklore) *Let Ω be an open connected subset of $\mathbb{S}^{d-1}$. Consider the Perron–Frobenius eigenvalue and eigenfunction $\lambda_\Omega, \psi_\Omega$ of the sphere-Laplacian with zero boundary condition on Ω. Then, the function*

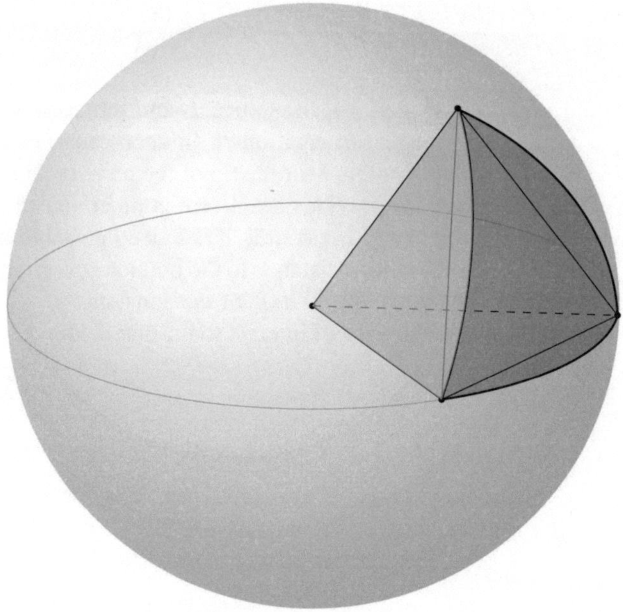

Fig. 5.4 The harmonic profile of a continuous cone is computed using the Perron–Frobenius eigenvalue and eigenfunction of the spherical Laplacian with Dirichlet boundary condition on the spherical base of the cone. In this figure, the cone is associated with the tetrahedron and the base is a particular equilateral triangle on the sphere, the equilateral triangle corresponding to the vertices of the tetrahedron

$$x = r\theta \mapsto h_V(x) = r^{\alpha_V}\psi_\Omega(\theta), \quad \alpha_V = \sqrt{((d/2)-1)^2 + \lambda_\Omega} - ((d/2)-1)$$

is a harmonic profile for $V = V_\Omega$.

The works [2, 3, 8, 12] study the harmonic profiles of cones (and more general sets in the case of [8, 12]) and provide a very useful comparison between the harmonic profile of a discrete cones and that of the associated continuous cone. For simplicity, in d-dimensional Euclidean space, $\mathbb{R}^d$, consider a co-compact lattice (discrete subgroup) L with the property that simple random walk on L has covariance matrix the identity (or proportional to the identity).

Let $V = V_\Omega = \{x : x = r\theta, r > 0, \theta \in \Omega \subseteq \mathbb{S}^{d-1}\}$ be an open half-cone and assume that Ω is a convex polygonal subset of the sphere. Let $V_L^* = L \cap V$ be the associated lattice-cone. A (discrete) harmonic profile for that lattice-cone is a function $h_{V,L}$ on the lattice L which vanishes on $L \setminus V_L^*$ and non-negative harmonic in V_L^* and not identically 0. Here, harmonic is with respect to the Markov operator associated with simple random walk on L. Among the results of [3, 12] is the following fact of importance to us. There are positive constants c, C, and D such that, for any fix x_0 in V_L^* at distance at least D from $\mathbb{R}^d \setminus V$, we have $h_{V,L}(x_0) > 0$ and, for all $x \in V_L^*$ at distance at least D from $\mathbb{R}^d \setminus V$,

$$c\frac{h_V(x)}{h_V(x_0)} \le \frac{h_{V,L}(x)}{h_{V,L}(x_0)} \le C\frac{h_V(x)}{h_V(x_0)}. \tag{5.6}$$

The reader should note that, in general, the lattice L and the cone V do not have to be neatly positioned with respect to each other. In such cases V_L^*, viewed as a subgraph of L, can have isolated points near the tip of the cone (at such point, $h_{V,L}$ has to vanish). The estimate (5.6) provides a uniform comparison of h_V and $h_{V,L}$ a few steps away from the boundary. In general, it is indeed possible that there is a sequence of points $(x_j)_1^\infty$ in V_L^* whose distance to the boundary is positive but tends to 0 as j tends to infinity. This explains the role of the constant D.

We now return to the particular case of interest to us, that is, the continuous open half-cone

$$C = \left\{ \mathbf{x} = (t\xi_1, \dots, t\xi_{k-1}, N(1-t)) : t > 0, \xi_i > 0, 1 \le i \le k-1, \sum_1^{k-1} \xi_i = N \right\}$$

and the lattice cone

$$\mathfrak{C}_{\mathfrak{L}} = C \cap \left\{ \mathbf{x} = (x_1, \dots, x_k) : \mathbf{x} - Ne_k = \sum_{1 \le i < j \le k} m_{ij}(e_i - e_j), m_{ij} \in \mathbb{Z} \right\},$$

both defined earlier. Recall that the tip of this cone is the point $\mathbf{t}_N$ with coordinates $t_1 = \cdots = t_{k-1} = 0$, $t_k = N$. In this case, the continuous cone and the lattice are neatly arranged in the sense that all the lattice boundary points of $\mathfrak{C}_{\mathfrak{L}}$ lie exactly on the topological boundary of the continuous cone C and no points in $\mathfrak{C}_{\mathfrak{L}}$ (viewed as a subgraph of $\mathfrak{L}$) are isolated. Moreover, there exists $\epsilon_k > 0$ such that any point in $\mathfrak{C}_{\mathfrak{L}}$ is at a distance at least ϵ_k from the boundary of C. This allows us to rephrase (5.6) in the following simplified form. Set

$$\mathbf{t}'_N = (1, \dots, 1, N - k + 1) \in \mathfrak{C}_{\mathfrak{L}}.$$

This point stands in the lattice cone, closest to the tip $\mathbf{t}_N$.

Proposition 5.2 *Let h and $h_{\mathfrak{L}}$ be, respectively, the (continuous) harmonic profile of the continuous cone C and the (lattice) harmonic profile of the lattice-cone $\mathfrak{C}_{\mathfrak{L}}$, both normalized by the condition*

$$h(\mathbf{t}'_N) = h_{\mathfrak{L}}(\mathbf{t}'_N) = 1.$$

There are constants $c, C \in (0, +\infty)$ such that, for all $\mathbf{x} \in \mathfrak{C}_{\mathfrak{L}}$,

$$ch(\mathbf{x}) \le h_{\mathfrak{L}}(\mathbf{x}) \le Ch(\mathbf{x}).$$

See [2, 3] for more precise results in the same spirit.

5.3.3 *Estimating ϕ_0 in Terms of the Profile h of the Continuous Cone*

Let $\lambda_k > 0$ be the Perron–Frobenius eigenvalue (i.e., the lowest eigenvalue) of the (positive) sphere-Laplacian with Dirichlet boundary condition in the regular spherical simplex Ω_k of dimension $k-2$ cut by the cone C on the unit sphere $\mathbb{S}^{k-2}$ of the $k-1$-space $\{\mathbf{x} : \sum_1^k x_i = N\}$ viewed has a vector space with origin at $\mathbf{t}_N = (0, \dots, 0, N)$. This number is defined by the variational formula

$$\lambda_k = \inf \left\{ \frac{\int_{\Omega_k} |\nabla f|^2 d\sigma}{\int_{\Omega_k} |f|^2 d\sigma} : f \in C_c^\infty(\Omega_k) \right\}. \tag{5.7}$$

Definition 5.3 Set

$$\alpha_k = \sqrt{(((k-1)/2)-1)^2 + \lambda_k} - (((k-1)/2)-1).$$

Recall that ϕ_0 is the (positive, normalized) Perron-Frobenious eigenfunction of K on the lattice simplex $\mathfrak{S}_N$. See (5.1)–(5.2). Because of the uniqueness of ϕ_0 and the obvious symmetry of the problem under permutation of the coordinates of a point $\mathbf{x} = (x_1, \dots, x_k)$, the function ϕ_0 is symmetric under any permutation of these coordinates. Because $\sum_1^k x_i = N$ for any point in $\mathfrak{S}_N$, at least one of the x_i's, $1 \le i \le k$, is larger than or equal to N/k and we can assume without loss of generality that $x_k \ge N/k$. See Fig. 5.5.

Proposition 5.3 *There are constants $c_k \le C_k \in (0, +\infty)$ such that, in $\mathfrak{S}_N \cap \{\mathbf{x} : x_k \ge N/k\}$,*

$$c_k h(\mathbf{x}) \le N^{((k-1)/2)+\alpha_k} \phi_0(\mathbf{x}) \le C_k h(\mathbf{x}).$$

Proof By Proposition 5.2 and Corollary 5.1 with $u = h_{\mathfrak{Q}}$, we have

$$\phi_0(\mathbf{x}) \asymp_k \phi_0(\mathbf{t}'_N) h(\mathbf{x}).$$

According to [7, (5.7)], we also have $\phi_0(\mathbf{o}_N)^2 \asymp_k N^{-(k-1)}$, where $\mathbf{o}_N$ is a lattice point nearest to the center $(N/k, \dots, N/k)$ of the simplex $\{\mathbf{x} = (x_1, \dots, x_k) : x_i > 0, \sum_1^k x_i = N\}$. This gives

$$N^{-(k-1)/2} \asymp_k \phi_0(\mathbf{o}_N) \asymp_k \phi_0(\mathbf{t}'_N) h(\mathbf{o}_N) \asymp_k \phi_0(\mathbf{t}'_N) N^{\alpha_k},$$

so that, as desired,

$$\phi_0(\mathbf{x}) \asymp_k N^{-(\alpha_k + ((k-1)/2))} h(\mathbf{x}).$$

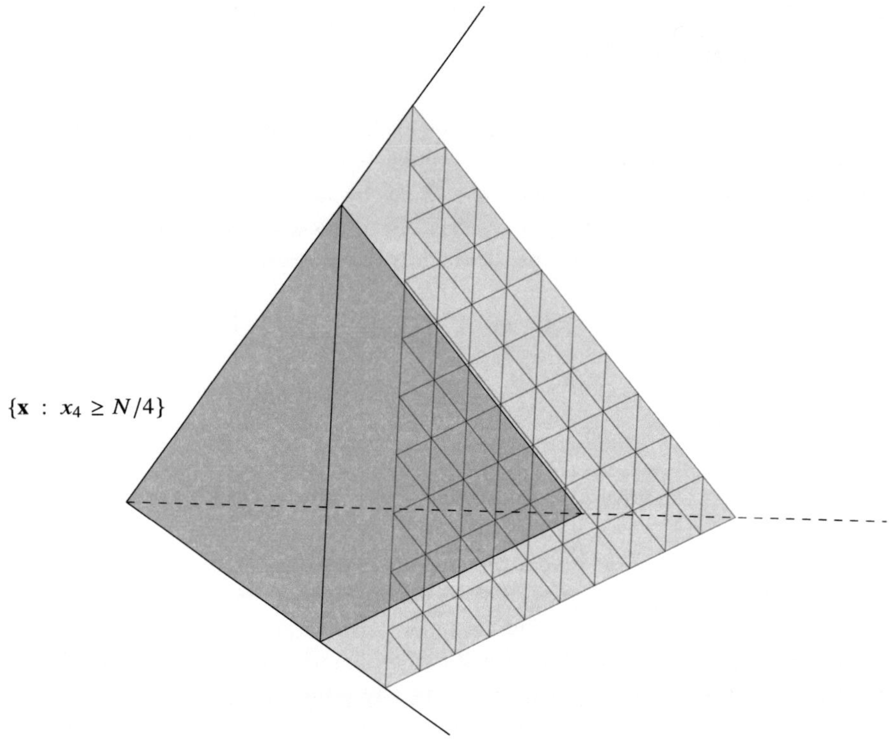

Fig. 5.5 The dark gray area represents $\mathfrak{S}_N \cap \{\mathbf{x} : x_k \geq N/k\}$. Here, $k = 4$ (4-player game) and $N = 10$. Note that the union of the zones corresponding to each corner covers the entire tetrahedron

Note that, in [4], Denisov and Wachtel prove a limit theorem related to Proposition 5.3 in the case $k = 3$ (this natural result is complicated by the presence of corners which makes classical approximation theorems inapplicable).

5.4 Harmonic Profile and ϕ_0 in Coordinates

5.4.1 *The Continuous Harmonic Profile h in Coordinates When k = 4*

The goal of this section is to provide (explicit) estimates for the continuous harmonic profile h of the continuous cone

$$C = \{\mathbf{x} = (t\xi_1, t\xi_2, t\xi_3, N(1-t)) : t > 0, \xi_i > 0, 1 \leq i \leq 3, \xi_1 + \xi_2 + \xi_3 = N\}$$

in terms of the (free) coordinates $(x_1, \ldots, x_3)$ of the point $\mathbf{x} = (x_1, x_2, x_3, x_4) \in C$.
By proposition 5.1, we know that h has the form

$$h(\mathbf{x}) = r^{\alpha_4} \psi_4(\theta), \tag{5.8}$$

where (r, θ) are the polar coordinates of a point in the hyperplane $\sum_1^4 x_i = N$ with respect to the origin $\mathbf{t}_4$ (the tip or our cone C). We postpone the discussion of the value of the real α_4 but note that $\alpha_4 \approx 5.68$. We note that the radius r satisfies

$$r \asymp x_1 + x_2 + x_3 \tag{5.9}$$

for any point $\mathbf{x} = (x_1, x_2, x_3, x_4)$ contained in C. So our goal is to understand the function ψ_4. By definition, $\psi = \psi_4$ is the Perron–Frobenius eigenfunction of the spherical Laplacian in the spherical domain Ω_4 cut by our cone C on the two-dimensional unit sphere centered at $\mathbf{t}_4$ in the 3-space $\{\mathbf{x} : \sum_1^4 x_i = N\}$. The final result we want to prove reads as follows.

Proposition 5.4 *The harmonic profile h of the cone C satisfies*

$$h(\mathbf{x}) \asymp (x_1 + x_2 + x_3)^{\alpha - 3\beta + 3}[(x_1 + x_2)(x_1 + x_3)(x_2 + x_3)]^{\beta - 2} x_1 x_2 x_3,$$

where $\alpha = \alpha_4 \approx 5.68$ *is as defined above and* $\beta = \pi/\arccos(1/3) \approx 2.55$.

Before embarking with the proof, we make the following observations. Working in Euclidean space $\mathbb{R}^m$ with canonical coordinates $(y_1, \ldots, y_m)$, few cones have harmonic profile whose expression in coordinates is simple and explicit.
In dimension 2, $\{\mathbf{y} = (y_1, y_2) : y_1 \in \mathbb{R}, y_2 > 0\}$ has $h_\pi(\mathbf{y}) = y_2$, while the cone $\{\mathbf{y} = (y_1, y_2) : y_1 > 0, y_2 > 0\}$ has $h_{\pi/2}(\mathbf{y}) = y_1 y_2$. For any $\eta \in (0, 2\pi)$, the cone

$$V_\eta = \{\mathbf{y} = re^{i\theta} : r > 0, 0 < \theta < \eta\}$$

of aperture η has

$$h_\eta(\mathbf{y}) = r^{\pi/\eta} \sin(\pi\theta/\eta).$$

Even so there is no easy formula for h in terms of (y_1, y_2), it is helpful to observe that

$$h(\mathbf{y}) \asymp (y_1^2 + y_2^2)^{(\pi/2\eta)-1} d(\mathbf{y}, L_0) d(\mathbf{y}, L_\eta),$$

where L_θ is the half-line $\{\mathbf{z} = re^{i\theta} : r > 0\}$ and $d(\mathbf{y}, L)$ is the distance from $\mathbf{y}$ to L. In such an estimate, the terms of the form $d(\mathbf{y}, L)$ can be replaced by any equivalent quantity. For instance, for $\eta \in (0, 3\pi/4]$, we can write

$$h_\eta(\mathbf{y}) \asymp (y_1 + y_2)^{(\pi/\eta)-2}(y_1 \sin\theta - y_2 \cos\theta) y_2.$$

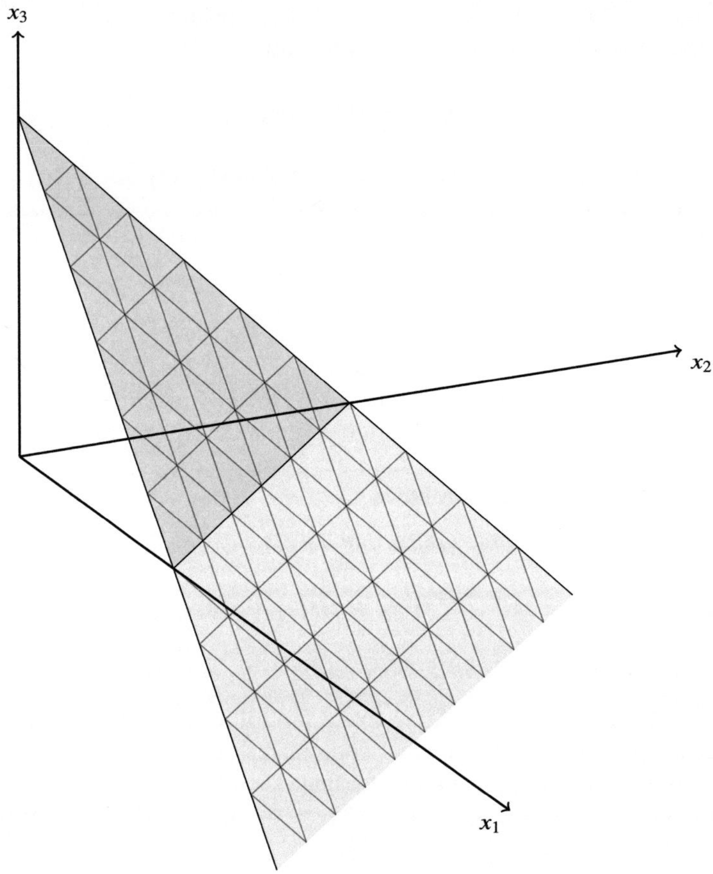

Fig. 5.6 The triangle (dark gray), cone, and lattice for the 3-player game: the harmonic profile satisfies $h(\mathbf{x}) \asymp (x_1 + x_2)x_1x_2$, $\mathbf{x} = (x_1, x_2, x_3)$, $\sum_1^3 x_i = N$. Here, $N = 6$

In the two-dimensional cone,

$$C = \{\mathbf{x} = (t\xi_1, t\xi_2, N(1-t)) : t > 0, \xi_1, \xi_2 > 0, \xi_1 + \xi_2 = N\},$$

which has aperture $\pi/3$, with $\mathbf{x} = (x_1, x_2, x_3)$, $x_1 + x_2 + x_3 = N$, this gives

$$h_{\pi/3}(\mathbf{x}) \asymp (x_1 + x_2)x_1x_2,$$

where $\asymp$ is uniform in N. See Fig. 5.6.

Such two-dimensional computations apply to a wedge determined by two half-planes meeting along their edge (a copy of $\mathbb{R}$) in three dimensions after choosing coordinates wisely. Namely, one can represent such a wedge as $W = \mathbb{R} \times V_\eta$, where V_η lies in a plane orthogonal to the mentioned copy of $\mathbb{R}$. If we call

(z, y_1, y_2) the coordinate of $\mathbb{R}^3$ corresponding to $\mathbb{R} \times V_\eta$, the profile of this wedge is $h_W(z, y_1, y_2) = r^{\pi/\eta} \sin(\pi\theta/\eta)$, where r and θ are the polar coordinates of the point (y_1, y_2). Hence, we have

$$h_W(z, y_1, y_2) \asymp (y_1 + y_2)^{(\pi/\eta)-2}(y_1 \sin\theta - y_2 \cos\theta)y_2.$$

Proof (Proof of Proposition 5.4) We now explain how to use this information to express h in (5.8) purely in terms of x_1, x_2, x_3. Consider a region in the cone C centered around one of the tips of the spherical triangle $C \cap \mathbb{S}^2(\mathbf{t})$, $\mathbf{t} = \mathbf{t}_4 = (0, 0, 0, N)$, say $\mathbf{q}_1 = (1, 0, 0, N - 1)$). In an Euclidean ball A of radius $1/2$ around this point, h is a positive harmonic function in a wedge W with aperture equal to the angle between the normal vectors of the two planes Q_2 and Q_3 meeting along the edge equal to the line L_1 passing through $\mathbf{t}$ and $\mathbf{q}_1$. The plane Q_i is the plane determined by the three points $\mathbf{t}, \mathbf{q}_1, \mathbf{q}_i$, parallel to the plane spanned by the vectors $\overrightarrow{\mathbf{tq}_1}, \overrightarrow{\mathbf{tq}_i}$, with $\overrightarrow{\mathbf{tq}_1} = (1, 0, 0, -1)$, $\overrightarrow{\mathbf{tq}_2} = (0, 1, 0, -1)$, $\overrightarrow{\mathbf{tq}_3} = (0, 0, 1, -1)$. It follows that the vectors $(1, 1, -3, 1)$ and $(1, -3, 1, 1)$ are, respectively, normal to Q_2 and Q_3 and contained in $\{\sum_1^4 x_i = 0\}$. The cosine of their angle $\eta \in [0, \pi)$ is $\cos\eta = 4/12 = 1/3$. Set

$$\beta = \frac{\pi}{\arccos(1/3)} \approx \frac{\pi}{1.23} \approx 2.55.$$

For a point $\boldsymbol{\zeta} = (\zeta_1, \zeta_2, \zeta_3, \zeta_4) \in C \cap A$, $\zeta_1 \asymp 1$, $\zeta_4 \asymp N$, $\zeta_2, \zeta_3 \in (0, 1/2)$, and

$$h_W(\boldsymbol{\zeta}) \asymp (\zeta_2 + \zeta_3)^{\beta-2}\zeta_2\zeta_3.$$

This gives that, in the ball $\frac{1}{2}A$,

$$\frac{h(\mathbf{x})}{h(\mathbf{y})} \leq C\frac{h_W(\mathbf{x})}{h_W(\mathbf{y})}.$$

Picking $\mathbf{y} \asymp (1/\sqrt{3}, 1/\sqrt{3}, 1/\sqrt{3}, N - 3/\sqrt{3})$ and $\mathbf{x} = r\theta$, $r = 1$, gives

$$h(\mathbf{x}) = \psi(\theta) \leq C(x_2 + x_3)^{\beta-2}x_2x_3.$$

Exchanging the role of $\mathbf{x}$ and $\mathbf{y}$ gives a matching lower bound so that

$$h(\mathbf{x}) = \psi(\theta) \asymp C(x_2 + x_3)^{\beta-2}x_2x_3.$$

By symmetry, this gives, on $C \cap \mathbb{S}^2(\mathbf{t})$, $\mathbf{x} = r\theta$, $r = 1$,

$$\psi(\theta) \asymp (x_1 + x_2)^{\beta-2}(x_1 + x_3)^{\beta-2}(x_2 + x_3)^{\beta-2}x_1x_2x_3.$$

Finally, for any $\mathbf{x} \in C$,

$$h(\mathbf{x}) \asymp (x_1 + x_2 + x_3)^{\alpha-3\beta+3}(x_1 + x_2)^{\beta-2}(x_1 + x_3)^{\beta-2}(x_2 + x_3)^{\beta-2}x_1x_2x_3.$$

5.4.2 The Function ϕ_0 in Coordinates

With Proposition 5.4 at hands, it is an easy matter to estimate the Perron–Frobenius function ϕ_0 of the simplex $\mathfrak{S}_n$ in the 4-player case ($k = 4$). For this purpose, we define the symmetric functions in the variables x_1, x_2, x_3, x_4:

$$\begin{aligned}
\tau_1 &= x_1x_2x_3x_4,\\
\tau_2 &= (x_1 + x_2)(x_1 + x_3)(x_1 + x_4)(x_2 + x_3)(x_2 + x_4)(x_3 + x_4),\\
\tau_3 &= (x_1 + x_2 + x_3)(x_1 + x_2 + x_4)(x_1 + x_3 + x_4)(x_2 + x_3 + x_4).
\end{aligned}$$

Theorem 5.4 *In the 4-player case, the Perron–Frobenius function ϕ_0 of the simplex $\mathfrak{S}_N$ satisfies*

$$\forall \mathbf{x} \in \mathfrak{S}_N, \quad \phi_0(\mathbf{x}) \asymp N^{-((3/2)+4\alpha-6\beta+4)}\tau_3^{\alpha-3\beta+3}\tau_2^{\beta-2}\tau_1 \tag{5.10}$$

uniformly in N. If $\mathbf{x} = (x_1, \ldots, x_4)$ satisfies $x_4 \geq N/4$, this simplifies to

$$\phi_0(\mathbf{x}) \asymp N^{-((3/2)+\alpha)}(x_1+x_2+x_3)^{\alpha-3\beta+3}(x_1+x_2)^{\beta-2}(x_1+x_3)^{\beta-2}$$
$$(x_2+x_3)^{\beta-2}x_1x_2x_3.$$

Proof We simply need to check that (5.10) is compatible with Propositions 5.4 and 5.3. Let us start with the central part of $\mathfrak{S}_N$ where each x_i satisfies $x_i \asymp N$. By the basic Harnack inequality and the normalization $\|\phi_0\|_2^2 = 1$, we know that $\phi_0(\mathbf{x}) \asymp N^{-3/2}$ there. In that region, the expression

$$T = \tau_3^{\alpha-3\beta+3}\tau_2^{\beta-2}\tau_1$$

is of order

$$N^{4(\alpha-3\beta+3)+6(\beta-2)+4} = N^{4\alpha-6\beta+4}.$$

This shows that (5.10) amounts to $\phi_0(\mathbf{x}) \asymp N^{-3/2}$ in this middle part, as desired. Then, we focus on each of the four corners of $\mathfrak{S}_N$. By symmetry, it suffices to consider one of this corner, say, the corner where $x_4 \geq N/4$. In that corner, the expression T satisfies

$$T \asymp N^{3\alpha-6\beta+4}(x_1+x_2+x_3)^{\alpha-3\beta+3}(x_1+x_2)^{\beta-2}(x_1+x_3)^{\beta-2}(x_2+x_3)^{\beta-2}x_1x_2x_3.$$

Hence, the right-hand side of (5.10) becomes

$$N^{-((3/2)+\alpha)}(x_1+x_2+x_3)^{\alpha-3\beta+3}(x_1+x_2)^{\beta-2}(x_1+x_3)^{\beta-2}(x_2+x_3)^{\beta-2}x_1x_2x_3.$$

This is indeed what Propositions 5.4 and 5.3 entail. It also gives the announced approximation for ϕ_0 in the corner $x_4 \geq N/4$. □

5.4.3 The Approximate Computation of α

As explained above (Definition 5.3), computing $\alpha = \alpha_4$ is equivalent to computing the Dirichlet eigenvalue λ_4 at (5.7) for the equilateral spherical triangle obtained on the unit sphere $\mathbb{S}^2$ by drawing a unit simplex in $\mathbb{R}^3$ with one vertex at the origin of $\mathbb{R}^3$. The exact value for this eigenvalue is not known. Grady Wright has constructed a numerical algorithm that approximates λ by computing the eigenvalue of a finite matrix corresponding to a radial basis function (RBF) finite difference approximation using a carefully selected grid [11, 13]. One of the difficulties is associated with dealing with corners of the spherical triangle, and the grid is selected to be more clustered near these corners.

When discretizing the continuous problem to the matrix eigenvalue problem, the "symmetry" of the original problem is lost and one is led to the computation of the spectrum (for us, just the lowest eigenvalue) of a non-symmetric matrix. In general, such computations are known to be difficult as the spectrum of the non-symmetric matrix can be more sensitive to perturbations than the underlying continuous problem. This is one of the reasons for which, although one should be confident that the value given by the algorithm is a good approximation, there is no proof of it and no error estimate. One way to "check" the algorithm is by testing its result on the equilateral spherical triangle associated with the first quadrant, which has eigenvalue 12. The algorithm gives 11.99. For the triangle associated with the tetrahedron, Grady Wright's algorithm gives $\lambda_4 \approx 38.447$ and produces Fig. 5.7.

5.5 Applications

In this section, we describe various explicit estimates obtained by applying the knowledge of ϕ_0. Recall that $\alpha \approx 5.68$ and $\beta = (\pi/\arccos 1/3) \approx 2.55$.

5.5.1 General Estimate

Putting together Theorem 5.2 and formula (5.10) yields the following general estimate. In order to simplify formula, we use the natural symmetries to focus on three cases depending on the positions of the starting point $\mathbf{s} \in \mathfrak{S}_N$ and the target point $\mathbf{z} \in \partial\mathfrak{S}_N$. Define three cases as follows:

- [Case 1] Assume $s_4 \geq N/4$, $z_4 = 0$, and $z_3 \geq N/3$ (one of the above average players ends up losing). In this case, the distance between $\mathbf{s}$ and $\mathbf{z}$ is of order N.
- [Case 2] Assume $s_4 \geq N/4 > \max\{s_1, s_2, s_3\}$, and $z_3 = 0$, $z_2 \geq N/3$, and $d(\mathbf{s}, \mathbf{z}) \asymp N$.
- [Case 3] Assume $s_4 \geq N/4 > \max\{s_1, s_2, s_3\}$, and $z_3 = 0$, $z_4 \geq N/3$ (one of the below average players ends up losing, while one of the above average players remains above average). The distance between $\mathbf{s}$ and $\mathbf{z}$ may be small.

Note that there is some overlap between cases 2 and 3.

Theorem 5.5 *In the 4-player game, consider points $\mathbf{s} \in \mathfrak{S}_N$ and $\mathbf{z} \in \partial\mathfrak{S}_N$ with $s_4 \geq N/4$. Assume Case 1 or Case 2 above. In Case 1, set $z_* = z_2$ and, in Case 2, set $z_* = z_4$. Then, we have*

$$\begin{aligned}\mathbf{P}(\mathbf{X}_\tau = \mathbf{z}|\mathbf{X}_0 = \mathbf{s}) &\asymp \\ &N^{-1-2\alpha}(s_1+s_2+s_3)^{\alpha-3\beta+3}[(s_1+s_2)(s_1+s_3)(s_2+s_3)]^{\beta-2}s_1s_2s_3 \\ &\times (z_1+z_*)^{\alpha-2\beta+1}(z_1z_*)^{\beta-1}.\end{aligned}$$

In particular,

$$\begin{aligned}&\mathbf{P}(\mathbf{X}_\tau \in \mathfrak{T}_{N,4}|\mathbf{X}_0 = \mathbf{s}) \\ &\asymp N^{-\alpha}(s_1+s_2+s_3)^{\alpha-3\beta+3}[(s_1+s_2)(s_1+s_3)(s_2+s_3)]^{\beta-2}s_1s_2s_3,\end{aligned}$$

and even

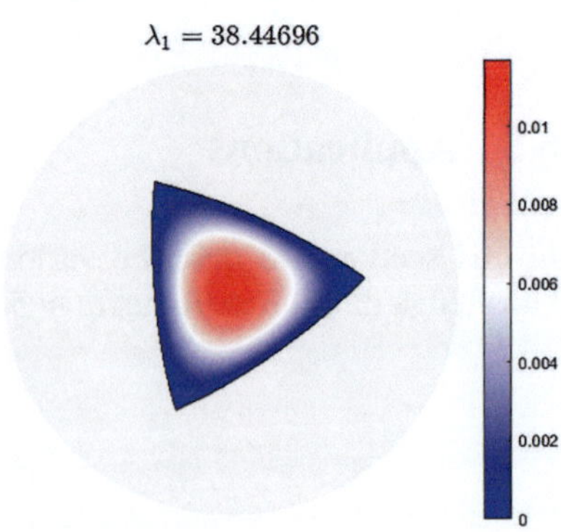

Fig. 5.7 The eigenvalue λ_4 (for the 4-player game; called λ_1 in the figure) and the associated eigenfunction

$$\mathbf{P}(\mathbf{X}_\tau \in \{\mathbf{z} : z_4 \le N/5\}|\mathbf{X}_0 = \mathbf{s})$$
$$\asymp N^{-\alpha}(s_1 + s_2 + s_3)^{\alpha-3\beta+3}[(s_1 + s_2)(s_1 + s_3)(s_2 + s_3)]^{\beta-2}s_1s_2s_3.$$

Proof In Cases 1 and 2 where we are sure that $d(\mathbf{s}, \mathbf{z}) \asymp N$, Theorem 5.2 and (5.3) give

$$\mathbf{P}(\mathbf{X}_\tau = \mathbf{z}|\mathbf{X}_0 = \mathbf{s}) \asymp N^2\phi_0(\mathbf{s})\phi_0(\mathbf{y_z}),$$

where $\mathbf{y_z}$ is an interior neighbor of the boundary point $\mathbf{z}$. Theorem 5.4 gives

$$\phi_0(\mathbf{s}) \asymp N^{-((3/2)+\alpha)}(s_1 + s_2 + s_3)^{\alpha-3\beta+3}[(s_1 + s_2)(s_1 + s_3)(s_2 + s_3)]^{\beta-2}s_1s_2s_3.$$

Similarly, the point $\mathbf{y}_z$ has the third coordinate greater than $(N/3) - 1$ in Case 1 and the second coordinate greater than $(N/3) - 1$ in Case 2 which yields

$$\phi_0(\mathbf{z}) \asymp N^{-(3/2)+\alpha}(z_1 + z_*)^{\alpha-2\beta+1}(z_1z_*)^{\beta-1}.$$

This gives the desired result for $\mathbf{P}(\mathbf{X}_\tau = \mathbf{z}|\mathbf{X}_0 = \mathbf{s})$. Summing over the free variables z_1, z_* gives the other statements.

Example: Players distributed on a power scale
As an illustration, assume that

$$s_4 \asymp N, \text{ and } s_i \asymp N^{\epsilon_i}, i \in \{1, 2, 3\} \text{ with } 0 \le \epsilon_1 \le \epsilon_2 \le \epsilon_3 < 1.$$

What is the (order of magnitude of) probability that the fourth player, which is currently the dominant player because $\epsilon_i \in [0, 1)$, $i \in \{1, 2, 3\}$, ends up losing first? The answer is

$$\mathbf{P}(\mathbf{X}_\tau \in \mathfrak{T}_{N,4}|\mathbf{X}_0 = \mathbf{s}) \asymp N^{-\alpha(1-\epsilon_3)-\beta(\epsilon_3-\epsilon_2)-(\epsilon_2-\epsilon_1)}.$$

When $\epsilon_1 = \epsilon_2 = \epsilon_3$,

$$\mathbf{P}(\mathbf{X}_\tau \in \mathfrak{T}_{N,4}|\mathbf{X}_0 = \mathbf{s}) \asymp N^{-\alpha(1-\epsilon_3)}.$$

When $\epsilon_2 = \epsilon_1 = \epsilon_3/2$,

$$\mathbf{P}(\mathbf{X}_\tau \in \mathfrak{T}_{N,4}|\mathbf{X}_0 = \mathbf{s}) \asymp N^{-\alpha(1-\epsilon_3)-\epsilon_3\beta/2}.$$

Theorem 5.6 *In the 4-player game, consider points $\mathbf{s} \in \mathfrak{S}_N$ and $\mathbf{z} \in \partial\mathfrak{S}_N$ with $s_4 \ge N/4$. Assume Case 3 above with $d = d(\mathbf{s}, \mathbf{z})$, that is, $z_3 = 0$, $z_4 \ge N/3$. Then,* *we have*

$$\mathbf{P}(\mathbf{X}_\tau = \mathbf{z}|\mathbf{X}_0 = \mathbf{s}) \asymp \frac{(s_1+s_2+s_3)^{\alpha-3\beta+3}}{d(s_1+s_2+s_3+d)^{2(\alpha-3\beta+3)}} \times \frac{[(s_1+s_2)(s_1+s_3)(s_2+s_3)]^{\beta-2}}{[(s_1+s_2+d)(s_1+s_3+d)(s_2+s_3+d)]^{2(\beta-2)}} \times \frac{s_1s_2s_3}{[(s_1+d)(s_2+d)(s_3+d)]^2} \times (z_1+z_2)^{\alpha-2\beta+1}(z_1z_2)^{\beta-1}.$$

Example: Most likely outcome in the very dominant player case
To illustrate this result, assume $\mathbf{s} = (1, 1, 1, N-3)$ and $\mathbf{z} = (z_1, z_2, 0, z_4)$ with $z_4 \geq N/3$. In this situation, the result simplifies to

$$\mathbf{P}(\mathbf{X}_\tau = \mathbf{z}|\mathbf{X}_0 = \mathbf{s}) \asymp \frac{(z_1+z_2)^{\alpha-2\beta+1}(z_1z_2)^{\beta-1}}{d^{1+2\alpha}} \asymp \frac{1}{d^{1+\alpha}}\left(\frac{z_1+z_2}{d}\right)^{\alpha}\left(\frac{z_1z_2}{(z_1+z_2)^2}\right)^{\beta}\left(\frac{z_1+z_2}{z_1z_2}\right).$$

In the last expression, every factor is bounded above and this shows that most of the mass is obtained when $d \asymp 1$ which forces $z_1 \asymp z_2 \asymp 1$ and $|z_4 - N| \asymp 1$.

Example: Probability that the second dominant player loses first
We saw that the probability that a very dominant player (say, $s_4 = N-3$) loses first is of order $N^{-\alpha}$. If there is a dominant player, i.e., $s_4 \sim N$, and a subdominant player, i.e., $s_3 \asymp N^{\epsilon}$, $\epsilon \in (0, 1)$, while $s_1 \asymp s_2 \asymp 1$, the probability that the dominant player loses first is of order

$$N^{-\alpha(1-\epsilon)-\beta\epsilon} = N^{-\alpha+(\alpha-\beta)\epsilon}.$$

Theorems 5.5 and 5.6 allow us to estimate the probability that the subdominant player ends up losing first in this situation. For this, we assume that $s_3 \asymp N^{\epsilon}$ with $\epsilon \in (0, 1)$, and $s_1, s_2 \asymp 1$. Of course, this implies that $s_4 \sim N$. We want to compute the probability that $\mathbf{X}_\tau \in \mathfrak{T}_{N,3} = \{\mathbf{z} \in \partial\mathfrak{S}_N \cap \{z_3 = 0\}\}$.

Given $\mathbf{z} \in \partial\mathfrak{S}_N$ with $z_3 = 0$ and $z_1 + z_2 \geq 2N/3$, hence $d(\mathbf{s}, \mathbf{z}) \asymp N$ and $z_4 \leq N/3$, we use Theorem 5.5 to see that

(continued)

$$\mathbf{P}(\mathbf{X}_\tau = \mathbf{z}|\mathbf{X}_0 = \mathbf{s}) \asymp N^{-\alpha+\epsilon(\alpha-\beta)-\beta-1}\min\{z_1, z_2\}^{\beta-1}.$$

The contribution of this to $\mathbf{P}(\mathbf{X}_\tau \in \mathfrak{T}_{N,3}|\mathbf{X}_0 = \mathbf{s})$ is of order $N^{-\alpha-\epsilon(\alpha-\beta)}$.

To estimate the contribution of those $\mathbf{z} \in \partial\mathfrak{S}_N$ with $z_3 = 0$ and $z_4 \geq N/3$, we use Theorem 5.6 to find

$$\mathbf{P}(\mathbf{X}_\tau = \mathbf{z}|\mathbf{X}_0 = \mathbf{s}) \asymp \frac{N^{\epsilon(\alpha-\beta)}(z_1+z_2)^{\alpha-2\beta+1}(z_1z_2)^{\beta-1}}{d^{1+2\beta}(N^\epsilon+d)^{2(\alpha-\beta)}}.$$

To sum the formula above over all $(z_1, z_2, 0, z_4)$ in $\mathfrak{T}_{N,3}$ with $z_4 \geq N/3$, we consider two cases. When $\max\{z_1, z_2\} \leq N^\epsilon$, which implies $N - z_4 \leq 2N^\epsilon$, we have $d \asymp N^\epsilon$. Otherwise, $\max\{z_1, z_2\} \geq N^\epsilon$, and $d \asymp (z_1 + z_2) \geq N^\epsilon$. Call the corresponding regions U_1 and U_2, respectively. In the first case,

$$\sum_{U_1}\mathbf{P}(\mathbf{X}_\tau = \mathbf{z}|\mathbf{X}_0 = \mathbf{s}) \asymp N^{-\epsilon\beta}.$$

In the second case (using symmetry, we can assume $d \asymp z_2 \geq z_1$),

$$\begin{aligned}\sum_{U_1}\mathbf{P}(\mathbf{X}_\tau = \mathbf{z}|\mathbf{X}_0 = \mathbf{s}) &\asymp N^{\epsilon(\alpha-\beta)}\sum_{N^\epsilon\leq d\leq N}\left(d^{-1-\alpha-\beta}\sum_1^d z_1^{\beta-1}\right)\\ &\asymp N^{\epsilon(\alpha-\beta)}\sum_{N^\epsilon\leq d\leq N} d^{-1-\alpha} \asymp N^{-\epsilon\beta}.\end{aligned}$$

It follows that, for $\mathbf{s} = (1, 1, s_3, N - s_3)$ and $s_3 \asymp N^\epsilon$, we have

$$\mathbf{P}(\mathbf{X}_\tau \in \mathfrak{T}_{N,3}|\mathbf{X}_0 = \mathbf{s}) \asymp N^{-\epsilon\beta}.$$

It is useful to compare this with the following heuristic: if player 4 has $x_4 \sim N$ chips, player 3 $x_3 \sim N^\epsilon$ chips and players 2 and 1 have just one chip each, to estimate the probability that player 3 loses, we can ignore player 4 and imagine we are watching a game with 3 players, N^ϵ chips, and with player 2 and 1 having only one chip each. From the 3-player game results, the probability that the very dominant player 3 loses such a game is of order $N^{-3\epsilon}$. Hence, this heuristic is off since the correct order of magnitude is $N^{-\epsilon\beta}$ and $\beta \approx 2.55$

Acknowledgments We thank Grady Wright from Boise State University for providing an algorithm that computes the approximate value of α given in the above theorem (see Sect. 5.4.3). We thank Alex Townsend for paying attention to our questions and connecting us with Grady.

We thank Persi Diaconis for encouraging us in our efforts and the anonymous reviewers for their helpful remarks. K. O'Connor's research was supported in part by NSF grant DMS-1645643. L. Saloff-Coste's research was partially supported by NSF grant DMS-2054593.

References

1. Thomas M. Cover. Gambler's ruin: A random walk on the simplex. In Thomas M. Cover and B. Gopinath, editors, *Open Problems in Communication and Computation*, pages 155–155. Springer New York, New York, NY, 1987.
2. Denis Denisov and Vitali Wachtel. Random walks in cones. *Ann. Probab.*, 43(3):992–1044, 2015.
3. Denis Denisov and Vitali Wachtel. Alternative constructions of a harmonic function for a random walk in a cone. *Electron. J. Probab.*, 24:Paper No. 92, 26, 2019. Author name corrected by publisher.
4. Denis Denisov and Vitali Wachtel. Harmonic measure in a multidimensional gambler's problem, 2022.
5. Persi Diaconis and Stewart N. Ethier. Gambler's ruin and the ICM. *Statist. Sci.*, 37(3):289–305, 2022.
6. Persi Diaconis, Kelsey Houston-Edwards, and Laurent Saloff-Coste. Analytic-geometric methods for finite Markov chains with applications to quasi-stationarity. *ALEA Lat. Am. J. Probab. Math. Stat.*, 17(2):901–991, 2020.
7. Persi Diaconis, Kelsey Houston-Edwards, and Laurent Saloff-Coste. Gambler's ruin estimates on finite inner uniform domains. *Ann. Appl. Probab.*, 31(2):865–895, 2021.
8. Pavel Gyrya and Laurent Saloff-Coste. Neumann and Dirichlet heat kernels in inner uniform domains. *Astérisque*, (336):viii+144, 2011.
9. Janna Lierl. Parabolic Harnack inequality for time-dependent non-symmetric Dirichlet forms. *J. Math. Pures Appl. (9)*, 140:1–66, 2020.
10. Janna Lierl and Laurent Saloff-Coste. The Dirichlet heat kernel in inner uniform domains: local results, compact domains and non-symmetric forms. *J. Funct. Anal.*, 266(7):4189–4235, 2014.
11. Varun Shankar, Grady B. Wright, Robert M. Kirby, and Aaron L. Fogelson. A radial basis function (RBF)-finite difference (FD) method for diffusion and reaction-diffusion equations on surfaces. *J. Sci. Comput.*, 63(3):745–768, 2015.
12. N. Th. Varopoulos. The central limit theorem in Lipschitz domains. *Boll. Unione Mat. Ital.*, 7(2):103–156, 2014.
13. Grady Wright. private communication, Sep. 2021.

Part III
Analysis on Manifolds

Chapter 6
Product Manifolds with Improved Spectral Cluster and Weyl Remainder Estimates

Xiaoqi Huang, Christopher D. Sogge, and Michael E. Taylor

6.1 Introduction

Spectral cluster estimates are operator norm estimates from L^2 to L^q of spectral projectors for the Laplace operator on a compact Riemannian manifold. In more detail, if X is a compact Riemannian manifold of dimension d_X and Laplace–Beltrami operator Δ_X, the universal estimate of [18] has the form

$$\| \mathbb{1}_{[\lambda-1,\lambda+1]}(P_X) \|_{L^2(X)\to L^q(X)} = O(\lambda^{\alpha(q)}), \quad P_X = \sqrt{-\Delta_X}, \tag{6.1}$$

where, with $n = d_X$,

$$\begin{aligned}\alpha(q) = \alpha(q,n) &= \max\left(n(\tfrac{1}{2}-\tfrac{1}{q})-\tfrac{1}{2},\ \tfrac{n-1}{2}(\tfrac{1}{2}-\tfrac{1}{q})\right) \\ &= \begin{cases} n(\frac{1}{2}-\frac{1}{q})-\frac{1}{2}, & \text{if } q \ge q_c(n) = \frac{2(n+1)}{n-1}, \\ \frac{n-1}{2}(\frac{1}{2}-\frac{1}{q}), & \text{if } 2 \le q \le q_c(n). \end{cases}\end{aligned} \tag{6.2}$$

X. Huang
Department of Mathematics, University of Maryland, College Park, MD, Maryland
e-mail: xhuang49@umd.edu

C. D. Sogge (✉)
Department of Mathematics, Johns Hopkins University, Baltimore, MD, Maryland
e-mail: sogge@jhu.edu

M. E. Taylor
Department of Mathematics, University of North Carolina, Chapel Hill, NC, USA
e-mail: met@math.unc.edu

P. Alonso Ruiz et al. (eds.), *From Classical Analysis to Analysis on Fractals*, Applied and Numerical Harmonic Analysis, https://doi.org/10.1007/978-3-031-37800-3_6

Much work has been done on the study of special classes of compact Riemannian manifolds for which stronger estimates hold. One of our goals here is to show that if Y is a compact Riemannian manifold, of dimension d_Y, with Laplace–Beltrami operator Δ_Y, for which such stronger results (cf. (6.3)) hold, and if X is an arbitrary compact Riemannian manifold, as described above, then, under broad circumstances, the product manifold $X \times Y$, of dimension $d = d_X + d_Y$, with product metric tensor and Laplace operator $\Delta = \Delta_X + \Delta_Y$, has also improved spectral cluster estimates.

We formulate the improved spectral cluster estimates on Y as follows:

$$\| \mathbb{1}_{[\lambda-\varepsilon(\lambda),\lambda+\varepsilon(\lambda)]}(P_Y) \|_{L^2(Y)\to L^q(Y)} \lesssim \sqrt{\varepsilon(\lambda)}\, B(\lambda), \tag{6.3}$$

where $P_Y = \sqrt{-\Delta_Y}$, $B(\lambda) = B(\lambda, q, Y)$, and $2 < q \le \infty$. Typically, $B(\lambda)$ is λ raised to a power (see, e.g., [5, 11, 22], and [14]) and possibly also involving $\log\lambda$-powers (see, e.g., [2, 4, 7, 10, 20]). So, it is natural to assume that

$$B(\theta\lambda) \le C_0\, B(\lambda) \quad \text{if } \lambda^{-1} \le \theta \le 2. \tag{6.4}$$

As we shall see later in (6.26), the improved bound in (6.3) requires

$$B(\lambda) \gtrsim \lambda^{\alpha(q)}, \quad \forall\, \lambda \ge 1, \tag{6.4'}$$

with $\alpha(q) = \alpha(q, d_Y)$ being the exponent in (6.2) depending on the dimension d_Y and q.

If $B(\lambda) = \lambda^{\alpha(q)}$, with $\alpha(q)$ being the exponent in the universal bounds that are due to the second author [16, 18] (see (6.2)), then (6.3) would represent a $\sqrt{\varepsilon(\lambda)}$ improvement over these bounds by projecting onto bands of size $\approx \varepsilon(\lambda)$ (as opposed to unit sized bands). We shall always assume that $\varepsilon(\lambda)$ decreases to 0 as $\lambda \to \infty$.

We prefer to express improvements including the factor $\sqrt{\varepsilon(\lambda)}$ since this would match with the bounds for the Stein–Tomas extension operators for $q \ge q_c$ (see Sect. 6.4). If $B(\lambda) = \lambda^{\alpha(q)}$ as in (6.2) and if $\varepsilon(\lambda)$ is very small, then they are only possible for certain q larger than the critical exponent $q_c = \frac{2(d_Y+1)}{d_Y-1}$. For instance, if $\varepsilon(\lambda) = \lambda^{-1}$, then one must have $q \ge \frac{2d_Y}{d_Y-2}$, since otherwise $\sqrt{\varepsilon(\lambda)}\,\lambda^{\alpha(q)} \to 0$ as $\lambda \to \infty$.

We shall also assume that $\varepsilon(\lambda)$ does not go to zero faster than the wavelength of eigenfunctions of frequency λ. More precisely, we shall assume that

$$t \to t\varepsilon(t) \quad \text{is non-decreasing for } t \ge 1. \tag{6.5}$$

We note that this is equivalent to the condition that

$$\theta\,\varepsilon(\theta\lambda) \le \varepsilon(\lambda) \quad \text{if } \lambda \ge 1 \text{ and } \lambda^{-1} \lesssim \theta \le 1. \tag{6.5'}$$

Typically, $\varepsilon(\lambda) = \lambda^{-\delta}$ for some $\delta \in (0, 1]$ or $\varepsilon(\lambda) = (\log\lambda)^{-\delta}$ for some $\delta > 0$.

Here is our first main result.

Theorem 6.1 *Assume that* (6.3) *is valid with* $B(\lambda)$ *and* $\varepsilon(\lambda)$ *satisfying* (6.4)–(6.4') *and* (6.5), *respectively. Then, if* $\Delta = \Delta_X + \Delta_Y$ *is the Laplace–Beltrami operator on the product manifold and* $P = \sqrt{-\Delta}$, *we have for all* $q > 2$

$$\left\| \mathbb{1}_{[\lambda-\varepsilon(\lambda),\lambda+\varepsilon(\lambda)]}(P) \right\|_{L^2(X\times Y)\to L^q(X\times Y)} \lesssim \sqrt{\varepsilon(\lambda)}B(\lambda)\lambda^{\alpha(q,d_X)+1/2}, \quad \forall\lambda \geq 1. \tag{6.6}$$

As we shall see at the end of the next section, if (6.3) is an improvement on Y over the universal bounds in [18], then, at least for sufficiently large exponents, (6.6) says that there are improved L^q spectral projection estimates on $X \times Y$.

Our second main result deals with the Weyl law, for the spectral counting function of the Laplace operator. Recall that the universal Weyl formula of Avakumovic [1], Levitan [15], and Hörmander [12] states that if $N(X, \lambda)$ denotes the number of eigenvalues, counted with multiplicity, of P_X which are $\leq \lambda$, then

$$N(X,\lambda) = (2\pi)^{-d_X}\omega_{d_X}(\text{Vol } X)\,\lambda^{d_X} + O(\lambda^{d_X-1}), \tag{6.7}$$

with ω_n denoting the volume of the unit ball in $\mathbb{R}^n$.

This result cannot be improved if X is the round sphere of dimension d_X. On the other hand, there are a number of results that do yield improved Weyl remainder estimates. In [8], it is shown that one can improve $O(\lambda^{d_X-1})$ to $o(\lambda^{d_X-1})$ in case the set of periodic geodesics has measure zero. The paper [2] shows that under certain geometrical conditions, such as nonpositive curvature, one can improve the remainder estimate to $O(\lambda^{d_X-1}/\log\lambda)$. Recently, Canzani and Galkowski [6] obtained such an improved remainder estimate for a much broader class of Riemannian manifolds. Among the results obtained there is that one gets this $(\log\lambda)^{-1}$ improvement on each Cartesian product manifold, with the product metric. Iosevich and Wyman [14] showed there are power improvements for products of spheres.

As with the L^q improvements in Theorem 6.1, we shall assume that there are $\varepsilon(\lambda)$ improvements, with $\varepsilon(\lambda)$ as in (6.5)–(6.5'), for Y, and show that these carry over for $X \times Y$. So, we shall assume that

$$N(Y,\lambda) = (2\pi)^{-d_Y}\omega_{d_Y}(\text{Vol } Y)\,\lambda^{d_Y} + R_Y(\lambda), \quad \text{with} \quad R_Y(\lambda) = O(\varepsilon(\lambda)\,\lambda^{d_Y-1}). \tag{6.8}$$

Here is our second main result.

Theorem 6.2 *Assume that* (6.8) *is valid. Then, for each compact Riemannian manifold, of dimension* d_X,

$$N(X\times Y,\lambda) = (2\pi)^{-d}\omega_d\,\text{Vol}(X\times Y)\,\lambda^d + O(\varepsilon(\lambda)\,\lambda^{d-1}), \quad d = d_X+d_Y. \tag{6.9}$$

The structure of the rest of this chapter is the following. In Sect. 6.2, we prove Theorem 6.1 and give applications. In Sect. 6.3, we prove Theorem 6.2. Section 6.4 presents some further results, including a study of multiple products of spheres.

6.2 Proof of L^q-Improvements and Some Applications

Let us now turn to the proof of Theorem 6.1. We first choose an orthonormal basis, $\{e^X_{\mu_i}\}$, of eigenfunctions of P_X with eigenvalues μ_i and $\{e^Y_{\nu_j}\}$ of P_Y with eigenvalues ν_j. Thus,

$$-\Delta_X e^X_{\mu_i} = \mu_i^2 e^X_{\mu_i} \quad \text{and} \quad -\Delta_Y e^Y_{\nu_j} = \nu_j^2 e^Y_{\nu_j}. \tag{6.10}$$

Then, $\{e^X_{\mu_i}(x) \cdot e^Y_{\nu_j}(y)\}_{i,j}$ is an orthonormal basis of eigenfunctions of $P = \sqrt{-\Delta}$, where $\Delta = \Delta_X + \Delta_Y$, with eigenvalues $\sqrt{\mu_i^2 + \nu_j^2}$, i.e.,

$$-\Delta(e^X_{\mu_i} e^Y_{\nu_j}) = (\mu_i^2 + \nu_j^2) e^X_{\mu_i} e^Y_{\nu_j}. \tag{6.11}$$

Consequently, the first inequality (6.6) is equivalent to the following:

$$\Big\| \sum_{(\mu_i,\nu_j)\in A_{\lambda,\varepsilon}} a_{ij}\, e^X_{\mu_i} e^Y_{\nu_j} \Big\|_{L^q(X\times Y)} \lesssim B(\lambda)\lambda^{\alpha(q)} \sqrt{\lambda\varepsilon(\lambda)}\, \|a\|_{\ell^2} \tag{6.12}$$

with $\|a\|_{\ell^2} = (\sum_{i,j} |a_{ij}|^2)^{1/2}$ and $A_{\lambda,\varepsilon}$ denoting the $\varepsilon(\lambda)$-annulus about $\lambda \cdot S^2$, i.e.,

$$A_{\lambda,\varepsilon} = \{\, (\mu, \nu) : |\lambda - \sqrt{\mu^2 + \nu^2}| \le \varepsilon(\lambda)\}. \tag{6.13}$$

To be able to use our assumptions (6.3) and (6.5') and the universal bounds (6.1), it is natural to break up the annulus into several pieces. Specifically, let

$$\Omega_{high} = \{(\mu, \nu) \in A_{\lambda,\varepsilon} : |\nu| \ge \lambda/2\} \tag{6.14}$$

denote the portion of $A_{\lambda,\varepsilon}$ where $|\nu|$ is relatively large and

$$\Omega_{low} = \{(\mu, \nu) \in A_{\lambda,\varepsilon} : |\nu| \le 1\} \tag{6.15}$$

be the portion where it is relatively small. We shall also break up the remaining region into the following disjoint dyadic pieces:

$$\Omega_\ell = \{(\mu, \nu) \in A_{\lambda,\varepsilon} \backslash \Omega_{high} : |\nu| \in (\lambda 2^{-\ell}, \lambda 2^{-\ell+1}]\}. \tag{6.16}$$

Thus,

$$A_{\lambda,\varepsilon} = \Omega_{high} \cup \Omega_{low} \cup \Big(\bigcup_{2^{-\ell}\in[\lambda^{-1}, \frac{1}{4}]} \Omega_\ell \Big). \tag{6.17}$$

We shall use the following simple lemma describing the geometry of each of these pieces.

Lemma 6.1 *There is a uniform constant C_0 so that*

$$|\sqrt{\lambda^2-\mu^2}-\nu| \le C_0\varepsilon(\lambda) \quad \text{if} \ \ (\mu,\nu)\in\Omega_{high} \ \text{and} \ \ \mu,\nu\ge 0. \tag{6.18}$$

Also, if $\Omega_\ell \neq \emptyset$, then

$$|\sqrt{\lambda^2-\mu^2}-\nu| \le C_0 2^\ell\varepsilon(\lambda) \quad \text{if} \ \ (\mu,\nu)\in\Omega_\ell \ \text{and} \ \ \mu,\nu\ge 0, \tag{6.19}$$

and if

$$I_\ell = \{\mu : \ (\mu,\nu)\in\Omega_\ell, \ \mu,\nu\ge 0\}, \tag{6.20}$$

then for fixed ℓ with $2^{-\ell}\ge\lambda^{-\frac12}$

$$I_\ell \quad \text{is an interval in} \ \ [0,\lambda] \quad \text{of length} \ \ |I_\ell|\le C_0\lambda 2^{-2\ell}, \tag{6.21}$$

and also

$$\begin{aligned}|\sqrt{\lambda^2-\nu^2}-\mu| \le C_0 \quad &\text{if} \ \ \mu,\nu\ge 0,\\ &\text{and} \ \ (\mu,\nu)\in\Omega_\ell \ \text{with} \ 2^{-\ell}\le\lambda^{-\frac12}, \ \text{or} \ \ (\mu,\nu)\in\Omega_{low}.\end{aligned} \tag{6.22}$$

The bound in (6.18) is straightforward since $\nu\ge\lambda/2$ for the (μ,ν) there. One obtains (6.22) similarly and indeed can replace C_0 there by $C_0\varepsilon(\lambda)$, although this will not be needed. One obtains (6.19) and (6.21) by noting that the (μ,ν) there must be of the form

$$(\mu,\nu) = r(\cos\theta,\sin\theta) \quad \text{with} \ \ r\in[\lambda-\varepsilon(\lambda),\lambda+\varepsilon(\lambda)] \quad \text{and} \ \ \theta\approx 2^{-\ell}.$$

We also require the following estimates that are a simple consequence of our main assumption (6.3) and the universal bounds (6.1).

Lemma 6.2 *There is a universal constant C_0 so that*

$$\left\| \mathbb{1}_{[\lambda-\rho,\lambda+\rho]}(P_X) \right\|_{L^2(X)\to L^q(X)} \le C_0\rho^{1/2}\lambda^{\alpha(q)}, \quad \text{if} \ \ \rho\in[1,\lambda], \tag{6.23}$$

and

$$\left\| \mathbb{1}_{[\lambda-\rho,\lambda+\rho]}(P_Y) \right\|_{L^2(Y)\to L^q(Y)} \le C_0\rho^{1/2}B(\lambda), \quad \text{if} \ \ \rho\in[\varepsilon(\lambda),\lambda], \tag{6.24}$$

and also

$$\| \mathbb{1}_{[2^k,2^{k+1}]}(P_Y) \|_{L^2(Y)\to L^q(Y)} \le C_0 2^{d_Y(\frac{1}{2}-\frac{1}{q})k}. \tag{6.25}$$

Note that if we let $\rho = 1$ in (6.24), we have

$$\begin{aligned} B(\lambda) &\gtrsim \| \mathbb{1}_{[\lambda-1,\lambda+1]}(P_Y) \|_{L^2(Y)\to L^q(Y)} \\ &\gtrsim \lambda^{\alpha(q)}, \end{aligned} \tag{6.26}$$

where in the second inequality we used the lower bounds on the spectral projection operator, which holds in the general case (see [19]).

Proof The proofs are well known. One obtains (6.23) from (6.1) by writing $[\lambda - \rho, \lambda+\rho]$ as the union of $O(\rho)$ disjoint intervals I_k of length 1 or less, each contained in $[0, 2\lambda]$. By the Cauchy–Schwarz inequality, one then has

$$\begin{aligned} \| \mathbb{1}_{[\lambda-\rho,\lambda+\rho]}(P_X) f \|_{L^q(X)} &\lesssim \rho^{1/2} \Big(\sum_k \| \mathbb{1}_{I_k}(P_X) f \|^2_{L^q(X)} \Big)^{1/2} \\ &\lesssim \rho^{1/2} \lambda^{\alpha(q)} \Big(\sum_k \| \mathbb{1}_{I_k}(P_X) f \|^2_{L^2(X)} \Big)^{1/2} \\ &\lesssim \rho^{1/2} \lambda^{\alpha(q)} \| f \|_{L^2(X)}, \end{aligned}$$

using (6.1) in the second inequality and orthogonality in the last one.

To prove (6.24), we note that our assumptions (6.4) and (6.5') imply that $B(\lambda_1) \approx B(\lambda_2)$ and $\varepsilon(\lambda_1) \approx \varepsilon(\lambda_2)$ if $\lambda_1 \approx \lambda_2$. Taking this into account, if $\rho \in [\varepsilon(\lambda), 1]$, one proves (6.24) by a similar argument used for (6.23) if one covers $[\lambda - \rho, \lambda + \rho]$ by $O(\rho/\varepsilon(\lambda))$ disjoint intervals of length $\varepsilon(\lambda)$ or less and uses the fact that (6.3) includes a $(\varepsilon(\lambda))^{1/2}$ factor in the right. Similarly, if $I_k = [k, k-1)$ with $k \le \lambda$, then one concludes that

$$\| \mathbb{1}_{I_k} \|_{L^2(Y)\to L^q(Y)} \lesssim B(k) \lesssim B(\lambda),$$

which can be used with the above argument involving the use of the Cauchy–Schwarz inequality to handle the case where $\rho \in [1, \lambda]$.

The remaining inequality (6.25) is a standard Bernstein estimate (see [19]). □

Having collected the tools we need, let us state the bounds associated with the various regions (6.14)–(6.16) of $A_{\lambda,\varepsilon}$ that will give us (6.12).

First, for all $\varepsilon(\lambda)$ satisfying the conditions in Theorem 6.1, we claim that we have

$$\Big\| \sum_{(\mu_i,\nu_j)\in\Omega_{high}} a_{ij}\, e^X_{\mu_i} e^Y_{\nu_j} \Big\|_{L^q(X\times Y)} \lesssim \sqrt{\varepsilon(\lambda)}\, B(\lambda) \cdot \lambda^{\alpha(q)} \sqrt{\lambda}\, \|a\|_{\ell^2}, \tag{6.27}$$

$$\Big\| \sum_{(\mu_i,\nu_j)\in\Omega_{low}} a_{ij}\, e^X_{\mu_i} e^Y_{\nu_j} \Big\|_{L^q(X\times Y)} \lesssim \lambda^{\alpha(q)}\, \|a\|_{\ell^2}, \tag{6.28}$$

as well as

$$\Big\| \sum_{(\mu_i,\nu_j)\in\Omega_\ell} a_{ij}\, e^X_{\mu_i} e^Y_{\nu_j} \Big\|_{L^q(X\times Y)} \lesssim \sqrt{2^\ell\varepsilon(\lambda)}\, B(2^{-\ell}\lambda)\cdot\lambda^{\alpha(q)}\,\sqrt{\lambda 2^{-2\ell}}\;\|a\|_{\ell^2}, \tag{6.29}$$

if $2^{-\ell}\ge\lambda^{-\frac{1}{2}}$.

For remaining pieces $2^{-\ell}\le\lambda^{-\frac{1}{2}}$, we need to obtain estimates to handle the two cases where $\varepsilon(\lambda)\le\frac{1}{4}\cdot\lambda 2^{-2\ell}$ and $\varepsilon(\lambda)\ge\frac{1}{4}\cdot\lambda 2^{-2\ell}$. For the first case, we shall use the following estimate which is valid for all $\varepsilon(\lambda)$ satisfying (6.5'):

$$\Big\| \sum_{(\mu_i,\nu_j)\in\Omega_\ell} a_{ij}\, e^X_{\mu_i} e^Y_{\nu_j} \Big\|_{L^q(X\times Y)} \lesssim \sqrt{2^\ell\varepsilon(\lambda)}\, B(2^{-\ell}\lambda)\cdot\lambda^{\alpha(q)}\;\|a\|_{\ell^2}, \tag{6.30}$$

while for the other case we shall use the fact that we also always have

$$\Big\| \sum_{(\mu_i,\nu_j)\in\Omega_\ell} a_{ij}\, e^X_{\mu_i} e^Y_{\nu_j} \Big\|_{L^q(X\times Y)} \lesssim \lambda^{\alpha(q)}\,(\lambda 2^{-\ell})^{d_Y(\frac{1}{2}-\frac{1}{q})}\|a\|_{\ell^2}. \tag{6.31}$$

Let us now see how the bounds in (6.27)–(6.31) yield those in Theorem 6.1. To use (6.29), we note that by (6.4) with $\theta=2^{-\ell}$

$$\sqrt{2^\ell\varepsilon(\lambda)}\, B(2^{-\ell}\lambda)\,\lambda^{\alpha(q)}\,\sqrt{\lambda 2^{-2\ell}} \lesssim 2^{-\ell/2}\, B(\lambda)\,\lambda^{\alpha(q)}\,\sqrt{\lambda\varepsilon(\lambda)}$$

and so

$$\sum_{2^{-\ell}\in[\lambda^{-\frac{1}{2}},\frac{1}{4}]} \Big\| \sum_{(\mu_i,\nu_j)\in\Omega_\ell} a_{ij} e^X_{\mu_i} e^Y_{\nu_j} \Big\|_{L^q(X\times Y)} \lesssim B(\lambda)\,\lambda^{\alpha(q)}\,\sqrt{\lambda\varepsilon(\lambda)}\,\|a\|_{\ell^2}. \tag{6.32}$$

To use (6.30), we note that by (6.4) with $\theta=2^{-\ell}$,

$$\sqrt{2^\ell\varepsilon(\lambda)}\, B(2^{-\ell}\lambda)\,\lambda^{\alpha(q)} \lesssim 2^{\ell/2}\, B(\lambda)\,\lambda^{\alpha(q)}\,\sqrt{\varepsilon(\lambda)}$$

and so

$$\sum_{2^{-\ell}\in[2\lambda^{-\frac{1}{2}}\varepsilon(\lambda)^{\frac{1}{2}},\lambda^{-\frac{1}{2}}]} \Big\| \sum_{(\mu_i,\nu_j)\in\Omega_\ell} a_{ij} e^X_{\mu_i} e^Y_{\nu_j} \Big\|_{L^q(X\times Y)} \lesssim B(\lambda)\,\lambda^{\alpha(q)}\,(\lambda\varepsilon(\lambda))^{1/4}\,\|a\|_{\ell^2}, \tag{6.33}$$

which is better than desired since we are assuming $\varepsilon(\lambda)\ge\lambda^{-1}$.

Finally, if $2^{-\ell}\in[\lambda^{-1}, 2\lambda^{-\frac{1}{2}}\varepsilon(\lambda)^{\frac{1}{2}}]$, by (6.31), we have

$$\sum_{2^{-\ell}\in[\lambda^{-1},2\lambda^{-\frac{1}{2}}\varepsilon(\lambda)^{\frac{1}{2}}]} \Big\| \sum_{(\mu_i,\nu_j)\in\Omega_\ell} a_{ij} e^X_{\mu_i} e^Y_{\nu_j} \Big\|_{L^q(X\times Y)} \lesssim \lambda^{\alpha(q)} (\lambda\varepsilon(\lambda))^{\frac{d_Y}{2}(\frac{1}{2}-\frac{1}{q})} \|a\|_{\ell^2}. \tag{6.34}$$

It is straightforward to check that for all $q \geq 2$,

$$(\lambda\varepsilon(\lambda))^{\frac{d_Y}{2}(\frac{1}{2}-\frac{1}{q})} \lesssim B(\lambda)\sqrt{\lambda\varepsilon(\lambda)},$$

given the fact that $B(\lambda) \geq \lambda^{\alpha(q)}$ and $\lambda^{-1} \leq \varepsilon(\lambda) \leq 1$. Thus, our proof would be complete if we could establish (6.27)–(6.31).

To prove the first one (6.27), we note that if $y \in Y$ is fixed, since $\mu_i \leq \lambda$ if $(\mu_i, \nu_j) \in \Omega_{high}$, by (6.23) with $\rho = \lambda$ and orthogonality

$$\begin{aligned}
&\Big\| \sum_{(\mu_i,\nu_j)\in\Omega_{high}} a_{ij} e^X_{\mu_i}(\,\cdot\,)\, e^Y_{\nu_j}(y) \Big\|_{L^q(X)} \\
&\lesssim \lambda^{\alpha(q)+1/2} \Big\| \sum_{(\mu_i,\nu_j)\in\Omega_{high}} a_{ij} e^X_{\mu_i}(\,\cdot\,)\, e^Y_{\nu_j}(y) \Big\|_{L^2(X)} \\
&= \lambda^{\alpha(q)+1/2} \Big(\sum_i \Big| \sum_{\{j:\,(\mu_i,\nu_j)\in\Omega_{high}\}} a_{ij} e^Y_{\nu_j}(y) \Big|^2 \Big)^{1/2}.
\end{aligned}$$

If we take the $L^q(Y)$ norm of the left side and use this inequality along with Minkowski's inequality, we conclude that

$$\begin{aligned}
&\Big\| \sum_{(\mu_i,\nu_j)\in\Omega_{high}} a_{ij} e^X_{\mu_i} e^Y_{\nu_j} \Big\|_{L^q(X\times Y)} \\
&\lesssim \lambda^{\alpha(q)+1/2} \Big(\sum_i \Big\| \sum_{\{j:\,(\mu_i,\nu_j)\in\Omega_{high}\}} a_{ij} e^Y_{\nu_j} \Big\|^2_{L^q(Y)} \Big)^{1/2}.
\end{aligned} \tag{6.35}$$

Since $\nu_j \approx \lambda$ if $(\mu_i, \nu_j) \in \Omega_{high}$, by (6.18) and (6.24) with $\rho = C_0\varepsilon(\lambda)$, we have for each fixed i

$$\begin{aligned}
\Big\| \sum_{\{j:\,(\mu_i,\nu_j)\in\Omega_{high}\}} a_{ij} e^Y_{\nu_j} \Big\|_{L^q(Y)} &\lesssim \sqrt{\varepsilon(\lambda)}B(\lambda) \Big\| \sum_{\{j:\,(\mu_i,\nu_j)\in\Omega_{high}\}} a_{ij} e^Y_{\nu_j} \Big\|_{L^2(Y)} \\
&= \sqrt{\varepsilon(\lambda)}B(\lambda) \Big(\sum_{\{j:\,(\mu_i,\nu_j)\in\Omega_{high}\}} |a_{ij}|^2 \Big)^{1/2}.
\end{aligned} \tag{6.36}$$

Clearly, (6.35) and (6.36) imply (6.27).

The proof of (6.29) is similar. Recall that the nonzero terms involve $\mu_i \in I_\ell$, and if $2^{-\ell} \geq \lambda^{-\frac{1}{2}}$, then I_ℓ is an interval of length $\rho \leq C_0\lambda 2^{-2\ell}$ as in (6.20). So, if we use the analog of (6.23) with this value of ρ and with λ replaced by the center of I_ℓ, we can repeat the proof of (6.35) to conclude that

$$\Big\| \sum_{(\mu_i,\nu_j)\in\Omega_\ell} a_{ij} e^X_{\mu_i} e^Y_{\nu_j} \Big\|_{L^q(X\times Y)}$$
$$\lesssim \lambda^{\alpha(q)} \sqrt{\lambda 2^{-2\ell}} \Big(\sum_i \Big\| \sum_{\{j:\,(\mu_i,\nu_j)\in\Omega_\ell\}} a_{ij} e^Y_{\nu_j} \Big\|^2_{L^q(Y)} \Big)^{1/2}. \tag{6.37}$$

Since $\nu_j \approx 2^{-\ell}\lambda$ if $(\mu_i,\nu_j)\in\Omega_\ell$ and by (6.5') together with the fact that $2^{-\ell} \geq \lambda^{-\frac{1}{2}}$, we have

$$\varepsilon(2^{-\ell}\lambda) \leq 2^\ell \varepsilon(\lambda) \leq 2^{-\ell}\lambda. \tag{6.38}$$

By (6.19) and (6.24) with $\rho = 2^\ell\varepsilon(\lambda)$, we can argue as above to see that for each fixed i we have

$$\Big\| \sum_{\{j:\,(\mu_i,\nu_j)\in\Omega_\ell\}} a_{ij} e^Y_{\nu_j} \Big\|_{L^q(Y)} \lesssim \big(2^\ell\varepsilon(\lambda)\big)^{1/2} B(2^{-\ell}\lambda) \Big(\sum_{\{j:\,(\mu_i,\nu_j)\in\Omega_\ell\}} |a_{ij}|^2 \Big)^{1/2}. \tag{6.39}$$

By combining (6.37) and (6.39), we obtain (6.29).

Next, we turn to (6.30). We note that if $2^{-\ell} \leq \lambda^{-\frac{1}{2}}$, there is a uniform constant C_0 so that

$$\mu_i \in [\lambda - C_0, \lambda + C_0], \quad \text{if } (\mu_i,\nu_j)\in\Omega_\ell \text{ for some } j. \tag{6.40}$$

This just follows from the fact that if $(\mu_i,\nu_j)\in\Omega_{low}$, then we can write $(\mu_i,\nu_j) = r(\cos\theta,\sin\theta)$ with $0 \leq \theta \lesssim \lambda^{-1/2}$ and $r \in [\lambda-\varepsilon(\lambda), \lambda+\varepsilon(\lambda)] \subset [\lambda-1, \lambda+1]$. If we use (6.40) and (6.1), we can argue as above to see that

$$\Big\| \sum_{(\mu_i,\nu_j)\in\Omega_\ell} a_{ij} e^X_{\mu_i} e^Y_{\nu_j} \Big\|_{L^q(X\times Y)} \lesssim \lambda^{\alpha(q)} \Big(\sum_i \Big\| \sum_{\{j:\,(\mu_i,\nu_j)\in\Omega_\ell\}} a_{ij} e^Y_{\nu_j} \Big\|^2_{L^q(Y)} \Big)^{1/2}. \tag{6.41}$$

For fixed μ_i, if $(\mu_i,\nu_j)\in\Omega_\ell$, we have $\nu_j \approx 2^{-\ell}\lambda$, and by (6.19), ν_j lie in an interval of length $\varepsilon(\lambda)2^\ell$. Now, using (6.5') again together with the fact that $\varepsilon(\lambda) \leq \frac{1}{4}\lambda 2^{-2\ell}$, one can see that (6.38) still holds in this case. Thus, by (6.24) with $\rho = 2^\ell\varepsilon(\lambda)$, we can argue as above to see that for each fixed i, we have the analogous inequality as in (6.39), which, combined with (6.41), implies (6.30).

To prove (6.31), if one uses (6.25), then we find we can replace (6.39) with

$$\Big\| \sum_{\{j:\,(\mu_i,\nu_j)\in\Omega_\ell\}} a_{ij} e^Y_{\nu_j} \Big\|_{L^q(Y)} \lesssim (\lambda 2^{-\ell})^{d_Y(\frac{1}{2}-\frac{1}{q})} \Big(\sum_{\{j:\,(\mu_i,\nu_j)\in\Omega_\ell\}} |a_{ij}|^2 \Big)^{1/2},$$

and this along with (6.41) yields (6.31).

The proof of (6.28) is similar. Since in this case, there is a uniform constant C_0 so that

$$\mu_i \in [\lambda - C_0, \lambda + C_0], \quad \text{if } (\mu_i, \nu_j) \in \Omega_{low} \text{ for some } j. \tag{6.42}$$

Thus, (6.41) still holds in this case, and by (6.25), we can replace (6.39) with

$$\Big\| \sum_{\{j:\,(\mu_i,\nu_j)\in\Omega_{low}\}} a_{ij} e^Y_{\nu_j} \Big\|_{L^q(Y)} \lesssim \Big(\sum_{\{j:\,(\mu_i,\nu_j)\in\Omega_{low}\}} |a_{ij}|^2 \Big)^{1/2}. \tag{6.43}$$

This along with (6.41) yields (6.28).

6.2.1 Some Applications

Let us now show that for products of round spheres $S^{d_1} \times S^{d_2}$, one can obtain power improvements over the universal bounds in [18] for *all* exponents $2 < q \le \infty$. This generalizes the L^∞ improvements of Iosevich and Wyman [14]. Using our improved L^q-estimates, we can also obtain improved bounds for large exponents for products of the form $S^{d_1} \times S^{d_2} \times M^n$, where M^n is an arbitrary compact manifold of dimension n. If M^n is a product of spheres and $q = \infty$, the bounds agree with the ones that are implicit in Iosevich and Wyman [14].

Theorem 6.3 *Suppose that $d_1, d_2 \ge 1$. Then, for all $\varepsilon > 0$, we have the following estimates for eigenfunctions on $S^{d_1} \times S^{d_2}$:*

$$\|e_\lambda\|_{L^q(S^{d_1}\times S^{d_2})} \le C_\varepsilon \lambda^{\alpha(q,d_1)+\alpha(q,d_2)+\varepsilon} \|e_\lambda\|_{L^2(S^{d_1}\times S^{d_2})}, \quad 2 < q \le \infty, \tag{6.44}$$

where

$$\alpha(q,d) = \max\big(d(\tfrac{1}{2} - \tfrac{1}{q}) - \tfrac{1}{2},\ \tfrac{d-1}{2}(\tfrac{1}{2} - \tfrac{1}{q}) \big) \tag{6.45}$$

is the λ-exponent in the d-dimensional universal bounds.

In order to use Theorem 6.1 to obtain bounds for product manifolds involving $S^{d_1} \times S^{d_2}$, we note that the distinct eigenvalues of $P = \sqrt{-\Delta_{S^{d_1}\times S^{d_2}}}$ are of the form

$$\sqrt{(k + (d_1-1)/2)^2 + (\ell + (d_2-1)/2)^2} \quad \text{with } k, \ell = 0, 1, 2, \ldots.$$

Consequently, the gap between successive distinct eigenvalues which are comparable to λ must be larger than a fixed multiple of λ^{-1}. So, (6.44) implies that we also have the corresponding bounds for the spectral projection operators onto windows of length $\varepsilon(\lambda) = \lambda^{-1}$:

$$\left\| \mathbb{1}_{[\lambda-\lambda^{-1},\lambda+\lambda^{-1}]}(P) \right\|_{L^2(S^{d_1}\times S^{d_2})\to L^q(S^{d_1}\times S^{d_2})} \le C_\varepsilon \lambda^{\alpha(q,d_1)+\alpha(q,d_2)+\varepsilon}, \ \forall\, \varepsilon > 0,$$

$$\text{if } 2 < q \le \infty \text{ and } P = \sqrt{-\Delta_{S^{d_1}\times S^{d_2}}}. \tag{6.44'}$$

A calculation shows that for *all* $q > 2$

$$\alpha(q, d_1) + \alpha(q, d_2) < \alpha(q, d_1 + d_2).$$

Thus, (6.44') says that on $S^{d_1} \times S^{d_2}$ one has power improvements over the universal bounds for manifolds of dimension $d = d_1 + d_2$ (but with $\varepsilon(\lambda) \equiv 1$). Also, by considering tensor products of spherical harmonics that saturate the $L^q(S^{d_j})$, $j = 1, 2$, bounds (see, e.g., [16]), one sees that (6.44) and hence (6.44') are optimal (up to possibly the λ^ε factor).

Let us now single out a couple of special cases of (6.44').

First, we have

$$\left\| \mathbb{1}_{[\lambda-\lambda^{-1},\lambda+\lambda^{-1}]}(P) \right\|_{L^2(S^{d_1}\times S^{d_2})\to L^q(S^{d_1}\times S^{d_2})} \le C_\varepsilon \lambda^{d(\frac12-\frac1q)-\frac12}\, \lambda^{-\frac12+\varepsilon}, \ \forall\, \varepsilon > 0,$$

$$\text{if } d = d_1 + d_2 \text{ and } q \ge \max\left(\tfrac{2(d_1+1)}{d_1-1}, \tfrac{2(d_2+1)}{d_2-1} \right), \tag{6.46}$$

which is a $\lambda^{-\frac12+\varepsilon}$ improvement for this range of exponents in dimension d versus the universal bounds. This is optimal in the sense that no bounds of this type may hold on *any* manifold of dimension d with $\lambda^{-\frac12+\varepsilon}$ replaced by $\lambda^{-\frac12-\delta}$ for some $\delta > 0$. For, by Bernstein inequalities, such an estimate would imply that the above spectral projection operators map $L^2 \to L^\infty$ with norm $O(\lambda^{\frac d2-1-\delta})$. This cannot hold in dimension d since it would imply that the number of eigenvalues of the square root of minus the Laplacian counted with multiplicity which are in subintervals of length λ^{-1} in $[\lambda/2, \lambda]$ would be $O(\lambda^{d-2-\delta})$, and this would contradict the Weyl formula for P.

Second, we note that if $d_1 = d_2 = 1$ and $q = \infty$, then (6.46) is just

$$\left\| \mathbb{1}_{[\lambda-\lambda^{-1},\lambda+\lambda^{-1}]}(\sqrt{-\Delta_{\mathbb{T}^2}}) \right\|_{L^2(\mathbb{T}^2)\to L^\infty(\mathbb{T}^2)} = O(\lambda^\varepsilon), \ \forall \varepsilon > 0. \tag{6.47}$$

This is equivalent to the classical fact that the number of integer lattice points on $\lambda \cdot S^1$ is $O(\lambda^\varepsilon)$, i.e.,

$$\#\{j \in \mathbb{Z}^2 : |j| = \lambda\} = O(\lambda^\varepsilon) \quad \forall\, \varepsilon > 0. \tag{6.47'}$$

We shall use this bound in our proof of Theorem 6.3.

Before proving this result, let us show how we can use the bounds in Theorem 6.1 to obtain a couple of corollaries.

The first says that sufficiently large exponents we can obtain power improvements of the universal bounds for products involving $S^{d_1} \times S^{d_2}$.

Corollary 6.1 *Let M^n be a compact manifold of dimension $n \geq 1$ and consider the product manifold $S^{d_1} \times S^{d_2} \times M^n$, where $d_1, d_2 \geq 1$. Then, if*

$$P = \sqrt{-(\Delta_{S^{d_1}} + \Delta_{S^{d_2}} + \Delta_{M^n})},$$

we have

$$\begin{aligned}&\| \mathbb{1}_{[\lambda-\lambda^{-1},\lambda+\lambda^{-1}]}(P) \|_{L^2(S^{d_1}\times S^{d_2}\times M^n)\to L^q(S^{d_1}\times S^{d_2}\times M^n)} \leq C_\varepsilon \lambda^{d(\frac{1}{2}-\frac{1}{q})-\frac{1}{2}} \lambda^{-\frac{1}{2}+\varepsilon},\\ &\forall \varepsilon > 0, \ \ \text{if} \ \ d = d_1 + d_2 + n \ \ \text{and} \ \ q \geq \max\big(\tfrac{2(d_1+1)}{d_1-1}, \tfrac{2(d_2+1)}{d_2-1}, \tfrac{2(n+1)}{n-1}\big).\end{aligned} \tag{6.48}$$

Furthermore, we have

$$\| \mathbb{1}_{[\lambda-\lambda^{-1},\lambda+\lambda^{-1}]}(P) \|_{L^2(S^{d_1}\times S^{d_2}\times M^n)\to L^q(S^{d_1}\times S^{d_2}\times M^n)} \leq C_q \lambda^{d(\frac{1}{2}-\frac{1}{q})-\frac{1}{2}-\delta}, \tag{6.49}$$

for some $\delta = \delta(q, d_1, d_2, n) > 0$ if $q > \frac{2(d+1)}{d-1}$, $d = d_1 + d_2 + n$.

To prove these to bounds, we note that for q as in (6.48) we have

$$\alpha(q, d_1) + \alpha(q, d_2) + \alpha(q, n) + \tfrac{1}{2} = d(\tfrac{1}{2} - \tfrac{1}{q}) - 1.$$

Consequently, (6.48) follows immediately from (6.44) and (6.6) with $\varepsilon(\lambda) = \lambda^{-1}$ and

$$B(\lambda) = \lambda^{(d_1+d_2)(\frac{1}{2}-\frac{1}{q})-\frac{1}{2}+\varepsilon}.$$

Since (6.48) is a power improvement over $O(\lambda^{d(\frac{1}{2}-\frac{1}{q})-\frac{1}{2}})$ bounds for large exponents and the universal bounds imply that (6.49) is valid when $\delta = 0$ and $q = q_c(d) = \frac{2(d+1)}{d-1}$, one obtains (6.49) via a simple interpolation argument.

A calculation shows that we cannot use Theorem 6.1 to obtain improvements over the universal bounds when $q \in (2, q_c(d)]$ with $q_c(d)$ as above being the critical exponent. We should point out that Canzani and Galkowski [7] recently obtained $\sqrt{\log \lambda}$-improvements over the universal bounds (with $\varepsilon(\lambda) = (\log \lambda)^{-1}$) for arbitrary products of manifolds and $q > q_c(d)$. They as well as Iosevich and Wyman [14] conjectured that for such manifolds appropriate power improvements over the universal bounds should always be possible. Obtaining any improvements for $q \in (2, q_c(d)]$, though, appears difficult except in special cases such as for products involving products of spheres as above. Perhaps, though, the Kakeya–Nikodym approach that was used in [3] and [4] to obtain log-power improvements of eigenfunction estimates for manifolds of nonpositive sectional curvature could be used to handle critical and subcritical exponents.

Let us also state one more corollary which generalizes the well-known higher dimensional version of (6.47'):

$$\#\{j \in \mathbb{Z}^n : |j| = \lambda\} = O(\lambda^{n-2+\varepsilon}) \ \forall \varepsilon > 0 \text{ if } n \geq 3. \tag{6.50}$$

Just as for the special case where $n = 2$ discussed above, this is easily seen to be equivalent to the following sup-norm bounds:

$$\left\| \mathbb{1}_{[\lambda-\lambda^{-1},\lambda+\lambda^{-1}]}(\sqrt{-\Delta_{\mathbb{T}^n}}) \right\|_{L^2(\mathbb{T}^n)\to L^\infty(\mathbb{T}^n)} = O(\lambda^{\frac{n-1}{2}} \lambda^{-\frac{1}{2}+\varepsilon}) \ \forall \varepsilon > 0 \text{ if } n \geq 3. \tag{6.50'}$$

If we use Theorem 6.1 for $q = \infty$ with $\varepsilon(\lambda) = \lambda^{-1}$ and $B(\lambda) = \lambda^{\frac{n-1}{2}+\varepsilon}$, we can argue as above to obtain the following generalization of (6.50').

Corollary 6.2 *Let M^{n-2} be a compact Riemannian manifold of dimension $n-2$, where $n \geq 3$. Then, if*

$$P = \sqrt{-(\Delta_{\mathbb{T}^2} + \Delta_{M^{n-2}})}$$

is the square root of minus the Laplacian on the n-dimensional product manifold $\mathbb{T}^2 \times M^{n-2}$, we have

$$\left\| \mathbb{1}_{[\lambda-\lambda^{-1},\lambda+\lambda^{-1}]}(P) \right\|_{L^2(\mathbb{T}^2\times M^{n-2})\to L^\infty(\mathbb{T}^2\times M^{n-2})} = O(\lambda^{\frac{n-1}{2}} \lambda^{-\frac{1}{2}+\varepsilon}) \ \forall \varepsilon > 0. \tag{6.51}$$

Consequently, if $0 = \lambda_0 \leq \lambda_1 \leq \lambda_2 \leq \ldots$ are the eigenvalues of P,

$$\#\{\lambda_j \in [\lambda - \lambda^{-1}, \lambda + \lambda^{-1}]\} = O(\lambda^{n-2+\varepsilon}), \ \forall \varepsilon > 0. \tag{6.52}$$

The first estimate (6.51) follows from (6.47) and Theorem 6.1. As is well known (see, e.g., [20]), it implies the counting bounds (6.52), since (6.51) is equivalent to the statement that the kernel of the projection operators there are $O(\lambda^{n-2+\varepsilon})$, $\forall \varepsilon > 0$, on the diagonal, and the left side of (6.52) is the trace of this kernel. The bounds in (6.52) are optimal, since, as we discussed before, $O(\lambda^{n-2-\delta})$ with $\delta > 0$ bounds cannot hold due to the Weyl formula.

One can also obtain power improvements for products $X \times \mathbb{T}^n$ using the following "discrete restriction theorem" of Bourgain and Demeter [5] (total eigenfunction bounds):

$$\left\| \mathbb{1}_{[\lambda-\lambda^{-1},\lambda+\lambda^{-1}]} (\sqrt{-\Delta_{\mathbb{T}^n}}) \right\|_{L^2(\mathbb{T}^n)\to L^{\frac{2(n+1)}{n-1}}(\mathbb{T}^n)} \lesssim \lambda^\varepsilon, \ \forall \varepsilon > 0. \tag{6.53}$$

This represents a $1/q_c$–power improvement over the universal estimates [18] with $q_c = \frac{2(n+1)}{n-1}$. Similar to the case above in Corollary 6.1, if we use (6.6) with $B(\lambda) = \lambda^{n(\frac{1}{2}-\frac{1}{q})-\frac{1}{q_c}+\varepsilon}$, we obtain from (6.53) that if $P = \sqrt{-(\Delta_X + \Delta_{\mathbb{T}^n})}$ then

$$
\begin{aligned}
&\| \mathbb{1}_{[\lambda-\lambda^{-1},\,\lambda+\lambda^{-1}]}(P) \|_{L^2(X\times\mathbb{T}^n)\to L^q(X\times\mathbb{T}^n)} \lesssim \lambda^{d(\frac{1}{2}-\frac{1}{q})-1/2-1/q_c+\varepsilon}, \ \forall \varepsilon>0,\\
&\text{if } q_c=\frac{2(n+1)}{n-1}, \ \text{ and } d=d_X+n,
\end{aligned}
\tag{6.54}
$$

and

$$
q \ge \max\Big(\frac{2(n+1)}{n-1}, \frac{2(d_X+1)}{d_X-1}\Big). \tag{6.55}
$$

It is conjectured that (6.53) should also be valid when $\frac{2(n+1)}{n-1}$ is replaced by the larger exponent $\frac{2n}{n-2}$, which would represent the optimal $\lambda^{-1/2+\varepsilon}$ improvement of the universal bounds in [18]. If this result held, then one would obtain the optimal bounds where in the exponent in (6.54) $-1/q_c$ is replaced by $-1/2$, which would be optimal, as well as the range of exponents in (6.55).

Also, using the results of Hickman [11] and Germain and Myerson [9], one can also obtain improved spectral projection bounds when $\varepsilon(\lambda)=\lambda^{-\sigma}$ with $\sigma\in(0,1)$.

Let us now present the proof of Theorem 6.3 which in the case of $q=\infty$ strengthens the bounds that are implicit in Iosevich and Wyman [14].

Proof of Theorem 6.3 Let $\{e_k^\mu\}_\mu$ be an orthonormal basis for spherical harmonics of degree k on S^{d_1} and $\{e_\ell^\nu\}_\nu$ be an orthonormal basis of spherical harmonics of degree ℓ on S^{d_2}. Then, an orthonormal basis of eigenfunctions on $S^{d_1}\times S^{d_2}$ is of the form

$$
e_k\, e_\ell, \tag{6.56}
$$

where $e_k=e_k^\nu$ for some ν and $e_\ell=e_\ell^\mu$ for some μ. So,

$$
(-\Delta_{d_1}+(\tfrac{d_1-1}{2})^2)e_k=(k+\tfrac{d_1-1}{2})^2 e_k
$$

and

$$
(-\Delta_{d_2}+(\tfrac{d_2-1}{2})^2)e_\ell=(\ell+\tfrac{d_1-1}{2})^2 e_\ell
$$

so that for the Laplacian on $S^{d_1}\times S^{d_2}$, $\Delta=\Delta_{d_1}+\Delta_{d_2}$, we have

$$
(-\Delta+(\tfrac{d_1-1}{2})^2+(\tfrac{d_2-1}{2})^2)e_ke_\ell=((k+\tfrac{d_1-1}{2})^2+(\ell+\tfrac{d_1-1}{2})^2)e_ke_\ell.
$$

Thus, if $P=\sqrt{(-\Delta+(\frac{d_1-1}{2})^2+(\frac{d_2-1}{2})^2)}$, its eigenvalues are

$$
\lambda=\lambda_{k,\ell}=\sqrt{((k+\tfrac{d_1-1}{2})^2+(\ell+\tfrac{d_2-1}{2})^2)}. \tag{6.57}
$$

Thus, if e_λ as in Theorem 6.3 is an eigenfunction of P with this eigenvalue, we must have

$$e_\lambda(x,y) = \sum_{\{(k,\ell):\,\lambda_{k,\ell}=\lambda\}} \Bigl(\sum_{\mu,\nu} a_{k,\ell}^{\mu,\nu}\, e_k^\mu(x)e_\ell^\nu(y)\Bigr), \quad (x,y)\in S^{d_1}\times S^{d_2}. \tag{6.58}$$

Let us now prove (6.44). We note that if (k,ℓ) are fixed, then, for every fixed $y\in S^{d_2}$, the function on S^{d_1}

$$x\to \sum_{\mu,\nu} a_{k,\ell}^{\mu,\nu}\, e_k^\mu(x)e_\ell^\nu(y)$$

is a spherical harmonic of degree k. Thus, by [16] or [18],

$$\Bigl\|\sum_{\mu,\nu} a_{k,\ell}^{\mu,\nu}\, e_k^\mu(\,\cdot\,)e_\ell^\nu(y)\Bigr\|_{L^q(S^{d_1})} \le C\lambda^{\alpha(q,d_1)}\Bigl\|\sum_{\mu,\nu} a_{k,\ell}^{\mu,\nu}\, e_k^\mu(\,\cdot\,)e_\ell^\nu(y)\Bigr\|_{L^2(S^{d_1})}.$$

Next, by Minkowski's inequality and another application of the universal bounds, we obtain from this

$$\begin{aligned}\Bigl\|\sum_{\mu,\nu} a_{k,\ell}^{\mu,\nu}\, e_k^\mu e_\ell^\nu\Bigr\|_{L^q(S^{d_1}\times S^{d_2})} &\le C\lambda^{\alpha(q,d_1)}\Bigl\|\sum_{\mu,\nu} a_{k,\ell}^{\mu,\nu}\, e_k^\mu(x)e_\ell^\nu(y)\Bigr\|_{L^2_xL^q_y(S^{d_1}\times S^{d_2})}\\ &\lesssim \lambda^{\alpha(q,d_1)+\alpha(q,d_2)}\Bigl\|\sum_{\mu,\nu} a_{k,\ell}^{\mu,\nu}\, e_k^\mu e_\ell^\nu\Bigr\|_{L^2(S^{d_1}\times S^{d_2})}.\end{aligned} \tag{6.59}$$

Since, by (6.47'), the number of $\{(k,\ell):\lambda_{k,\ell}=\lambda\}$ is $O(\lambda^\varepsilon)$, we also have by the Cauchy–Schwarz inequality that if e_λ is as in (6.58),

$$|e_\lambda(x,y)| \lesssim \lambda^\varepsilon\Bigl(\sum_{\{(k,\ell):\,\lambda_{k,\ell}=\lambda\}}\Bigl|\Bigl(\sum_{\mu,\nu} a_{k,\ell}^{\mu,\nu}\, e_k^\mu(x)e_\ell^\nu(y)\Bigr)\Bigr|^2\Bigr)^{1/2}. \tag{6.60}$$

Thus, by (6.59)–(6.60),

$$\begin{aligned}&\|e_\lambda\|_{L^q(S^{d_1}\times S^{d_2})}\\ &\lesssim \lambda^{\alpha(q,d_1)+\alpha(q,d_2)+\varepsilon}\cdot\Bigl(\sum_{\{(k,\ell):\,\lambda_{k,\ell}=\lambda\}}\Bigl\|\sum_{\mu,\nu} a_{k,\ell}^{\mu,\nu}\, e_k^\mu e_\ell^\nu\Bigr\|^2_{L^2(S^{d_1}\times S^{d_2})}\Bigr)^{1/2},\end{aligned} \tag{6.61}$$

which leads to (6.44) since, by orthogonality, the last factor in (6.61) is $\|e_\lambda\|_{L^2}$. □

6.3 Improved Weyl Formulae

To prove Theorem 6.2, we first observe that if as above μ_i^2 are the eigenvalues of $-\Delta_X$, then by (6.8) we have

$$N(X \times Y, \lambda) = \sum_{\mu_i \le \lambda} \Big[(2\pi)^{-d_Y} \omega_{d_Y} (\mathrm{Vol}\, Y)(\lambda^2 - \mu_i^2)^{d_Y/2} + R_Y\big(\sqrt{\lambda^2 - \mu_i^2}\big) \Big]. \tag{6.62}$$

We can estimate the last sum using (6.8) and (6.5'):

$$\begin{aligned} R_\lambda &= \sum_{\mu_i \le \lambda} R_Y\big(\sqrt{\lambda^2 - \mu_i^2}\big) \\ &\lesssim \sum_{\mu_i \le \lambda} \varepsilon\big(\sqrt{\lambda^2 - \mu_i^2}\big) \cdot \big(\lambda^2 - \mu_i^2\big)^{\frac{d_Y - 1}{2}} \\ &\lesssim \lambda^{d_Y - 1} \sum_{\mu_i \le \lambda} \varepsilon\big(\lambda \cdot \sqrt{1 - \mu_i^2/\lambda^2}\big) \cdot \sqrt{1 - \mu_i^2/\lambda^2} \times \big(1 - \mu_i^2/\lambda^2\big)^{\frac{d_Y - 2}{2}}. \end{aligned}$$

Since $d_Y \ge 2$, we have

$$\big(1 - \mu_i^2/\lambda^2\big)^{\frac{d_Y - 2}{2}} \le 1, \quad \text{if} \quad \mu_i \le \lambda.$$

Thus, if we use (6.5') with $\theta = \sqrt{1 - \mu_i^2/\lambda^2}$ to estimate the terms with $\sqrt{\lambda^2 - \mu_i^2} \ge 1$, we get

$$\varepsilon\big(\lambda \cdot \sqrt{1 - \mu_i^2/\lambda^2}\big) \cdot \sqrt{1 - \mu_i^2/\lambda^2} \times \big(1 - \mu_i^2/\lambda^2\big)^{\frac{d_Y - 2}{2}} \le \varepsilon(\lambda) \quad \text{if} \quad \sqrt{\lambda^2 - \mu_i^2} \ge 1.$$

Thus, by (6.7),

$$\sum_{\mu_i \le \lambda,\, \sqrt{\lambda^2 - \mu_i^2} \ge 1} R_Y\big(\sqrt{\lambda^2 - \mu_i^2}\big) \lesssim \lambda^{d_Y - 1} \varepsilon(\lambda) \sum_{\mu_i \le \lambda} 1 \lesssim \varepsilon(\lambda) \lambda^{d_Y - 1} \cdot \lambda^{d_X} = \varepsilon(\lambda) \lambda^{d-1},$$

with, as before, $d = d_X + d_Y$. We also need to estimate the terms where $\sqrt{\lambda^2 - \mu_i^2} \le 1$. In this case, we just use that $R_Y(\theta) = O(1)$ if $\theta \le 1$ and so

$$\sum_{\mu_i \le \lambda,\, \sqrt{\lambda^2 - \mu_i^2} \le 1} R_Y\big(\sqrt{\lambda^2 - \mu_i^2}\big) \le \sum_{\mu_i \le \lambda} 1 \lesssim \lambda^{d_X} = \lambda^{d-1} \cdot \lambda^{1 - d_Y},$$

and since $d_Y \ge 2$,

$$\lambda^{1-d_Y} \le \lambda^{-1} \le \varepsilon(\lambda).$$

By combining these two estimates, we conclude that if, as above, R_λ is the second sum in the right side of (6.62), then

$$R_\lambda = O(\varepsilon(\lambda)\,\lambda^{d-1}), \quad d = d_X + d_Y,$$

as desired.

Based on this, we conclude that the improved Weyl formula (6.9) would be a consequence of the following:

$$\begin{aligned}(2\pi)^{-d_Y}\omega_{d_Y}(\text{Vol}\,Y)\,\lambda^{d_Y}\cdot\sum_{\mu_i\le\lambda}\big(1-\mu_i^2/\lambda^2\big)^{d_Y/2} \\ = (2\pi)^{-d}\omega_d(\text{Vol}\,Y\cdot\text{Vol}\,X)\lambda^d + O(\lambda^{d-2}).\end{aligned} \tag{6.63}$$

Note that $\sum_{\mu_i\le\lambda}\big(1-\mu_i^2/\lambda^2\big)^{d_Y/2}$ is the trace of the kernel of $(1-(P_X)^2/\lambda^2)_+^{d_Y/2}$, i.e.,

$$\sum_{\mu_i\le\lambda}\big(1-\mu_i^2/\lambda^2\big)^{d_Y/2} = \int_M \sum_{\mu_i\le\lambda}\big(1-\mu_i^2/\lambda^2\big)^{d_Y/2} e^X_{\mu_i}(x)\overline{e^X_{\mu_i}(x)}\,dV(x), \tag{6.64}$$

with dV denoting the volume element on X. For $\delta\ge 0$,

$$S^\delta_\lambda(x,y) = \sum_{\mu_i\le\lambda}\big(1-\mu_i^2/\lambda^2\big)^\delta e^X_{\mu_i}(x)\overline{e^X_{\mu_i}(y)} \tag{6.65}$$

denotes the kernel of the Bochner–Riesz operators $(1-(P_X)^2/\lambda^2)^\delta_+$ (see, e.g., [19]). Keeping (6.64) in mind, we claim that (6.63) (and hence Theorem 6.2) would be a consequence of the following pointwise estimates for these kernels restricted to the diagonal in $X\times X$.

Proposition 6.1 *Let S^δ_λ be as in* (6.65)*. Then, if $\delta\ge 1$, we have*

$$S^\delta_\lambda(x,x) = (2\pi)^{-d_X}|S^{d_X-1}|\times\frac{1}{2}B(\delta+1, d_X/2)\,\lambda^{d_X} + O(\lambda^{d_X-2}), \tag{6.66}$$

with $|S^{d_X-1}|$ denoting the area of the unit sphere in $\mathbb{R}^{d_X}$ and

$$B(s,t) = \int_0^1 (1-u)^{s-1}u^{t-1}\,du$$

being the beta function.

To see that (6.64)–(6.66) imply (6.63), we recall the formulae

$$\omega_n = \frac{1}{n}\,|S^{n-1}| = \frac{\pi^{n/2}}{\Gamma(n/2+1)},$$

and

$$B(s,t) = \frac{\Gamma(s)\Gamma(t)}{\Gamma(s+t)}.$$

Thus,

$$\begin{aligned}|S^{d_X-1}|\cdot\frac{1}{2}B(d_Y/2+1, d_X/2) &= \frac{d_X\pi^{d_X/2}}{\Gamma(d_X/2+1)}\cdot\frac{\Gamma(d_Y/2+1)\Gamma(d_X/2)}{2\Gamma(d/2+1)}\\ &= \frac{\pi^{d_X/2}\Gamma(d_Y/2+1)}{\Gamma(d/2+1)}\end{aligned}$$

and so

$$\begin{aligned}\omega_{d_Y}\cdot|S^{d_X-1}|\cdot\frac{1}{2}B(d_Y/2+1, d_X/2) &= \frac{\pi^{d_Y/2}}{\Gamma(d_Y/2+1)}\cdot\frac{\pi^{d_X/2}\Gamma(d_Y/2+1)}{\Gamma(d/2+1)},\\ &= \frac{\pi^{d/2}}{\Gamma(d/2+1)} = \omega_d.\end{aligned} \tag{6.67}$$

Thus, since $d_Y/2 \geq 1$, if (6.66) were valid, we would have

$$\begin{aligned}&(2\pi)^{-d_Y} w_{d_Y}(\mathrm{Vol}\,Y)\lambda^{d_Y} \times \int_X S_\lambda^{d_Y/2}(x,x)\,dV(x)\\ &= (2\pi)^{-d}\lambda^{d_Y}\omega_d(\mathrm{Vol}\,Y\cdot\mathrm{Vol}\,X)\cdot\lambda^{d_X} + O(\lambda^{d_Y}\cdot\lambda^{d_X-2})\\ &= (2\pi)^{-d}\omega_d\,\mathrm{Vol}(X\times Y)\lambda^d + O(\lambda^{d-2}).\end{aligned}$$

Since, by (6.64) and (6.65), this yields (6.63), we conclude that the proof of Theorem 6.2 would be complete if we could establish Proposition 6.1.

Proof of Proposition 6.1 The proof of kernel estimates for Bochner–Riesz estimates is well known. See, e.g., [13, 17, 19] and [20]. We shall adapt the argument in the latter reference, which is based on the arguments that exploit the Hadamard parametrix and go back to Avakumovic [1] and Levitan [15]. These dealt with the analog of (6.66) where $\delta = 0$, and then the error bounds in (6.66) must be replaced by $O(\lambda^{d_X-1})$.

To proceed, let $m_\delta(\tau)$, $\tau \in \mathbb{R}$, denote the even function

$$\tau \to m_\delta(\tau) = (1-\tau^2)_+^\delta.$$

Then, if $\hat{m}_\delta$ denotes its Fourier transform, we have by Fourier's inversion theorem

$$S_\lambda^\delta f(x) = m_\delta(P_X/\lambda) f(x) = \frac{1}{\pi}\int_0^\infty \lambda \hat{m}_\delta(\lambda t)\big(\cos t P_X\big) f(x)\, dt.$$

Thus,

$$\begin{aligned} S_\lambda^\delta(x,y) &= \frac{1}{\pi}\int_0^\infty \lambda \hat{m}_\delta(\lambda t)\big(\cos t P_X\big)(x,y)\, dt \\ &= \frac{1}{\pi}\sum_i \int_0^\infty \lambda \hat{m}_\delta(\lambda t)\, \cos t\mu_i\, e_{\mu_i}^X(x)\, \overline{e_{\mu_i}^X(y)}\, dt. \end{aligned} \tag{6.68}$$

To be able to exploit this, we require a couple of facts about m_δ. First, we can write its Fourier transform as follows:

$$\begin{aligned} &\hat{m}_\delta(t) = a_0^\delta(t) + a_+^\delta(t)e^{it} + a_-^\delta(t)e^{-it}, \quad \text{where} \\ &|\partial_t^j a(t)| \lesssim O((1+|t|)^{-1-\delta-j}) \;\forall\; j = 0,1,2,\ldots, \quad \text{if } a = a_0, a_+, a_-. \end{aligned} \tag{6.69}$$

Also,

$$\int_0^\infty m_\delta(r)\, r^{d_X-1}\, dr = \frac{1}{2}B(\delta+1, d_X/2). \tag{6.70}$$

Let us postpone the simple proofs of these two facts for a moment and see how they can be used, along with the Hadamard parametrix, to prove Proposition 6.1.

Let us first fix an even function $\rho(t) \in C_0^\infty(\mathbb{R})$ satisfying the following:

$$\operatorname{supp} \rho \subset (-c/2, c/2) \quad \text{and} \quad \rho \equiv 1 \text{ on } [-c/4, c/4], \tag{6.71}$$

where we assume that

$$c = \min\{1, \text{ Inj } X/2\},$$

with Inj X denoting the injectivity radius of (X, g_X). It follows from (6.69) that

$$\begin{aligned} r_\lambda^\delta(\mu) &= \frac{1}{\pi}\int_0^\infty (1-\rho(t))\, \lambda \hat{m}_\delta(\lambda t) \cos t\mu\, dt \\ &= O\big(\lambda^{-\delta}(1+|\lambda-\mu|)^{-N}\big), \;\forall N, \quad \text{if } \mu \ge 0. \end{aligned} \tag{6.72}$$

Thus, we modify the kernels in (6.68) as follows:

$$\begin{aligned}\widetilde{S}^\delta_\lambda(x,y) &= \frac{1}{\pi}\int_0^\infty \rho(t)\lambda\hat{m}_\delta(\lambda t)\big(\cos tP_X\big)(x,y)\,dt \\ &= \frac{1}{\pi}\sum_i\int_0^\infty \rho(t)\lambda\hat{m}_\delta(\lambda t)\,\cos t\mu_i\,e^X_{\mu_i}(x)\,\overline{e^X_{\mu_i}(y)}\,dt,\end{aligned} \tag{6.73}$$

and let

$$R^\delta_\lambda(x,y) = S^\delta_\lambda(x,y) - \widetilde{S}^\delta_\lambda(x,y).$$

It follows that

$$R^\delta_\lambda(x,y) = \sum_i r_\lambda(\mu_i)e^X_{\mu_i}(x)\overline{e^X_{\mu_i}(y)} \tag{6.74}$$

satisfies

$$|R^\delta_\lambda(x,y)| \lesssim \lambda^{-\delta}\sum_i \big(1+|\lambda-\mu_i|\big)^{-N}|e^X_{\mu_i}(x)|\,|e^X_{\mu_i}(y)|, \quad N=1,2,\dots. \tag{6.75}$$

As is well known, the pointwise Weyl formula of Avakumovic [1] and Levitan [15] (see also [12, 20]) yields the uniform bounds

$$\sum_{\mu_i\in[\tau,\tau+1)} |e^X_{\mu_i}(x)|^2 \lesssim (1+\tau)^{d_X-1}, \quad \tau\geq 0,$$

which in turn gives us

$$R^\delta_\lambda(x,y) = O(\lambda^{d_X-1-\delta}) = O(\lambda^{d_X-2}), \tag{6.76}$$

since we are assuming in Proposition 6.1 that $\delta\geq 1$.

Consequently, it suffices to show that $\widetilde{S}^\delta_\lambda(x,x)$ equals the first term in the right side of (6.66) up to error terms which are $O(\lambda^{d_X-2})$.

To do this, if $d_X\geq 2$, we recall that the Hadamard parametrix implies that for $|t|$ smaller than half the injectivity radius of X we can write

$$\begin{aligned}\big(\cos tP_X\big)(x,x) &= (2\pi)^{-d_X}\int_{\mathbb{R}^{d_X}} \cos t|\xi|\,d\xi + \alpha_0(x)\int_{\mathbb{R}^{d_X}} t\frac{\sin t|\xi|}{|\xi|}d\xi \\ &+ \int_{\mathbb{R}^{d_X}} \cos t|\xi|\,\alpha_1(t,x,\xi)\,d\xi + \int_{\mathbb{R}^{d_X}} \sin t|\xi|\,\alpha_2(t,x,\xi)\,d\xi + O(1),\end{aligned} \tag{6.77}$$

where α_0 is a smooth function, and the α_j are symbols of order -3, so, in particular,

$$|\partial^\gamma_\xi\alpha_j| \lesssim (1+|\xi|)^{-3-|\gamma|}.$$

We shall first deal with the second term on the right side. If we take $x = y$ in (6.73) and replace $\big(\cos tP_X\big)(x,x)$ by the second term in the right side of (6.77), our goal is to show that

$$\int_0^\infty \int_{\mathbb{R}^{d_X}} t\frac{\sin t|\xi|}{|\xi|}\lambda\hat{m}_\delta(\lambda t)\rho(t)d\xi dt \lesssim \lambda^{d_X-2}. \tag{6.78}$$

To handle the part of the integral where $|\xi| \geq 2\lambda$, note that since

$$\hat{m}_\delta(\lambda t) = \frac{1}{\pi}\int_{-1}^{1} e^{-it\lambda\tau} m_\delta(\tau)d\tau$$

is an even function in t, it suffices to show that

$$\int_{|\xi|\geq 2\lambda}\int_{-\infty}^{\infty}\int_{-1}^{1} t\frac{\sin t|\xi|}{|\xi|}\lambda e^{-it\lambda\tau} m_\delta(\tau)\rho(t)\, d\tau dt d\xi \lesssim \lambda^{d_X-2}. \tag{6.79}$$

However, by integrating by parts in t, if $|\xi| \geq 2\lambda$, it is easy to see that

$$\int_{-\infty}^{\infty}\int_{-1}^{1} t\frac{\sin t|\xi|}{|\xi|}\lambda e^{-it\lambda\tau} m_\delta(\tau)\rho(t)\, d\tau dt \lesssim O(1+|\xi|)^{-N}, \ \ \forall N, \tag{6.80}$$

which clearly implies (6.79).

For the part of integral where $|\xi| \leq 2\lambda$, let us fix $\eta \in C_0^\infty$ satisfying

$$\operatorname{supp}\eta \subset (1/2, \infty) \quad \text{and } \eta \equiv 1 \text{ on } [1, +\infty)$$

and write

$$\begin{aligned}
&\int_0^\infty \int_{|\xi|\leq 2\lambda} t\frac{\sin t|\xi|}{|\xi|}\lambda\hat{m}_\delta(\lambda t)\rho(t)d\xi dt \\
&= \int_0^\infty \int_{|\xi|\leq 2\lambda} \big(1-\eta(t|\xi|)\big)t\frac{\sin t|\xi|}{|\xi|}\lambda\hat{m}_\delta(\lambda t)\rho(t)d\xi dt \\
&\quad + \int_0^\infty \int_{|\xi|\leq 2\lambda} \eta(t|\xi|)t\frac{\sin t|\xi|}{|\xi|}\lambda\hat{m}_\delta(\lambda t)\rho(t)d\xi dt \\
&= I + II.
\end{aligned} \tag{6.81}$$

For the first term on the right, note that $|\xi| \leq \min\{t^{-1}, 2\lambda\}$ and so by (6.69)

$$
\begin{aligned}
I &\lesssim \int_{t\geq(2\lambda)^{-1}}\int_{|\xi|\leq t^{-1}} \frac{1}{(t\lambda)^{\delta}|\xi|}\,d\xi dt \\
&\quad + \int_{t\leq(2\lambda)^{-1}}\int_{|\xi|\leq 2\lambda} \frac{1}{|\xi|}\,d\xi dt \\
&\lesssim \int_{t\geq(2\lambda)^{-1}} t^{1-d_X-\delta}\lambda^{-\delta}dt + \int_{t\leq(2\lambda)^{-1}} \lambda^{d_X-1}dt \\
&\lesssim \lambda^{d_X-2}.
\end{aligned}
\tag{6.82}
$$

To bound II, by integrating by parts in t, we rewrite it as

$$
\begin{aligned}
&\int_{|\xi|\leq 2\lambda}\int_0^{\infty} \eta(t|\xi|)\frac{\sin t|\xi|}{|\xi|}\, t\lambda\hat{m}_{\delta}(\lambda t)\rho(t)dtd\xi \\
&= \int_{|\xi|\leq 2\lambda}\int_0^{\infty} \eta(t|\xi|)\frac{\cos t|\xi|}{|\xi|^2}\, t\lambda\hat{m}_{\delta}(\lambda t)\rho'(t)dtd\xi \\
&\quad + \int_{|\xi|\leq 2\lambda}\int_0^{\infty} \eta'(t|\xi|)\frac{\cos t|\xi|}{|\xi|}\, t\lambda\hat{m}_{\delta}(\lambda t)\rho(t)dtd\xi \\
&\quad + \int_{|\xi|\leq 2\lambda}\int_0^{\infty} \eta(t|\xi|)\frac{\cos t|\xi|}{|\xi|^2}\, \big(t\lambda\hat{m}_{\delta}(\lambda t)\big)'\rho(t)dtd\xi \\
&= I + II + III.
\end{aligned}
\tag{6.83}
$$

For the first term, since $\rho'(t)$ is supported where $t \approx 1$, by (6.69), we have

$$
\begin{aligned}
I &\lesssim \int_{|\xi|\leq 2\lambda} \frac{1}{|\xi|^2}\lambda^{-\delta}d\xi \\
&\lesssim \lambda^{d_X-2}, \text{ if } \delta > 0.
\end{aligned}
$$

For the second term, since $\eta'(t|\xi|)$ is supported where $t \approx |\xi|^{-1}$, by (6.69), we have

$$
\begin{aligned}
II &\lesssim \int_{|\xi|\leq 2\lambda}\int_{t\approx|\xi|^{-1}} \frac{1}{(t\lambda)^{\delta}|\xi|}\,dtd\xi \\
&\lesssim \int_{|\xi|\leq 2\lambda} \frac{1}{|\xi|^{2-\delta}}\lambda^{-\delta}d\xi \\
&\lesssim \lambda^{d_X-2}.
\end{aligned}
$$

For the third term, we use the fact that by (6.69), $\big(t\lambda\hat{m}_{\delta}(\lambda t)\big)' \lesssim \lambda^{-\delta}t^{-1-\delta}$, which implies

$$
\begin{aligned}
III &\lesssim \int_{|\xi|\le 2\lambda}\int \eta(t|\xi|)\frac{1}{|\xi|^2}\lambda^{-\delta}t^{-1-\delta}\rho(t)dtd\xi \\
&\lesssim \int_{|\xi|\le 2\lambda}\frac{1}{|\xi|^{2-\delta}}\lambda^{-\delta}d\xi \\
&\lesssim \lambda^{d_X-2}.
\end{aligned}
$$

Thus, the proof of (6.78) is complete.

On the other hand, if we take $x = y$ in (6.73) and replace $\big(\cos tP_X\big)(x,x)$ by the third or fourth term in the right side of (6.77), one can use (6.69) to see that the resulting expression must be bounded by

$$
\begin{aligned}
&\int_{\{\xi\in\mathbb{R}^{d_X}:\,|\xi|\le 2\lambda\}}(1+|\xi|)^{-3}\,d\xi + \int_{\{\xi\in\mathbb{R}^{d_X}:\,|\xi|\ge 2\lambda\}}(1+|\lambda-|\xi||)^{-N}\,d\xi \\
&\qquad = \begin{cases} O(\lambda^{d_X-3}), & d_X>3, \\ O(\log\lambda), & d_X=3, \\ O(1), & d_X=2, \end{cases}
\end{aligned}
$$

which is better than desired. Clearly, if we also do this for the last term in the right side of (6.77), the resulting term will be $O(1)$.

Based on this, we would have the bounds in Proposition 6.1 for $d_X \ge 2$ if we could show that

$$
\begin{aligned}
&(2\pi)^{-d_X}\int_0^\infty\int_{\mathbb{R}^{d_X}}\frac{1}{\pi}\rho(t)\lambda\hat{m}_\delta(\lambda t)\,\cos t|\xi|\,d\xi dt \\
&= (2\pi)^{-d_X}|S^{d_X-1}|\times\frac{1}{2}B(\delta+1,d_X/2)\,\lambda^{d_X}+O(\lambda^{d_X-2}).
\end{aligned}
\tag{6.84}
$$

If we repeat the argument that leads to (6.75), we find that if we replace $\rho(t)$ here by one, then the difference between this expression and the left side of (6.84) is bounded by

$$
\lambda^{-\delta}\int_{\mathbb{R}^{d_X}}(1+|\lambda-|\xi||)^{-N}\,d\xi
$$

for any N and hence $O(\lambda^{d_X-1-\delta}) = O(\lambda^{d_X-2})$.

Thus, by Fourier's inversion formula, up to these errors, the expression in the left side of (6.84) is

$$
(2\pi)^{-d_X}\int_{\mathbb{R}^{d_X}} m_\delta(|\xi|/\lambda)\,d\xi = (2\pi)^{-d_X}|S^{d_X-1}|\Big(\int_0^1 m_\delta(r)\,r^{d_X-1}\,dr\Big)\cdot\lambda^{d_X}.
\tag{6.85}
$$

Since by (6.70) the integral in the right side is equal to $\frac{1}{2}B(\delta+1, d_X/2)$, we obtain (6.84).

For the remaining case $d_X = 1$, we shall use the fact that for $|t|$ smaller than a fixed constant c, by choosing coordinates such that the metric equals dx^2, we have

$$\big(\cos tP_X\big)(x,y) = (2\pi)^{-1}\int_{-\infty}^{\infty} \cos t\tau e^{i\tau(x-y)}\,d\tau\,. \tag{6.86}$$

Moreover, one can simply repeat the argument in (6.78)–(6.79) to see that

$$(2\pi)^{-1}\int_0^{\infty}\int_{-\infty}^{\infty}\frac{1}{\pi}\rho(t)\lambda\hat{m}_\delta(t)\,\cos t\tau\,d\tau dt = (2\pi)^{-1}B(\delta+1,1/2)\,\lambda + O(\lambda^{-1}). \tag{6.87}$$

By (6.86), (6.87), and the arguments in (6.71)–(6.76), we obtain (6.66) when $d_X = 1$.

To finish, we still need to prove the facts (6.69) and (6.70) about m_δ that we used above. The latter just follows from the standard formula for the beta function stated above and a change of variables. To prove the former (6.69), we note that if ρ is as in (6.71), then $\hat{m}_\delta(t)$ can be written as

$$\int \rho(1-\tau)\,(1-\tau)_+^\delta(1+\tau)^\delta\,e^{-it\tau}\,d\tau + \int \rho(1+\tau)\,(1+\tau)_+^\delta(1-\tau)^\delta\,e^{-it\tau}\,d\tau + a_0^\delta(t),$$

where $a_0^\delta \in \mathcal{S}(\mathbb{R})$ and hence satisfies the bounds in (6.69). A simple argument shows that the first two terms in the right can be written as $a_+^\delta(t)e^{it}$ and $a_-^\delta(t)e^{-it}$, respectively, with $a_\pm^\delta$ as in (6.69), which finishes the proof. □

6.4 Further Results and Remarks

In our main results, Theorems 6.1, 6.2, and 6.3, we focused on products of length two, as was the case of some of the earlier results, e.g., [6] and [7]. On the other hand, Iosevich and Wyman [14] obtained further improved Weyl error bounds for products of spheres $S^{d_1}\times S^{d_2}\times\cdots\times S^{d_n}$ as the length $n = 2, 3, \ldots$ increased, and their $O(\lambda^{d-1-\delta_n})$ bounds, $d = d_1+\cdots+d_n$, have $\delta_n \to 1$ as $n \to \infty$. As we noted earlier, such bounds are impossible for $\delta_n > 1$. Iosevich and Wyman conjectured that for such products of length $n \geq 5$ one should be able to take $\delta_n = 1-\varepsilon$ for all $\varepsilon > 0$ or even $\delta_n = 1$, which would agree with the classical error term bounds for the n-torus (i.e., $d_1 = \cdots = d_n = 1$ and $n \geq 5$). See, e.g., Walfisz [21].

Let us now show that the proof of Theorem 6.3 yields optimal L^q estimates for such products with q large. The particular case where $q = \infty$ can be thought of as a weaker version of the conjecture of Iosevich and Wyman [14] in the sense that it would follow from the somewhat stronger pointwise Weyl remainder variant of their conjecture.

The improved variant of Theorem 6.3 that follows from its proof and the aforementioned optimal bounds for $\mathbb{T}^n$ for $n = 5$ is the following.

Theorem 6.4 *Let $Y = S^{d_1} \times \cdots \times S^{d_5}$ be a product of 5 round spheres, and let $-\Delta_Y = -(\Delta_{S^{d_1}} + \cdots + \Delta_{S^{d_5}})$ and $P_Y = \sqrt{-\Delta_Y}$. We then have for $\lambda \geq 1$*

$$\| \mathbb{1}_{[\lambda-\lambda^{-1},\lambda+\lambda^{-1}]}(P_Y) \|_{L^2(Y)\to L^q(Y)} = O(\lambda^{d(\frac{1}{2}-\frac{1}{q})-1}), \tag{6.88}$$

$$\text{if } d = d_1 + \cdots + d_5 \text{ and } q \geq \max\left\{\tfrac{2(d_j+1)}{d_j-1} : 1 \leq j \leq 5\right\}. \tag{6.89}$$

Additionally, if $X = M^n$ is an n-dimensional, $n \geq 1$, compact Riemannian manifold and $P = \sqrt{-(\Delta_Y + \Delta_X)}$, then for $\lambda \geq 1$

$$\| \mathbb{1}_{[\lambda-\lambda^{-1},\lambda+\lambda^{-1}]}(P) \|_{L^2(X\times Y)\to L^q(X\times Y)} = O(\lambda^{(d+n)(\frac{1}{2}-\frac{1}{q})-1}), \tag{6.90}$$

$$\text{if } q \geq \max\{\tfrac{2(d_1+1)}{d_1-1}, \tfrac{2(d_2+1)}{d_2-1}, \ldots, \tfrac{2(d_5+1)}{d_5-1}, \tfrac{2(n+1)}{n-1}\}. \tag{6.91}$$

Both estimates represent a $\lambda^{-1/2}$ improvement over the universal bounds in [18], and, as mentioned before, this is optimal.

To prove the theorem, we first note that the second estimate, (6.90), is a simple consequence of the first one (6.88) and Theorem 6.1 after noting that $\alpha(q, n) = n(\frac{1}{2} - \frac{1}{q}) - \frac{1}{2}$ for q as in (6.90).

Let us now see how we can use the proof of Theorem 6.3 and the classical improved lattice point counting bounds in dimension 5 to obtain (6.88).

Just as we did before for products of length 2, we first note that if $\{e^{\nu}_{j,k}\}_\nu$ is an orthonormal basis of spherical harmonics of degree k on S^{d_j}, then an orthonormal basis of eigenfunctions on $Y = S^{d_1} \times \cdots \times S^{d_5}$ is of the form

$$e_{1,k_1} \cdot e_{2,k_2} \cdot e_{3,k_3} \cdot e_{4,k_4} \cdot e_{5,k_5},$$

where $e_{j,k_j} = e^{\nu_j}_{j,k_j}$ for some ν_j, with $j = 1, 2, 3, 4, 5$. Consequently,

$$\begin{aligned}&\Big(-\Delta_Y + \big(\tfrac{d_1-1}{2}\big)^2 + \cdots + \big(\tfrac{d_5-1}{2}\big)^2\Big) \big[e_{1,k_1} \cdots e_{5,k_5}\big] \\ &= \Big(\big(k_1 + \tfrac{d_1-1}{2}\big)^2 + \cdots + \big(k_5 + \tfrac{d_5-1}{2}\big)^2\Big)\, e_{1,k_1} \cdots e_{5,k_5}\end{aligned}$$

and so, analogous to (6.57), the eigenvalues of P_Y are

$$\begin{aligned}\lambda = \lambda_{k_1,\ldots,k_5} = \sqrt{\big(k_1 + \tfrac{d_1-1}{2}\big)^2 + \cdots + \big(k_5 + \tfrac{d_5-1}{2}\big)^2}, \\ k_j = 0, 1, 2, \ldots, \quad 1 \leq j \leq 5.\end{aligned} \tag{6.92}$$

Also, analogous to before, an eigenfunction with this eigenvalue must be of the form

$$e_\lambda(x_1,\dots,x_5) = \sum_{\{(k_1,\dots,k_5):\,\lambda_{k_1,\dots,k_5}=\lambda\}} \Big(\sum_{\nu^1_{j_1},\dots,\nu^5_{j_5}} a^{\nu^1_{j_1},\dots,\nu^5_{j_5}}_{k_1,\dots,k_5} e^{\nu^1_{j_1}}_{k_1}(x_1)\cdots e^{\nu^5_{j_5}}_{k_5}(x_5)\Big). \tag{6.93}$$

Here, $\{e^{\nu^\ell_{j_\ell}}_{k_\ell}\}_{j_\ell}$ is the orthonormal basis of spherical harmonics of degree k_ℓ on S^{d_ℓ}, $\ell = 1,\dots,5$.

Next, we note that for q as in (6.88), we have that if, as in (6.2), $\alpha(q,d_j)$ denotes the λ-power in the universal L^q-estimates, then

$$\alpha(q,d_j) = d_j(\tfrac12 - \tfrac1q) - \tfrac12, \tag{6.94}$$

if q is as in (6.88). Thus, if we inductively use the universal bounds from [18] (or the earlier bounds for spherical harmonics [16]), we find that if $k_1,\dots,k_5$ are fixed and $\lambda_{k_1,\dots,k_5} = \lambda$

$$\Big\| \sum_{\nu^1_{j_1},\dots,\nu^5_{j_5}} a^{\nu^1_{j_1},\dots,\nu^5_{j_5}}_{k_1,\dots,k_5} e^{\nu^1_{j_1}}_{k_1}\cdots e^{\nu^5_{j_5}}_{k_5} \Big\|_{L^q(Y)} \tag{6.95}$$

$$\le C\Big(\prod_{j=1}^5 \lambda^{d_j(\frac12-\frac1q)-\frac12}\Big)\cdot\Big(\sum_{\nu^1_{j_1},\dots,\nu^5_{j_5}} |a^{\nu^1_{j_1},\dots,\nu^5_{j_5}}_{k_1,\dots,k_5}|^2\Big)^{1/2} \tag{6.96}$$

$$= C\lambda^{d(\frac12-\frac1q)-\frac52}\Big(\sum_{\nu^1_{j_1},\dots,\nu^5_{j_5}} |a^{\nu^1_{j_1},\dots,\nu^5_{j_5}}_{k_1,\dots,k_5}|^2\Big)^{1/2}. \tag{6.97}$$

To use this, we recall that when $n \ge 5$ we have the following improvement of (6.50):

$$\#\{j\in\mathbb{Z}^n : |j| = \lambda\} = O(\lambda^{n-2}), \quad \text{if } n \ge 5. \tag{6.98}$$

Indeed, this is a consequence of the stronger result for the problem of counting the number of integer lattice points inside λ-balls centered at the origin (e.g., [21, p. 45]).

If we use (6.98) along with the Cauchy–Schwarz inequality, we deduce that (6.95) implies That if e_λ is as in (6.93),

$$\|e_\lambda\|_{L^q(Y)} \le C\lambda^{d(\frac12-\frac1q)-\frac52}\sqrt{\lambda^3}\Big(\sum_{\nu^1_{j_1},\dots,\nu^5_{j_5}} |a^{\nu^1_{j_1},\dots,\nu^5_{j_5}}_{k_1,\dots,k_5}|^2\Big)^{1/2} \tag{6.88'}$$

$$= C\lambda^{d(\frac12-\frac1q)-1}\|e_\lambda\|_{L^2(Y)}.$$

As before, this estimate for eigenfunctions implies the spectral projection bounds due to the fact that successive distinct eigenvalues of P_Y which are comparable to λ have gaps that are bounded below by $c_0\lambda^{-1}$ for some uniform $c_0 > 0$. This completes the proof of Theorem 6.4.

Remark It would be interesting to investigate other situations involving product manifolds where one is able to obtain L^q estimates that improve ones that follow from Theorem 6.1. For instance, Canzani and Galkowski [7] showed that if Y is a product manifold, then one has $\sqrt{\log \lambda}$ improvements over the universal bounds for large q (i.e., $\varepsilon(\lambda) = (\log \lambda)^{-1}$ in Theorem 6.1). In this case, $X \times Y$ in Theorem 6.1 would be a product of three manifolds, yet our results do not give further improvements over the results coming from [7]. Similarly, if *both* X and Y have improved eigenfunction bounds, are there situations where $X \times Y$ can inherit both improvements, as opposed to the better of the two improvements for X and Y as guaranteed by Theorem 6.1? Our proof does not seem to yield such a result. Moreover, in many cases, one cannot obtain $\sqrt{\varepsilon_X(\lambda)} \cdot \sqrt{\varepsilon_Y(\lambda)}$ improvements if X and Y, respectively, have $\sqrt{\varepsilon_X(\lambda)}$ and $\sqrt{\varepsilon_Y(\lambda)}$ improvements. This is the case, for instance, when they both involve power improvements $\sqrt{\varepsilon_X(\lambda)} = \lambda^{-\delta_X}$ and $\sqrt{\varepsilon_Y(\lambda)} = \lambda^{-\delta_Y}$ with $\delta_X + \delta_Y > 1/2$, for reasons mentioned before.

Acknowledgments The second author was supported in part by the NSF (DMS-1665373). The authors would like to thank the referee for remarks and corrections that improved the exposition.

References

1. V. G. Avakumović. Über die Eigenfunktionen auf geschlossenen Riemannschen Mannigfaltigkeiten. *Math. Z.*, 65:327–344, 1956.
2. P. H. Bérard. On the wave equation on a compact Riemannian manifold without conjugate points. *Math. Z.*, 155(3):249–276, 1977.
3. M. D. Blair and C. D. Sogge. Concerning Toponogov's theorem and logarithmic improvement of estimates of eigenfunctions. *J. Differential Geom.*, 109(2):189–221, 2018.
4. M. D. Blair and C. D. Sogge. Logarithmic improvements in L^p bounds for eigenfunctions at the critical exponent in the presence of nonpositive curvature. *Invent. Math.*, 217(2):703–748, 2019.
5. J. Bourgain and C. Demeter. The proof of the ℓ^2 decoupling conjecture. *Ann. of Math. (2)*, 182(1):351–389, 2015.
6. Y. Canzani and J. Galkowski. Weyl remainders: an application of geodesic beams. *preprint, arXiv:2010.03969.*
7. Y. Canzani and J. Galkowski. Eigenfunction concentration via geodesic beams. *J. Reine Angew. Math.*, 775:197–257, 2021.
8. J. J. Duistermaat and V. W. Guillemin. The spectrum of positive elliptic operators and periodic bicharacteristics. *Invent. Math.*, 29(1):39–79, 1975.
9. P. Germain and S. L. R. Myerson. Bounds for spectral projectors on tori. *arXiv:2104.13274.*
10. A. Hassell and M. Tacy. Improvement of eigenfunction estimates on manifolds of nonpositive curvature. *Forum Mathematicum*, 27(3):1435–1451, 2015.
11. J. Hickman. Uniform L^p resolvent estimates on the torus. *Mathematics Research Reports*, 1:31–45, 2020.

12. L. Hörmander. The spectral function of an elliptic operator. *Acta Math.*, 121:193–218, 1968.
13. L. Hörmander. *The analysis of linear partial differential operators. III*. Classics in Mathematics. Springer, Berlin, 2007. Pseudo-differential operators, Reprint of the 1994 edition.
14. A. Iosevich and E. Wyman. Weyl law improvement for products of spheres. *Anal. Math.*, 47(3):593–612, 2021.
15. B. M. Levitan. On the asymptotic behavior of the spectral function of a self-adjoint differential equation of the second order. *Izvestiya Akad. Nauk SSSR. Ser. Mat.*, 16:325–352, 1952.
16. C. D. Sogge. Oscillatory integrals and spherical harmonics. *Duke Math. J.*, 53(1):43–65, 1986.
17. C. D. Sogge. On the convergence of Riesz means on compact manifolds. *Ann. of Math. (2)*, 126(2):439–447, 1987.
18. C. D. Sogge. Concerning the L^p norm of spectral clusters for second-order elliptic operators on compact manifolds. *J. Funct. Anal.*, 77(1):123–138, 1988.
19. C. D. Sogge. *Fourier integrals in classical analysis*, volume 105 of *Cambridge Tracts in Mathematics*. Cambridge University Press, Cambridge, 1993.
20. C. D. Sogge. *Hangzhou lectures on eigenfunctions of the Laplacian*, volume 188 of *Annals of Mathematics Studies*. Princeton University Press, Princeton, NJ, 2014.
21. A. Walfisz. *Gitterpunkte in mehrdimensionalen Kugeln*. Monografie Matematyczne, Vol. 33. Państwowe Wydawnictwo Naukowe, Warsaw, 1957.
22. A. Zygmund. On Fourier coefficients and transforms of functions of two variables. *Studia Math.*, 50:189–201, 1974.

Chapter 7
A Scalar-Valued Fourier Transform for the Heisenberg Group

Sundaram Thangavelu

Dedicated to the memory of Robert Strichartz

I think it was during the summer of 1987 when my friend Der-Chen Chang brought a preprint of Strichartz to my attention. In this work, which was published in 1989 [17], Strichartz developed harmonic analysis as spectral theory of Laplacians. In the context of the Heisenberg group $\mathbb{H}^n$, the relevant operator was the sub-Laplacian $\mathcal{L}$, and Strichartz developed the joint spectral theory of $\mathcal{L}$ and T. I read this paper with great zeal and learnt for the first time about special Hermite functions and their importance in the harmonic analysis on $\mathbb{H}^n$. This point of view was further developed in another influential paper [18] and both works have played a vital role in all my works related to the Heisenberg group. During the academic year 1991–92, as a visitor to Cornell University, I was fortunate to have the opportunity to discuss mathematics with Strichartz. Needless to say, I greatly benefitted by interacting with him and I am eternally grateful for his kindness and the genuine interest he showed in my works. As one can see, this article draws quite a lot from the two papers mentioned above.

7.1 Introduction

Fourier transform of a function f on a locally compact Lie group G is defined as a function on the unitary dual $\widehat{G}$ consisting of the equivalence classes of irreducible unitary representations of G. Given a representative π of an element from $\widehat{G}$, we define

$$\hat{f}(\pi) = \int_G f(g)\pi(g)dg, \tag{7.1}$$

S. Thangavelu (✉)
Department of Mathematics, Indian Institute of Science, Bangalore, India
e-mail: veluma@iisc.ac.in

P. Alonso Ruiz et al. (eds.), *From Classical Analysis to Analysis on Fractals*, Applied and Numerical Harmonic Analysis, https://doi.org/10.1007/978-3-031-37800-3_7

where dg is the left invariant Haar measure on G. It is also customary to use the notation $\pi(f)$ instead of $\hat{f}(\pi)$ especially when the group is non-abelian. Note that for $f \in L^1(G)$, $\hat{f}(\pi)$ is a bounded linear operator on the Hilbert space $\mathcal{H}_\pi$ on which the representation π is realised. When G is abelian, all the irreducible unitary representations of G are one-dimensional, and hence we can think of them as homomorphisms of G into the circle group S^1 consisting of all complex numbers of absolute value one. They are known as unitary characters, and for any given $\chi \in \widehat{G}$ the Fourier transform

$$\hat{f}(\chi) = \int_G f(g)\chi(g)dg \tag{7.2}$$

is a scalar. We can also turn $\widehat{G}$ into a multiplicative group, and hence $\hat{f}$ becomes a function on this dual group. This allows us to treat Fourier transform on locally compact abelian groups on par with the well-understood Euclidean Fourier transform on $\mathbb{R}^n$.

When the group G is non-abelian, it is no longer true that all the elements of $\widehat{G}$ are one-dimensional. If the group is compact, then $\widehat{G}$ consists only of finite-dimensional representations, and hence the Fourier transform becomes a matrix-valued function which is still manageable. When the group is non-compact and non-abelian, we need to deal with infinite-dimensional representations, and hence the Fourier transform becomes operator-valued. As we are mainly interested in studying the Fourier transform on the Heisenberg group, let us leave the generality and specialise to the case in hand.

The simplest example of a non-abelian nilpotent Lie group is provided by the Heisenberg group $\mathbb{H}^n$ whose underlying manifold is $\mathbb{C}^n \times \mathbb{R}$. The representation theory of $\mathbb{H}^n$ is very simple, and its unitary dual can be described explicitly. There are certain one-dimensional representations of $\mathbb{H}^n$ which are characters, parametrised by $\mathbb{C}^n$, but we discard them as they do not play any role in the Plancherel theorem. We let $\widehat{\mathbb{H}^n}$ to stand for the set of all infinite-dimensional irreducible unitary representations of $\mathbb{H}^n$. By a theorem of Stone and von Neumann, the members of $\widehat{\mathbb{H}^n}$ are parametrised by the set of non-zero real numbers $\mathbb{R}^*$. Thus, to each $\lambda \in \mathbb{R}^*$, we have an irreducible unitary representation π_λ realised on the same Hilbert space $L^2(\mathbb{R}^n)$. The Fourier transform of $f \in L^1(\mathbb{H}^n)$ is the operator-valued function

$$\hat{f}(\lambda) = \int_{\mathbb{H}^n} f(g)\pi_\lambda(g)dg$$

defined on $\mathbb{R}^*$. This operator-valued Fourier transform satisfies all the standard properties of a Fourier transform. Hence, for suitable functions, we have an inversion formula, and for $f \in L^1 \cap L^2(\mathbb{H}^n)$, we have the Plancherel formula

$$\int_{\mathbb{H}^n} |f(g)|^2 \, dg = (2\pi)^{-n-1} \int_{-\infty}^{\infty} \|\hat{f}(\lambda)\|_{HS}^2 \, |\lambda|^n \, d\lambda. \tag{7.3}$$

This allows us to extend the Fourier transform to the whole of $L^2(\mathbb{H}^n)$ as a unitary operator onto $L^2(\mathbb{R}^*, \mathcal{S}_2, d\mu)$, the L^2 space of Hilbert–Schmidt operator-valued functions on $\mathbb{R}^*$ taken with respect to the measure $d\mu(\lambda) = (2\pi)^{-n-1}|\lambda|^n d\lambda$.

Though the Fourier transform defined above shares many features with its abelian counterparts, it is unwieldy and not suitable for studying several standard problems in harmonic analysis. For example, there are no simple descriptions of the images either of the Schwartz space $\mathcal{S}(\mathbb{H}^n)$ or $C_0^\infty(\mathbb{H}^n)$ under the Fourier transform. Consequently, it is not clear how to define Fourier transforms of distributions on the Heisenberg group. Though there are versions of Paley–Wiener theorems for the so-called Fourier–Weyl transform, the results are not very satisfactory, see [22, 23].

Recently, the authors of the paper [6] have attempted with certain degree of success to define a scalar-valued Fourier transform on $\mathbb{H}^n$. The idea is very simple: as $\hat{f}(\lambda)$ for $f \in L^1(\mathbb{H}^n)$ is a bounded linear operator on $L^2(\mathbb{R}^n)$, it is natural to fix an orthonormal basis Φ_α^λ, $\alpha \in \mathbb{N}^n$ for $L^2(\mathbb{R}^n)$, and consider the map $f \to \langle \hat{f}(\lambda)\Phi_\alpha^\lambda, \Phi_\beta^\lambda\rangle$ as a candidate for a scalar-valued Fourier transform. In [6], the authors have introduced a metric d on the set $\widetilde{\mathbb{H}}^n = \mathbb{N}^n \times \mathbb{N}^n \times \mathbb{R}^*$ so that the above map is continuous from $L^1(\mathbb{H}^n)$ into $\widehat{\mathbb{H}}^n$, the completion of $\widetilde{\mathbb{H}}^n$ in the metric d.

In this chapter, we propose the following definition of a scalar-valued Fourier transform on $\mathbb{H}^n$ that shares several properties with the Helgason Fourier transform on non-compact rank one Riemannian symmetric spaces. In order to define this Fourier transform, which we call the Strichartz Fourier transform, we need to set up some notations. For any $\delta \geq -1/2$, we let $L_k^\delta(t), k \in \mathbb{N}, t \geq 0$, stand for the Laguerre polynomials of type δ. For any $\lambda \in \mathbb{R}^*$, we define the Laguerre functions by

$$\varphi_{k,\lambda}^{n-1}(z) = L_k^{n-1}(\frac{1}{2}|\lambda||z|^2)e^{-\frac{1}{4}|\lambda||z|^2}, \quad z \in \mathbb{C}^n.$$

Then, $e_{k,\lambda}^{n-1}(z,t) = e^{i\lambda t}\varphi_{k,\lambda}^{n-1}(z)$ are joint eigenfunctions of the sub-Laplacian $\mathcal{L}$ and $T = \frac{\partial}{\partial t}$:

$$\mathcal{L}e_{k,\lambda}^{n-1}(z,t) = (2k+n)|\lambda|\, e_{k,\lambda}^{n-1}(z,t), \quad -iT e_{k,\lambda}^{n-1}(z,t) = \lambda\, e_{k,\lambda}^{n-1}(z,t).$$

Let Ω stand for the Heisenberg fan which is the union of the rays $R_k = \{(\lambda,\tau) \in \mathbb{R}^* \times \mathbb{R} : \tau = (2k+n)|\lambda|\}$ for $k = 0, 1, 2, \ldots$ and the limiting ray $R_\infty = \{(0,\tau) : \tau \geq 0\}$. For each $a = (\lambda, (2k+n)|\lambda|) \in R_k$, we use the notation $e_a(z,t)$ in place of $e_{k,\lambda}^{n-1}(z,t)$. For any $f \in L^1 \cap L^2(\mathbb{H}^n)$,, we define its Strichartz Fourier transform $\widehat{f}(a,z)$ on $\Omega \times \mathbb{C}^n$ as follows. For $a \in R_k$, $z \in \mathbb{C}^n$, $\lambda \neq 0$ we define

$$\widehat{f}(a,z) = \int_{\mathbb{H}^n} f(w,s)e_a((w,s)^{-1}(z,0))dw\, ds. \tag{7.4}$$

Here, $(w, s)^{-1} = (-w, -s)$ and the group law in $\mathbb{H}^n$ is given in Section 2. For $a = (0, \tau)$ coming from the limiting ray R_∞, we set

$$\widehat{f}(0, \tau, z) = (n-1)!2^{n-1} \int_{\mathbb{H}^n} f(w, s) \frac{J_{n-1}(\sqrt{\tau}|z-w|)}{(\sqrt{\tau}|z-w|)^{n-1}} dw\, ds, \tag{7.5}$$

where J_{n-1} is the Bessel function of order $(n-1)$. As a subset of $\mathbb{R}^2$, Ω inherits the Euclidean metric and topology. We define the normalised Strichartz Fourier transform $\widetilde{f}(a, z)$ by

$$\widetilde{f}(a, z) = \frac{k!(n-1)!}{(k+n-1)!} \widehat{f}(a, z), \quad a \in R_k, \quad \widetilde{f}(0, \tau, z) = \widehat{f}(0, \tau, z).$$

For this transform, we can prove the following analogue of the Riemann–Lebesgue lemma.

Theorem 7.1 *For any $f \in L^1(\mathbb{H}^n)$, the normalised Strichartz Fourier transform $\widetilde{f}(a, z)$ is uniformly continuous on Ω for any $z \in \mathbb{C}^n$ fixed. Moreover, $\widetilde{f}(a, z)$ vanishes at infinity, i.e. $\widetilde{f}(a, z) \to 0$ as $|a| \to \infty$ for each $z \in \mathbb{C}^n$ fixed.*

On the Heisenberg fan Ω, we consider the measure ν which is defined by

$$\int_\Omega \varphi(a) d\nu(a) = (2\pi)^{-2n-1} \int_{-\infty}^{\infty} \Big(\sum_{k=0}^{\infty} \varphi(\lambda, (2k+n)|\lambda|) \Big) |\lambda|^{2n} d\lambda.$$

We can now state the inversion formula and the Plancherel theorem for our Strichartz Fourier transform.

Theorem 7.2 *For any Schwartz class function f on $\mathbb{H}^n$, we have the following inversion formula for the Strichartz Fourier transform:*

$$f(z, t) = \int_\Omega \int_{\mathbb{C}^n} \widehat{f}(a, w) e_a((-w, 0)(z, t)) dw\, d\nu(a). \tag{7.6}$$

Moreover, for any $f \in L^1 \cap L^2(\mathbb{H}^n)$, we have the Plancherel formula

$$\int_{\mathbb{H}^n} |f(z, t)|^2 dz\, dt = \int_\Omega \int_{\mathbb{C}^n} |\widehat{f}(a, w)|^2 dw\, d\nu(a). \tag{7.7}$$

It is not true that the Strichartz Fourier transform takes $L^2(\mathbb{H}^n)$ onto $L^2(\Omega \times \mathbb{C}^n, d\nu\, dw)$. This is because $\widehat{f}(a, z)$ has to satisfy a necessary condition (see (7.26)) which we write as

$$(2\pi)^{-n} |\lambda|^n \varphi_{k,\lambda}^{n-1} *_\lambda \widehat{f}(a, \cdot)(z) = \widehat{f}(a, z), \quad a \in R_k. \tag{7.8}$$

If we let $L_0^2(\Omega\times\mathbb{C}^n, d\nu\, dw)$ stand for the subspace of $L^2(\Omega\times\mathbb{C}^n, d\nu\, dw)$ consisting of functions $F(a,z)$ satisfying the above condition, we can prove the following result.

Theorem 7.3 *The Strichartz Fourier transform initially defined on $L^1 \cap L^2(\mathbb{H}^n)$ can be extended as a unitary operator from $L^2(\mathbb{H}^n)$ onto $L_0^2(\Omega \times \mathbb{C}^n, d\nu\, dw)$.*

Remark 7.1.1 For $f \in L^2(\mathbb{H}^n)$, the function f^λ is defined only for almost every $\lambda \in \mathbb{R}^*$, and hence we may not be able to define $\widehat{f}(a,z)$ at every $a \in \Omega$. In particular, $\widehat{f}(0,\tau,z)$ need not be defined. As the limiting ray R_∞ has zero Plancherel measure, this does not create any problem. □

As a consequence of the estimates satisfied by the Laguerre functions $\varphi_{k,\lambda}^{n-1}(z)$ and the Bessel function $J_{n-1}(t)$, the Strichartz Fourier transform of any $f \in L^1(\mathbb{H}^n)$ satisfies the estimate

$$\sup_{(a,z)\in\Omega\times\mathbb{C}^n} |\widetilde{f}(a,z)| \leq c_{n,1}\, \|f\|_1.$$

Restating the Plancherel theorem in terms of $\widetilde{f}$ and interpolating with change of measures, we can prove the following Hausdorff–Young inequality.

Theorem 7.4 *For $1 \leq p \leq 2$, there is a measure ν_p on Ω such that for all $f \in L^p(\mathbb{H}^n)$ we have*

$$\Big(\int_\Omega \int_{\mathbb{C}^n} |\widetilde{f}(a,w)|^{p'} dw\, d\nu_p(a)\Big)^{1/p'} \leq c_{n,p} \Big(\int_{\mathbb{H}^n} |f(z,t)|^p dz\, dt\Big)^{1/p}.$$

Let us compare the definition (7.4) with the Euclidean Fourier transform and the Helgason Fourier transform on non-compact rank one Riemannian symmetric spaces $X = G/K$. For functions f on $\mathbb{R}^n$, we can write the Fourier transform in the form

$$\widehat{f}(\lambda,\omega) = (2\pi)^{-n/2} \int_{\mathbb{R}^n} f(x) e^{-i\lambda x\cdot\omega} dx, \quad \lambda \in \mathbb{R}, \omega \in S^{n-1}. \tag{7.9}$$

Note that the functions $e_{\lambda,\omega}(x) = e^{i\lambda x\cdot\omega}$ are eigenfunctions of the Laplacian Δ on $\mathbb{R}^n$ and the inversion formula which reads as

$$f(x) = (2\pi)^{-n/2} \int_{-\infty}^{\infty} \int_{S^{n-1}} \widehat{f}(\lambda,\omega) e_{\lambda,\omega}(x) |\lambda|^{n-1} d\omega\, d\lambda \tag{7.10}$$

expresses f as a superposition of the eigenfunctions $e_{\lambda,\omega}$. As $e_a(z,t)$ are eigenfunctions of $\mathcal{L}$, the similarity between (7.6) and (7.10) is clear. It is even more illuminating to compare our Fourier transform with the Helgason Fourier transform.

Let $X = G/K$ be a rank one Riemannian symmetric space of non-compact type. The Helgason Fourier transform $\widetilde{f}(\lambda, b)$ of a function f on G/K is given by

$$\widetilde{f}(\lambda, b) = \int_{G/K} f(x) e^{(-i\lambda+\rho)A(x,b)} dx, \tag{7.11}$$

where $\lambda \in \mathbb{C}$ and $b \in K/M$, see Helgason [11, 12] for the notations. The inversion formula is given by

$$f(x) = c \int_{-\infty}^{\infty} \int_{K/M} \widetilde{f}(\lambda, b) e^{(i\lambda+\rho)A(x,b)} |c(\lambda)|^{-2} db\, d\lambda, \tag{7.12}$$

where db is the normalised measure on K/M and $c(\lambda)$ is the Harish-Chandra c-function. Here again, the functions $e_{\lambda,b}(x) = e^{(i\lambda+\rho)A(x,b)}$ are eigenfunctions of the Laplace–Beltrami operator Δ_X on the symmetric space X bringing out the analogy between (7.4) and (7.11).

Since functions f on G/K can be considered as right K-invariant functions on the Lie group G, we can consider the group Fourier transform of f. For any $\lambda \in \mathbb{C}$, if we let π_λ stand for the spherical principal series representations realized on $L^2(K/M)$, each of these representations has a unique K-fixed vector $Y_0 \in L^2(K/M)$, which is the constant function $Y_0(b) = 1$. The Helgason Fourier transform of f is then related to the group Fourier transform via the equation

$$\widetilde{f}(\lambda, b) = \pi_\lambda(f) Y_0(b), \quad \pi_\lambda(f) = \int_G f(g) \pi_\lambda(g) dg. \tag{7.13}$$

We will show that there is a similar relation in the case of the Strichartz Fourier transform. The Heisenberg group $\mathbb{H}^n$ can be considered as a subgroup of a bigger group $G_n := \mathbb{H}^n \ltimes U(n)$ known as the Heisenberg motion group, so that functions on $\mathbb{H}^n$ are precisely the right $U(n)$-invariant functions on G_n. For each $a = (\lambda, (2k+n)|\lambda|) \in R_k$, there is a class one representation ρ_k^λ of G_n so that

$$\widehat{f}(a, z) = \rho_k^\lambda(f) e_a(z, 0), \quad \rho_k^\lambda(f) = \int_{G_n} f(g) \rho_k^\lambda(g) dg. \tag{7.14}$$

We will demonstrate that the Strichartz Fourier transform shares several other properties with the Helgason Fourier transform.

We conclude this introduction with a brief description of the plan of the work. In Sect. 7.2, after recalling the representation theory of $\mathbb{H}^n$ and the Heisenberg group motion group and the spectral theory of the sublaplacian, we define the Strichartz Fourier transform and prove all its basic properties. In Sect. 7.3, we explore the connection between the Strichartz Fourier transform and the Gelfand transform on the commutative Banach algebra consisting of $L^1(\mathbb{H}^n)$ functions which are radial in the z variable. We prove an analogue of Hecke–Bochner formula for the Strichartz Fourier transform and use it to characterize the image of the Schwartz space under the Strichartz Fourier transform. In Sect. 7.4, we describe the connection between the operator-valued group Fourier transform and the scalar-valued Strichartz Fourier transform and restate the well-known Hardy and Ingham theorems which are originally stated in terms of the former, in terms of the new transform.

7.2 Fourier Transforms on the Heisenberg Group

7.2.1 *Schrödinger Representations and the Group Fourier Transform*

Let $\mathbb{H}^n := \mathbb{C}^n \times \mathbb{R}$ be the $(2n+1)$-dimensional Heisenberg group with the group law

$$(z,t)(w,s) := \big(z+w, t+s+\frac{1}{2}\,\mathrm{Im}(z.\bar{w})\big), \ \forall (z,t),(w,s) \in \mathbb{H}^n.$$

This is a step two nilpotent Lie group where the Lebesgue measure $dzdt$ on $\mathbb{C}^n \times \mathbb{R}$ serves as the Haar measure. The representation theory of $\mathbb{H}^n$ is well studied in the literature, see the monographs [8, 20], and [23]. In order to define the Fourier transform, we use the Schrödinger representations as described below.

For each non-zero real number λ, we have an infinite-dimensional representation π_λ realised on the Hilbert space $L^2(\mathbb{R}^n)$. These are explicitly given by

$$\pi_\lambda(z,t)\varphi(\xi) = e^{i\lambda t} e^{i\lambda(x\cdot\xi + \frac{1}{2}x\cdot y)}\varphi(\xi+y),$$

where $z = x+iy$ and $\varphi \in L^2(\mathbb{R}^n)$. These representations are known to be unitary and irreducible. Moreover, by a theorem of Stone and Von-Neumann (see, e.g., [8]) up to unitary equivalence, these account for all the infinite-dimensional irreducible unitary representations of $\mathbb{H}^n$ which act as $e^{i\lambda t}I$ on the centre. Also, there is another class of finite-dimensional irreducible representations. As they do not contribute to the Plancherel measure, we will not describe them here.

The Fourier transform of a function $f \in L^1(\mathbb{H}^n)$ is the operator-valued function on the set of all non-zero reals $\mathbb{R}^*$ given by

$$\hat{f}(\lambda) = \int_{\mathbb{H}^n} f(z,t)\pi_\lambda(z,t)dzdt.$$

Note that $\hat{f}(\lambda)$ is a bounded linear operator on $L^2(\mathbb{R}^n)$. It is known that when $f \in L^1 \cap L^2(\mathbb{H}^n)$, its Fourier transform is actually a Hilbert–Schmidt operator and one has

$$\int_{\mathbb{H}^n} |f(z,t)|^2 dzdt = (2\pi)^{-(n+1)} \int_{-\infty}^{\infty} \|\widehat{f}(\lambda)\|_{HS}^2 |\lambda|^n d\lambda,$$

where $\|.\|_{HS}$ denotes the Hilbert–Schmidt norm. The above allows us to extend the Fourier transform as a unitary operator between $L^2(\mathbb{H}^n)$ and the Hilbert space of Hilbert–Schmidt operator-valued functions on $\mathbb{R}$ which are square integrable with respect to the Plancherel measure $d\mu(\lambda) = (2\pi)^{-n-1}|\lambda|^n d\lambda$.

Let $\mathcal{S}_2$ stand for the Hilbert space of Hilbert–Schmidt operators on $L^2(\mathbb{R}^n)$ equipped with the inner product $(T, S) = \mathrm{tr}(S^*T)$. We then have the following Plancherel theorem:

Theorem 7.5 *The group Fourier transform is a unitary operator from $L^2(\mathbb{H}^n)$ onto $L^2(\mathbb{R}^*, \mathcal{S}_2, d\mu)$.*

By polarizing the Plancherel formula, we obtain the Parseval identity: for $f, g \in L^2(\mathbb{H}^n)$,

$$\int_{\mathbb{H}^n} f(z,t)\overline{g(z,t)}dzdt = \int_{-\infty}^{\infty} tr(\widehat{f}(\lambda)\widehat{g}(\lambda)^*)\, d\mu(\lambda).$$

Also, for suitable functions f on $\mathbb{H}^n$, we have the following inversion formula:

$$f(z,t) = \int_{-\infty}^{\infty} tr(\pi_\lambda(z,t)^*\widehat{f}(\lambda))d\mu(\lambda).$$

Now, from the definition of π_λ, it is easy to see that $\widehat{f}(\lambda) = \int_{\mathbb{C}^n} f^\lambda(z)\pi_\lambda(z,0)dz$, where f^λ stands for the inverse Fourier transform of f in the central variable:

$$f^\lambda(z) := \int_{-\infty}^{\infty} e^{i\lambda t} f(z,t)dt.$$

This suggests that for any $g \in L^1(\mathbb{C}^n)$, we consider the following operator-valued function:

$$W_\lambda(g) := \int_{\mathbb{C}^n} g(z)\pi_\lambda(z,0)dz.$$

With these notations, we note that $\hat{f}(\lambda) = W_\lambda(f^\lambda)$. These transforms are called the Weyl transforms. We have the following Plancherel formula for Weyl transform (see [24, (2.2.9), p. 49]):

$$\|W_\lambda(g)\|_{HS}^2|\lambda|^n = (2\pi)^n\|g\|_2^2, \; g \in L^2(\mathbb{C}^n). \tag{7.15}$$

Many results for the group Fourier transform are proved by studying the analogues for the Weyl transform.

7.2.2 *Joint Spectral Theory of $\mathcal{L}$ and T*

The Heisenberg Lie algebra $\mathfrak{h}_n$ has a basis consisting of the following left invariant vector fields:

$$X_j = \frac{\partial}{\partial x_j} + \frac{1}{2} y_j \frac{\partial}{\partial t}, \quad X_j = \frac{\partial}{\partial y_j} - \frac{1}{2} x_j \frac{\partial}{\partial t}, \quad T = \frac{\partial}{\partial t}, \quad j = 1, 2, \ldots, n.$$

The sub-Laplacian $\mathcal{L} = -\sum_{j=1}^{n} \left(X_j^2 + Y_j^2\right)$ plays the role of the Laplacian on $\mathbb{H}^n$. As $\mathcal{L}$ commutes with T, there is a well-defined joint spectral theory of these two operators. The functions $e_{k,\lambda}^{n-1}(z,t)$ in the introduction are joint eigenfunctions of $\mathcal{L}$ and T. The operators L_λ defined by the relation $\mathcal{L}(f(z)e^{i\lambda t}) = e^{i\lambda t} L_\lambda f(z)$ are known as special Hermite operators and they have explicit spectral decomposition which we describe now.

The convolution on $\mathbb{H}^n$ gives rise to the so-called λ-twisted convolutions by the relation

$$(f * g)^\lambda(z) = \int_{\mathbb{C}^n} f^\lambda(z-w) g^\lambda(w) e^{\frac{i}{2}\lambda \operatorname{Im}(z\cdot\bar{w})} dw =: f^\lambda *_\lambda g^\lambda(z).$$

The Laguerre functions $\varphi_{k,\lambda}^{n-1}$ are eigenfunctions of L_λ with eigenvalues $(2k+n)|\lambda|$, and every $g \in L^2(\mathbb{C}^n)$ has the L^2 convergent expansion

$$g(z) = (2\pi)^{-n} |\lambda|^n \sum_{k=0}^{\infty} g *_\lambda \varphi_{k,\lambda}^{n-1}(z). \tag{7.16}$$

The eigenspaces of L_λ corresponding to the eigenvalues $(2k+n)|\lambda|$ are infinite-dimensional, and an orthonormal basis is provided by

$$\Phi_{\alpha,\beta}^\lambda(z) = (2\pi)^{-n/2} (\pi_\lambda(z,0)\Phi_\alpha^\lambda, \Phi_\beta^\lambda), \quad \alpha, \beta \in \mathbb{N}^n, |\alpha| = k.$$

Here, Φ_α^λ are the scaled Hermite functions which are eigenfunctions of the scaled Hermite operator $H(\lambda) = -\Delta + \lambda^2|x|^2$ with eigenvalues $(2k+n)|\lambda|$. We refer to [21, 24] for the details. The functions $\Phi_{\alpha,\beta}^\lambda$, known as special Hermite functions, form an orthonormal basis for $L^2(\mathbb{C}^n)$ and (7.16) is the compact form of the special Hermite expansion of a function g.

The projections $(2\pi)^{-n}|\lambda|^n g *_\lambda \varphi_{k,\lambda}^{n-1}$ are mutually orthogonal, and the Plancherel theorem associated with the expansion (7.16) reads as

$$\int_{\mathbb{C}^n} |g(z)|^2 dz = (2\pi)^{-2n} |\lambda|^{2n} \sum_{k=0}^{\infty} \int_{\mathbb{C}^n} |g *_\lambda \varphi_{k,\lambda}^{n-1}(z)|^2 dz. \tag{7.17}$$

The Weyl transform converts the twisted convolution into products, and we have the formula

$$(2\pi)^{-n}|\lambda|^n W_\lambda(g *_\lambda \varphi_{k,\lambda}^{n-1}) = W_\lambda(g) P_k(\lambda). \tag{7.18}$$

Here, $P_k(\lambda)$ is the orthogonal projection of $L^2(\mathbb{R}^n)$ onto the k-th eigenspace of the Hermite operator $H(\lambda)$ spanned by Φ_α^λ, $|\alpha| = k$.

7.2.3 Heisenberg Motion Group and Some Class One Representations

Let $U(n)$ denote the group of all unitary matrices of order n. This acts on $\mathbb{H}^n$ by the automorphisms

$$\sigma.(z,t) = (\sigma z, t), \ \ \sigma \in U(n).$$

We consider the semi-direct product of $\mathbb{H}^n$ and $U(n)$, $G_n := \mathbb{H}^n \ltimes U(n)$, which acts on $\mathbb{H}^n$ by

$$(z,t,\sigma).(w,s) = \big(z + \sigma w, t + s + \frac{1}{2}\operatorname{Im}(z \cdot \overline{\sigma w})\big),$$

whence the group law in G_n is given by

$$(z,t,\sigma).(w,s,\tau) = \big(z + \sigma w, t + s + \frac{1}{2}\operatorname{Im}(z \cdot \overline{\sigma w}), \sigma\tau\big).$$

The group G_n is called the *Heisenberg motion group*, which contains $\mathbb{H}^n$ and $U(n)$ as subgroups. Also, $\mathbb{H}^n$ can be identified with the quotient space $G_n/U(n)$. As a matter of fact, functions on $\mathbb{H}^n$ can be viewed as right $U(n)$ invariant functions on G_n. The Haar measure on G_n is given by $d\sigma\, dz\, dt$, where $d\sigma$ denotes the normalised Haar measure on $U(n)$. To bring out the connection between the group Fourier transform on $\mathbb{H}^n$ and the Heisenberg motion group, we need to describe a family of class one representations of G_n. We start with recalling the definition of such representations.

Let G be a locally compact topological group and K be a compact subgroup of G. Suppose π is a representation of G realised on the Hilbert space H. Let H_K denote the set of all K-fixed vectors given by

$$H_K := \{v \in H : \pi(k)v = v, \ \forall k \in K\}.$$

It can be easily checked that H_K is a subspace of H. We say that π is a class one representation of the pair (G, K) if $H_K \neq \{0\}$. Moreover, when (G, K) is a Gelfand pair, it is well known that $\dim H_K = 1$. In the following, we describe a certain family of class one representations for the Gelfand pair $(G_n, U(n))$. For that, we need to set up some more notations.

For $\alpha \in \mathbb{N}^n$ and $\lambda \neq 0$, let $\Phi_\alpha^\lambda(x) := |\lambda|^{n/4}\Phi_\alpha(\sqrt{|\lambda|}x)$, $x \in \mathbb{R}^n$, where Φ_α denote the normalised Hermite functions on $\mathbb{R}^n$. We know that for each $\lambda \neq 0$,

$\{\Phi_\alpha^\lambda : \alpha \in \mathbb{N}^n\}$ forms an orthonormal basis for $L^2(\mathbb{R}^n)$. Suppose

$$E_{\alpha,\beta}^\lambda(z,t) := (\pi_\lambda(z,t)\Phi_\alpha^\lambda, \Phi_\beta^\lambda), \ (z,t) \in \mathbb{H}^n,$$

denote the matrix coefficients of the Schrödinger representation π_λ of $\mathbb{H}^n$. For each $\lambda \neq 0$ and $k \in \mathbb{N}$, we consider the Hilbert space $\mathcal{H}_k^\lambda$ spanned by $\{E_{\alpha,\beta}^\lambda : \alpha, \beta \in \mathbb{N}^n, |\beta| = k\}$ and equipped with the inner product

$$(f,g)_{\mathcal{H}_k^\lambda} := (2\pi)^{-n}|\lambda|^n \int_{\mathbb{C}^n} f(z,0)\overline{g(z,0)}dz.$$

We define a representation ρ_k^λ of G_n realised on $\mathcal{H}_k^\lambda$ by the prescription

$$\rho_k^\lambda(z,t,\sigma)\varphi(w,s) := \varphi((z,t,\sigma)^{-1}(w,s)), \ (w,s) \in \mathbb{H}^n.$$

It is well known that ρ_k^λ is an irreducible unitary representation of G_n for all $\lambda \neq 0$ and $k \in \mathbb{N}$. Also, for $\lambda \neq 0$ and $k \in \mathbb{N}$, we consider the function $e_{k,\lambda}^{n-1}$ on $\mathbb{H}^n$ defined by

$$e_{k,\lambda}^{n-1}(z,t) = \sum_{|\alpha|=k} (\pi_\lambda(z,t)\Phi_\alpha^\lambda, \Phi_\alpha^\lambda).$$

It is known that the above function can be expressed in terms of Laguerre functions as follows (see [24, p. 52]):

$$e_{k,\lambda}^{n-1}(z,t) = e^{i\lambda t}\varphi_{k,\lambda}^{n-1}(z).$$

It can be checked that $e_{k,\lambda}^{n-1}$ is a $U(n)$-fixed vector corresponding to the representation ρ_k^λ, and hence ρ_k^λ is a class one representation of the pair $(G_n, U(n))$. Moreover, $(G_n, U(n))$ being a Gelfand pair, $e_{k,\lambda}^{n-1}$ is unique up to scalar multiple.

The representations ρ_k^λ when restricted to $\mathbb{H}^n$ are not irreducible but split into finitely many irreducible unitary representations each one being equivalent to π_λ. Given $f \in L^1(\mathbb{H}^n)$, considering it as a $U(n)$-invariant function on G_n, we associate an operator-valued function $\rho_k^\lambda(f)$ acting on $\mathcal{H}_k^\lambda$ defined by

$$\rho_k^\lambda(f) := \int_{G_n} f(z,t)\rho_k^\lambda(z,t,\sigma)d\sigma\, dz\, dt.$$

Now, since ρ_k^λ is unitary, it can be easily checked that $\rho_k^\lambda(f)$ is a bounded operator and the operator norm is bounded above by $\|f\|_1$. From the definition of ρ_k^λ, the following can be easily checked:

$$\rho_k^\lambda(f)e_{k,\lambda}^{n-1}(z,t) = e^{i\lambda t} f^{-\lambda} *_{-\lambda} \varphi_{k,\lambda}^{n-1}(z). \tag{7.19}$$

This leads to the Plancherel formula, proved in [13, Proposition 2.1], for the representations ρ_k^λ :

$$\frac{(k+n-1)}{k!(n-1)!}\|\rho_k^\lambda(f)\|_{HS}^2 = (2\pi)^{-n}|\lambda|^n \int_{\mathbb{C}^n} |f^{-\lambda} *_{-\lambda} \varphi_{k,\lambda}^{n-1}(z)|^2 dz. \tag{7.20}$$

It is easy to see that $\rho_k^\lambda(f)e_{k,\lambda}^{n-1}(z,t)$ are eigenfunctions of the sub-Laplacian $\mathcal{L}$ with eigenvalues $(2k+n)|\lambda|$ and f can be recovered by the formula

$$f(z,t) = (2\pi)^{-n-1} \int_{-\infty}^{\infty} e^{i\lambda t} \Big(\sum_{k=0}^{\infty} \rho_k^\lambda(f)e_{k,\lambda}^{n-1}(z,0)\Big)|\lambda|^n d\lambda. \tag{7.21}$$

This is a consequence of the special Hermite expansion (7.16) applied to f^λ. We can thus view the above as the spectral decomposition of the sub-Laplacian.

7.2.4 Strichartz Fourier Transform on the Heisenberg Group

We propose the following definition as a scalar-valued Fourier transform for functions on the Heisenberg group. Let Ω stand for the Heisenberg fan which is the union of the rays $R_k = \{(\lambda,\tau) \in \mathbb{R}^* \times \mathbb{R} : \tau = (2k+n)|\lambda|\}$ for $k = 0, 1, 2, \ldots$ and the limiting ray $R_\infty = \{(0,\tau) : \tau \geq 0\}$. For any $f \in L^1 \cap L^2(\mathbb{H}^n)$, we define its Strichartz Fourier transform $\widehat{f}(a,z)$ for $a \in R_k$ and $z \in \mathbb{C}^n$ by the relation

$$\widehat{f}(a,z) = f^{-\lambda} *_{-\lambda} \varphi_{k,\lambda}^{n-1}(z), \quad a = (\lambda, (2k+n)|\lambda|).$$

In view of the relation (7.19), namely $\rho_k^\lambda(f)e_{k,\lambda}^{n-1}(z,t) = e^{i\lambda t} f^{-\lambda} *_{-\lambda} \varphi_{k,\lambda}^{n-1}(z)$, we see that

$$\widehat{f}(a,z) = e^{-i\lambda t}\, \rho_k^\lambda(f)e_{k,\lambda}^{n-1}(z,t) = \rho_k^\lambda(f)e_{k,\lambda}^{n-1}(z,0).$$

For $a = (0,\tau)$ coming from the limiting ray R_∞, we set

$$\widehat{f}(0,\tau,z) = (n-1)!2^{n-1} \int_{\mathbb{C}^n} f^0(z-w) \frac{J_{n-1}(\sqrt{\tau}|w|)}{(\sqrt{\tau}|w|)^{n-1}} dw,$$

where J_{n-1} is the Bessel function of order $(n-1)$. At $(0,0,z)$, we let $\widehat{f}(0,0,z) = \int_{\mathbb{H}^n} f(w,t)dwdt$. As a subset of $\mathbb{R}^2$, Ω inherits the Euclidean metric and topology.

We define the normalised Strichartz Fourier transform by $\widetilde{f}(a,z) = \frac{k!(n-1)!}{(k+n-1)!}\widehat{f}(a,z)$ for $a \in R_k$. On the limiting ray, we simply set $\widetilde{f}(0,\tau,z) = \widehat{f}(0,\tau,z)$. It then follows that $\widetilde{f}$ is a continuous function on Ω for $f \in L^1(\mathbb{H}^n)$.

Proof of Theorem 7.1 The only nontrivial part which requires a proof is the continuity at $(0, \tau, z)$ when a sequence $a_j \in \Omega$ converges to $(0, \tau)$ running through different rays R_k. Thus, we have $(\lambda_j, (2k_j + n)|\lambda_j|)$ with $\lambda_j \to 0$ and $(2k_j + n)|\lambda_j| \to \tau$. We need to show that

$$\lim_{j\to\infty} \frac{k_j!(n-1)!}{(k_j+n-1)!} f^{-\lambda_j} *_{-\lambda_j} \varphi^{n-1}_{k_j,\lambda_j}(z) = (n-1)!2^{n-1} \times \int_{\mathbb{C}^n} f^0(z-w) \frac{J_{n-1}(\sqrt{\tau}|w|)}{(\sqrt{\tau}|w|)^{n-1}} dw.$$

This can be proved by using asymptotic properties of the Laguerre functions $\varphi^{n-1}_{k,\lambda}(z)$. From [19, (8.22.4), p.193], we have

$$\frac{k!(n-1)!}{(k+n-1)!}\varphi^{n-1}_{k,\lambda}(z) = (n-1)!2^{n-1} \frac{J_{n-1}(\sqrt{(2k+n)|\lambda|}\,|z|)}{(\sqrt{(2k+n)|\lambda|}\,|z|)^{n-1}} + m_k(\sqrt{|\lambda|}|z|), \tag{7.22}$$

where the error term satisfies the uniform estimates

$$|m_k(t)| \leq C\,(2k+n)^{-(n-1)/2-3/4}, \quad 0 < t \leq b. \tag{7.23}$$

Thus, we see that $\widetilde{f}(a_j, z)$ is a sum of two terms of which the main term is given by

$$(n-1)!2^{n-1} \int_{\mathbb{C}^n} f^{\lambda_j}(z-w) \frac{J_{n-1}(\sqrt{(2k_j+n)|\lambda_j|}\,|w|)}{(\sqrt{(2k_j+n)|\lambda_j|}\,|w|)^{n-1}} e^{\frac{i}{2}\lambda_j \operatorname{Im}(z\cdot\bar{w})} dw.$$

As $\lambda_j \to 0$ and $(2k_j + n)|\lambda_j| \to \tau$, it is clear that the above converges to

$$(n-1)!2^{n-1} \int_{\mathbb{C}^n} f^0(z-w) \frac{J_{n-1}(\sqrt{\tau}|w|)}{(\sqrt{\tau}|w|)^{n-1}} dw.$$

As λ_j remains bounded, the estimates on $m_k(\sqrt{|\lambda|}\,|z-w|)$ show that the error term goes to zero proving our claim.

In order to prove that $\widetilde{f}(a, z)$ vanishes at infinity, we first observe that the Strichartz Fourier transform of any $f \in L^1(\mathbb{H}^n)$ satisfies the estimate

$$\sup_{(a,z)\in\Omega\times\mathbb{C}^n} |\widetilde{f}(a, z)| \leq c_{n,1}\, \|f\|_1. \tag{7.24}$$

This is a consequence of the well-known fact that the Laguerre functions $\varphi^{n-1}_{k,\lambda}(z)$ and the Bessel function $J_{n-1}(t)$ satisfy the uniform estimates (see [19, (8.22.4) p. 193, (1.71.6) p. 15])

$$\frac{k!(n-1)!}{(k+n-1)!}|\varphi_{k,\lambda}^{n-1}(z)| \leq 1, \quad |J_{n-1}(t)| \leq c_n \, t^{n-1}(1+t)^{-1/2}. \tag{7.25}$$

Therefore, it is enough to prove that $\widetilde{f}(a, z)$ vanishes at infinity whenever f is compactly supported. The case $a = (0, \tau) \to \infty$ is easy to handle. In this case,

$$\widehat{f}(0, \tau, z) = (n-1)!\, 2^{n-1} \int_{\mathbb{C}^n} f(z-w) \frac{J_{n-1}(\sqrt{\tau}|w|)}{(\sqrt{\tau}|w|)^{n-1}} \, dw$$

is a constant multiple of the Hankel transform of the radial function

$$f_z(w) = \int_{U(n)} f(z - \sigma w) d\sigma.$$

Therefore, by appealing to the Riemann–Lebesgue lemma for Hankel transforms, we can conclude that $\widehat{f}(0, \tau, z)$ vanishes at infinity.

Consider the case when $a = (\lambda, (2k+n)|\lambda|) \in R_k$ goes to infinity. In case $|\lambda| \to \infty$, we have

$$\frac{k!(n-1)!}{(k+n-1)}|\widehat{f}(a, z)| \leq \int_{\mathbb{C}^n} |f^{\lambda}(z)| \, dz$$

in view of (7.25). The above certainly vanishes as $|\lambda| \to \infty$ in view of the Riemann–Lebesgue lemma for the Euclidean Fourier transform. In case $(2k_j + n)|\lambda_j| \to \infty$ but λ_j remains bounded, the main term in $\widetilde{f}(a_j, z)$ is bounded by a constant times

$$\int_{\mathbb{C}^n} |f^{\lambda_j}(z-w)| \frac{|J_{n-1}(\sqrt{(2k_j+n)|\lambda_j|}\, |w|)|}{(\sqrt{(2k_j+n)|\lambda_j|}\, |w|)^{n-1}} \, dw.$$

Since $|f^{\lambda_j}(w)| \leq |f^0(w)|$ is integrable, the estimate $|J_{n-1}(t)| \leq c_n \, t^{n-1} \, (1+t)^{-1/2}$ shows that the main term goes to zero as j goes to infinity. The same is true of the error term in view of (7.23) since $f^0(w)$ is compactly supported. This completes the proof. □

Proof of Theorem 7.2 In view of the relation (see Corollary 2.3.4 in [24])

$$\varphi_{k,\lambda}^{n-1} *_\lambda \varphi_{k,\lambda}^{n-1}(z) = (2\pi)^n |\lambda|^{-n} \, \varphi_{k,\lambda}^{n-1}(z),$$

we see that the Fourier transform $\widehat{f}(a, z)$ satisfies

$$\begin{aligned} &(2\pi)^{-n}|\lambda|^n \int_{\mathbb{C}^n} \widehat{f}(a, w) \varphi_{k,\lambda}^{n-1}(z-w) e^{\frac{i}{2}\lambda \operatorname{Im}(z \cdot \bar{w})} dw \\ &= f^{-\lambda} *_{-\lambda} \varphi_{k,\lambda}^{n-1}(z) = \widehat{f}(a, z). \end{aligned} \tag{7.26}$$

Recalling the definition of $e_{k,\lambda}^{n-1}(z,t)$, we can rewrite the above as

$$(2\pi)^{-n}\,|\lambda|^n\int_{\mathbb{C}^n}\widehat{f}(a,w)e_{k,\lambda}^{n-1}((w,0)^{-1}(z,t))dw = e^{i\lambda t}\,f^{-\lambda}*_{-\lambda}\varphi_{k,\lambda}^{n-1}(z).$$

Integrating the above over Ω with respect to $d\nu$ and recalling (7.21), we obtain the inversion formula

$$f(z,t)=\int_\Omega\int_{\mathbb{C}^n}\widehat{f}(a,w)e_a((-w,0)(z,t))dw\,d\nu(a). \tag{7.27}$$

The Plancherel theorem for the special Hermite expansions (7.17) and the Euclidean Fourier transform gives us the identity

$$\int_{\mathbb{H}^n}|f(z,t)|^2dz\,dt=c_n\int_\Omega\int_{\mathbb{C}^n}|\widehat{f}(a,w)|^2dw\,d\nu(a). \tag{7.28}$$

This completes the proof of Theorem 7.2. □

Proof of Theorem 7.3 For a given $F\in L_0^2(\Omega\times\mathbb{C}^n,d\nu\,dw)$, we define

$$f(z,t)=\int_\Omega\int_{\mathbb{C}^n}F(a,w)e_a((-w,0)(z,t))dw\,d\nu(a). \tag{7.29}$$

As $F\in L_0^2(\Omega\times\mathbb{C}^n,d\nu\,dw)$, the function f defined by

$$f(z,t)=(2\pi)^{-n-1}\int_{-\infty}^{\infty}\Big(\sum_{k=0}^{\infty}e^{i\lambda t}F(a,z)\Big)|\lambda|^n d\lambda$$

belongs to $L^2(\mathbb{H}^n)$. But then as $f^{-\lambda}(z)=(2\pi)^{-n}|\lambda|^n\big(\sum_{k=0}^{\infty}F(a,z)\big)$, in view of the necessary condition (7.8) and the orthogonality relation $\varphi_{k,\lambda}^{n-1}*_{-\lambda}\varphi_{j,\lambda}^{n-1}=0,\ j\neq k$ we obtain

$$\widehat{f}(a,z)=f^{-\lambda}*_{-\lambda}\varphi_{k,\lambda}^{n-1}(z)=F(a,z),$$

which completes the proof. □

Proof of Theorem 7.4 Recall that in the course of the proof of Theorem 7.1, we have verified the estimate (7.24), namely

$$\sup_{(a,z)\in\Omega\times\mathbb{C}^n}|\widetilde{f}(a,z)|\le c_{n,1}\,\|f\|_1.$$

This simply means that $f\to\widetilde{f}$ is bounded from $L^1(\mathbb{H}^n)$ into $L^\infty(\Omega\times\mathbb{C}^n,d\nu_1(a)\,dz)$ where

$$\int_{\Omega} \varphi(a)d\nu_1(a) = (2\pi)^{-2n-1} \int_{-\infty}^{\infty} \Big(\sum_{k=0}^{\infty} \varphi(\lambda, (2k+n)|\lambda|)\Big) d\lambda.$$

Restating the Plancherel theorem for $\widetilde{f}$, we also have

$$\Big(\int_{\Omega}\int_{\mathbb{C}^n} |\widetilde{f}(a,w)|^2 dw\, d\nu_2(a)\Big)^{1/2} = \Big(\int_{\mathbb{H}^n} |f(z,t)|^2 dz\, dt\Big)^{1/2}$$

with another measure $d\nu_2(a)$ on Ω defined by

$$\int_{\Omega} \varphi(a)d\nu_2(a) = \int_{-\infty}^{\infty} \Big(\sum_{k=0}^{\infty} \Big(\frac{k+n-1)!}{k!(n-1)!}\Big)^2 \varphi(\lambda, (2k+n)|\lambda|)\Big)|\lambda|^{2n} d\lambda.$$

Using interpolation theorem with change of measures [15], we obtain the Hausdorff–Young inequality

$$\Big(\int_{\Omega}\int_{\mathbb{C}^n} |\widetilde{f}(a,w)|^{p'} dw\, d\nu_p(a)\Big)^{1/p'} \leq c_{n,p} \Big(\int_{\mathbb{H}^n} |f(z,t)|^p dz\, dt\Big)^{1/p}$$

for all $f \in L^p(\mathbb{H}^n)$, $1 \leq p \leq 2$ for some measure $\nu_p(a)$ on Ω. The measure $\nu_p(a)$ can be constructed explicitly. □

Remark 7.2.1 Since $f \to \widetilde{f}$ is also bounded from $L^1(\mathbb{H}^n)$ into $L^\infty(\Omega \times \mathbb{C}^n, d\nu_2(a)\, dz)$, interpolation without change of measures gives the inequality

$$\Big(\int_{\Omega}\int_{\mathbb{C}^n} |\widetilde{f}(a,w)|^{p'} dw\, d\nu_2(a)\Big)^{1/p'} \leq c'_{n,p} \Big(\int_{\mathbb{H}^n} |f(z,t)|^p dz\, dt\Big)^{1/p}.$$

However, the result proved above with change of measures is sharper than this inequality. □

7.3 Strichartz Fourier Transform vs. Gelfand Transform

We now investigate further properties of the Fourier transform $f \to \widehat{f}$ which justifies our claim that this transform is the analogue of the Helgason Fourier transform on Riemannian symmetric spaces of non-compact type. If f is a K-biinvariant function on $X = G/K$, it is known that the Helgason Fourier transform $\widetilde{f}(\lambda, b)$ is independent of b and reduces to the Jacobi transform. In fact, if $f(g) = f(ka_rk') = f_0(r)$, then $\widetilde{f}(\lambda, b) = J_{\alpha,\beta} f_0(\lambda) Y_0(b)$, where $Y_0(b) = 1$ is the unique K-fixed vector associated with the representation π_λ and

$$J_{\alpha,\beta} f_0(\lambda) = \int_0^\infty f_0(r)\varphi_\lambda^{\alpha,\beta}(r) w_{\alpha,\beta}(r) dr,$$

where $\varphi_\lambda^{\alpha,\beta}$ is the Jacobi function of type (α, β) which depends on the symmetric space. For the Helgason Fourier transform, it is not true that $\widetilde{(f * g)}(\lambda, b) = \widetilde{f}(\lambda, b)\widetilde{g}(\lambda, b)$. However, if g is K-biinvariant, then it is indeed true and we have

$$\widetilde{(f * g)}(\lambda, b) = J_{\alpha,\beta} g_0(\lambda)\, \widetilde{f}(\lambda, b), \quad g(ka_r k') = g_0(r).$$

We have a similar situation for the Strichartz Fourier transform on the Heisenberg group.

7.3.1 Gelfand Transform on the Heisenberg Group

In the case of $\mathbb{H}^n$, the role of K-biinvariant functions is played by $U(n)$-invariant functions of $\mathbb{H}^n$, also known as radial functions. Note that when f is such a function, $f(z, t) = f_0(|z|, t)$ for a unique function on $\mathbb{R}^+ \times \mathbb{R}$ and we know that the special Hermite expansion of the radial function $f^\lambda(z)$ reduces to the Laguerre expansion, see [24, (2.4.5),p. 61]. Thus,

$$f^\lambda *_\lambda \varphi_{k,\lambda}^{n-1}(z) = R_k^{n-1}(\lambda, f)\varphi_{k,\lambda}^{n-1}(z),$$

where $R_k^{n-1}(\lambda, f)$ is the k-th Laguerre coefficient of f^λ given by

$$R_k^{n-1}(\lambda, f) = (2\pi)^n\, 2^{-n+1} \frac{k!}{(k+n-1)!} \int_0^\infty f_0^\lambda(r)\varphi_{k,\lambda}^{n-1}(r)\, r^{2n-1} dr.$$

Since the surface measure of S^{2n-1} is $2\,\frac{\pi^n}{\Gamma(n)}$, we can write the above in the form

$$R_k^{n-1}(\lambda, f) = \frac{k!(n-1)!}{(k+n-1)!} \int_{\mathbb{C}^n} f^\lambda(z)\varphi_{k,\lambda}^{n-1}(z)\, dz.$$

Thus, we see that if f is a radial function on $\mathbb{H}^n$, then

$$\widehat{f}(a, z) = R_k^{n-1}(-\lambda, f) e_{k,\lambda}^{n-1}(z, 0) \tag{7.30}$$

so that $\widehat{f}(a, z)$ is proportional to the unique $U(n)$-fixed vector for the representation ρ_k^λ. For $f, g \in L^1(\mathbb{H}^n)$, we also have

$$\widehat{f * g}(a, z) = R_k^{n-1}(-\lambda, g_0)\, \widehat{f}(a, z).$$

As in the case of the Helgason Fourier transform, a more general result, known as the Hecke–Bochner formula, is true.

Though the algebra $L^1(\mathbb{H}^n)$ under convolution is noncommutative, the subalgebra $L^1(\mathbb{H}^n/U(n))$ consisting of radial functions is commutative. This follows from the fact that for any radial function f the special Hermite expansion of f^λ reduces to the Laguerre expansion as proved above:

$$f^\lambda(z) = (2\pi)^{-n} |\lambda|^n \sum_{k=0}^{\infty} R_k^{n-1}(\lambda, f) \, \varphi_{k,\lambda}^{n-1}(z). \tag{7.31}$$

As elements of $L^1(\mathbb{H}^n/U(n))$ are precisely the $U(n)$-biinvariant functions on the Heisenberg motion group, this simply means that $(G_n, U(n))$ is a Gelfand pair. The multiplicative linear functionals on this algebra are given by bounded spherical functions which come in two families as shown in [7]. These are given by

$$\widetilde{e}_a(z,t) = \frac{k!(n-1)!}{(k+n-1)!} e_{k,\lambda}^{n-1}(z,t), \;\; \chi_\tau(z,t) = (n-1)! 2^{n-1} \frac{J_{n-1}(\sqrt{\tau}|z|)}{(\sqrt{\tau}|z|)^{n-1}}, \;\; \tau \geq 0. \tag{7.32}$$

Thus, we see that the Gelfand spectrum of the algebra $L^1(\mathbb{H}^n/U(n))$ is precisely the Heisenberg fan Ω. The Gelfand transform on $L^1(\mathbb{H}^n/U(n))$ is defined as the map which takes f into $\mathcal{G}f$ given by

$$\mathcal{G}f(a) = \int_{\mathbb{H}^n} f(z,t) \, \widetilde{e}_a(z,t) \, dz \, dt, \;\; \mathcal{G}f(0,\tau) = \int_{\mathbb{H}^n} f(z,t) \, \chi_\tau(z,t) \, dz \, dt.$$

Thus, the relation (7.30) reads as

$$\widehat{f}(a,z) = \mathcal{G}f(a) \, e_{k,\lambda}^{n-1}(z,0), \;\; f \in L^1(\mathbb{H}^n/U(n)). \tag{7.33}$$

The same relation holds also for $a = (0, \tau)$ as can be easily verified.

From the expansion (7.31) for radial functions, we see that their (group) Fourier transforms are functions of the Hermite operator:

$$\widehat{f}(\lambda) = \sum_{k=0}^{\infty} R_k(\lambda, f) \, P_k(\lambda) = \sum_{k=0}^{\infty} \mathcal{G}f(a) \, P_k(\lambda).$$

This relation allows us to study the Fourier transform of radial functions in terms of their Gelfand transforms, also called spherical Fourier transforms for obvious reasons. As the Gelfand transform is scalar-valued, it has been used as an alternative for the group Fourier transform by several authors. For example, in the work [1], the authors have studied the image of radial Schwartz functions under the spherical Fourier transform. In what follows, we will further explore the connection between the Strichartz Fourier transform and Gelfand transforms.

7.3.2 Hecke–Bochner Formula for the Strichartz Fourier Transform

We begin by recalling briefly some basic facts about bigraded spherical harmonics, referring to [24, Section 2.5] for details. Given a pair (p, q) of non-negative integers, we say that a polynomial $P(z, \bar{z})$ on $\mathbb{C}^n$ is a bigraded solid harmonic if it is harmonic and has the form

$$P(z, \bar{z}) = \sum_{|\alpha|=p} \sum_{|\beta|=q} c_{\alpha,\beta}\, z^{\alpha}\, \bar{z}^{\beta}.$$

They are uniquely determined by their restrictions to the unit sphere S^{2n-1}. We denote by $\mathcal{S}_{p,q}$ the space of all bigraded spherical harmonics, i.e. restrictions of bigraded solid harmonics. It is then known that $L^2(S^{2n-1}) = \bigoplus_{p,q} \mathcal{S}_{p,q}$. Each $\mathcal{S}_{p,q}$ is finite-dimensional, and by choosing an orthonormal basis $S^j_{p,q}$, $j = 1, 2, \ldots, d(p, q)$ for each (p, q), we obtain an orthonormal basis for $L^2(S^{2n-1})$.

Let us return to the Euclidean Fourier transform (7.9) for a moment and consider the integral

$$\int_{S^{n-1}} \widehat{f}(\lambda, \omega)\, Y_m(\omega)\, d\omega,$$

where Y_m is a (standard) spherical harmonic of degree m. Then, it is known that (see for example [16])

$$\lambda^{-m} \int_{S^{n-1}} \widehat{f}(\lambda, \omega)\, Y_m(\omega)\, d\omega = c_{n,m} \int_0^{\infty} r^{-m}\, f_m(r)\, \frac{J_{n/2+m-1}(\lambda r)}{(\lambda r)^{n/2+m-1}} r^{n+2m-1} dr, \tag{7.34}$$

where $f_m(r)$ is the spherical harmonic coefficient $(f(r\cdot), Y_m)_{L^2(S^{n-1})}$. Suppose g is a radial function on $\mathbb{R}^n$ and $f(x) = g(x)P_m(x)$, where P_m is a solid harmonic of degree m. Then, from the above formula, we easily see that the spherical harmonic expansion of $\widehat{f}(\lambda, \omega)$ has only one term and hence

$$\widehat{f}(\lambda, \omega) = c_{n,m}\, P_m(\lambda\omega) \int_0^{\infty} g(r)\, \frac{J_{n/2+m-1}(\lambda r)}{(\lambda r)^{n/2+m-1}} r^{n+2m-1} dr. \tag{7.35}$$

We also observe that if we consider g as a radial function on $\mathbb{R}^{n+2m}$, then the above can be rewritten as $\mathcal{F}_n(g P_m) = c_{n,m} P_m\, \mathcal{F}_{n+2m}(g)$, where $\mathcal{F}_k$ stands for the Fourier transform on $\mathbb{R}^k$. These formulas are known as Hecke–Bochner identities in the literature.

There is a Hecke–Bochner formula for the Helgason Fourier transform too, but we do not intend to state it here as it requires quite a bit of preparation, see [11, 12]. Given below is the Hecke–Bochner formula for the Strichartz Fourier transform which is the exact analogue of (7.34). In what follows, we use the following convention. A radial function g on $\mathbb{H}^n$ will be simultaneously considered as a radial function on any $\mathbb{H}^m$. When k is a negative integer, we set $\varphi_{k,\lambda}^m(z) = 0$ for any m and $z \in \mathbb{C}^m$. Moreover, the Gelfand transform for the algebra $L^1(\mathbb{H}^m/U(m))$ will be denoted by $\mathcal{G}_m$. With these notations, we have the following theorem:

Theorem 7.6 *For any $f \in L^1(\mathbb{H}^n)$ and $S_{p,q} \in \mathcal{S}_{p,q}$, let us define the spherical harmonic coefficient of $f(r\omega, t)$ by $f_{p,q}(r,t) = (f(r\cdot, t), S_{p,q})_{L^2(S^{2n-1})}$. For any $a \in R_k$, $\lambda > 0$, we have the following:*

$$\int_{S^{2n-1}} \widehat{f}(a, r\omega)\, S_{p,q}(\omega)\, d\omega =$$

$$(2\pi)^{-p-q}(|\lambda| r)^{p+q}\, \mathcal{G}_{n+p+q}(g_{p,q})(a(p,q))\, e_{k-q,\lambda}^{n+p+q-1}(r,0),$$

where $g_{p,q}(z,t) = |z|^{-p-q} f_{p,q}(|z|, t)$ and $a(p,q) = (\lambda, (2k+p-q+n)|\lambda|)$. When $\lambda < 0$, we have

$$\int_{S^{2n-1}} \widehat{f}(a, r\omega)\, S_{p,q}(\omega)\, d\omega =$$

$$(2\pi)^{-p-q}(|\lambda| r)^{p+q}\, \mathcal{G}_{n+p+q}(g_{p,q})(a(q,p))\, e_{k-p,\lambda}^{n+p+q-1}(r,0).$$

Proof The proof is indeed a rewriting of [24, Theorem 2.6.1] where the following result is proved. Let $f \in L^1(\mathbb{H}^n)$ be of the form $f(z,t) = g(z,t)P(z)$, where g is radial and P is a bigraded solid harmonic of degree (p,q). Then, for any $\lambda > 0$, we have

$$f^\lambda *_\lambda \varphi_{k,\lambda}^{n-1}(z) = (2\pi)^{-p-q}\lambda^{p+q}\, P(z)\, g^\lambda *_\lambda \varphi_{k-p,\lambda}^{n+p+q-1}(z), \tag{7.36}$$

where the convolution on the right-hand side is taken over $\mathbb{C}^{n+p+q}$. There is a similar formula when $\lambda < 0$ where the roles of p and q are interchanged. This result, stated in terms of Weyl transform, is due to Geller [10]. It is enough to prove the result when $S_{p,q} = S_{p,q}^j$ is a member of the orthonormal basis for $\mathcal{S}_{p,q}$ which we have described earlier.

Expanding $f^\lambda(z)$ in terms of $S_{p,q}^j$ and recalling the definitions of $f_{p,q}$ and $g_{p,q}$, we have

$$f^\lambda(z) = \sum_{p,q} \sum_{j=1}^{d(p,q)} P_{p,q}^j(z)\, (g_{p,q}^j)^\lambda(z).$$

As $\widehat{f}(a,z) = f^{-\lambda} *_{-\lambda} \varphi_{k,\lambda}^{n-1}(z)$, assuming $\lambda < 0$ (so that $-\lambda > 0$), the above formula gives

$$\widehat{f}(a,z) = \sum_{p,q} (2\pi)^{-p-q} |\lambda|^{p+q} \Big(\sum_{j=1}^{d(p,q)} P^j_{p,q}(z)\, (g^j_{p,q})^{|\lambda|} *_{|\lambda|} \varphi^{n+p+q-1}_{k-p,\lambda}(z) \Big).$$

As $g^j_{p,q}(z,t)$ is radial on $\mathbb{H}^{n+p+q}$ using (7.33), we can rewrite the above expansion as

$$\widehat{f}(a,z) = \sum\nolimits_{p,q} (2\pi)^{-p-q} |\lambda|^{p+q} \Big(\sum\nolimits_{j=1}^{d(p,q)} P^j_{p,q}(z)\, \mathcal{G}_{n+p+q}(g^j_{p,q})(a(q,p)) \Big)$$
$$e^{n+p+q-1}_{k-p,\lambda}(z,0).$$

By calculating the spherical harmonic coefficients of $\widehat{f}(a, r\omega)$, we obtain the stated formula. □

The identity (7.36) which was used in the proof of the above theorem can be restated as the following Hecke–Bochner formula for the Strichartz Fourier transform.

Corollary 7.1 *Suppose $f \in L^1(\mathbb{H}^n)$ is of the form $f(z,t) = P(z)g(z,t)$, where g is radial and P is a solid harmonic of degree (p,q). Then, for any $\lambda > 0$,*

$$\widehat{f}(a,z) = (2\pi)^{-p-q} |\lambda|^{p+q}\, P(z)\, \mathcal{G}_{n+p+q} g(a(p,q))\, e^{n+p+q-1}_{k-q,\lambda}(z,0).$$

A similar formula holds for $\lambda < 0$.

We also have a representation theoretic interpretation of the Hecke–Bochner formula. The spherical harmonic expansion of a function $f(z,t)$ on $\mathbb{H}^n$ can be written as

$$f(z,t) = \sum_{(p,q)\in\mathbb{N}^2} \sum_{j=1}^{d(p,q)} g_{(p,q),j}(r,t)\, P^j_{p,q}(z), \quad z = r\omega, \tag{7.37}$$

where $P^j_{p,q}$ is the solid harmonic corresponding to $S^j_{p,q}$. For each pair (p,q), the space $\mathcal{S}_{p,q}$ supports an irreducible unitary representation $R_{(p,q)}$ of the unitary group $K = U(n)$. The action of $R_{(p,q)}$ on $\mathcal{S}_{p,q}$ is given by $R_{(p,q)}(\sigma)S(\omega) = S(\sigma^{-1}\omega)$, $\sigma \in K$. Let M be the isotropic subgroup of K which fixes the coordinate vector $e_1 \in \mathbb{C}^n$ which can be identified with $U(n-1)$. It is known that each $R_{(p,q)}$ has a unique M-fixed vector in $\mathcal{S}_{p,q}$. Such representations are called class one representations and it is known that any such irreducible unitary representation of K is unitarily equivalent to one and only one of $R_{(p,q)}$. Let $\widehat{K_0}$ stand for the set of all equivalence classes of such representations of K. We can rewrite (7.37) as

$$f(z,t)=\sum_{\delta\in\widehat{K}_0} f_\delta(z,t), \qquad f_\delta(z,t)=\sum_{j=1}^{d(p,q)} g_{(p,q),j}(r,t)\,P^j_{p,q}(z), \quad \delta=(p,q). \tag{7.38}$$

We can view the functions f_δ as radial functions on $\mathbb{H}^n$ taking values in the finite-dimensional Hilbert space $\mathcal{S}_\delta$.

Let $L^2(\mathbb{H}^n/U(n),\mathcal{H}_\delta)$ stand for the space of all such radial functions taking values in $\mathcal{H}_\delta$, so that

$$L^2(\mathbb{H}^n)=\bigoplus_{\delta\in\widehat{K}_0} L^2(\mathbb{H}^n/U(n),\mathcal{S}_\delta). \tag{7.39}$$

It then follows that each of the spaces $L^2(\mathbb{H}^n/U(n),\mathcal{H}_\delta)$ is invariant under the Strichartz Fourier transform. Indeed, by the Hecke–Bochner formula, it is clear that, for $\lambda>0$,

$$\widehat{f_\delta}(a,z)=(2\pi)^{-p-q}|\lambda|^{p+q}\Big(\sum_{j=1}^{d(p,q)} P^j_{p,q}(z)\,\mathcal{G}_{n+p+q}(g_{\delta,j})(a(p,q))\Big)\,\varphi^{n+p+q-1}_{k-q,\lambda}(z). \tag{7.40}$$

Thus, we can consider the Strichartz Fourier transform as a family of Gelfand transforms $\mathcal{G}_\delta$ indexed by the class one representations δ of the pair $(U(n),U(n-1))$. If f_δ is thus identified with the vector $(g_{\delta,j})_{j=1}^{d(\delta)}$, then $\widehat{f_\delta}(a,z)$ is given by the vector with components

$$(2\pi)^{-p-q}|\lambda|^{p+q}\,r^{p+q}\,\mathcal{G}_{n+p+q}(g_{\delta,j})(a(p,q))\,\varphi^{n+p+q-1}_{k-q,\lambda}(r). \tag{7.41}$$

We will make use of this view in the next section where we study the image of the Schwartz space under the Strichartz Fourier transform.

7.3.3 *The Image of $\mathcal{S}(\mathbb{H}^n)$ Under the Strichartz Fourier Transform*

Let $\mathcal{S}(\mathbb{H}^n)$ stand for the space of all Schwartz class functions on the Heisenberg group. As the underlying manifold of $\mathbb{H}^n$ is just $\mathbb{R}^{2n+1}$, this space is nothing but $\mathcal{S}(\mathbb{R}^{2n+1})$. In his pioneering work, Geller [10] has proved a characterisation of the image of $\mathcal{S}(\mathbb{H}^n)$ under the group Fourier transform in terms of certain asymptotic series. In 1998, Benson et al. [7] studied the image of Schwartz functions on the Heisenberg group under the spherical Fourier transform associated with Gelfand pairs. In particular, for radial functions on $\mathbb{H}^n$, they have described the image as a space of rapidly decreasing functions on the Gelfand spectrum identified with the

Heisenberg fan Ω. In their work, rapidly decreasing functions on Ω are defined in terms of certain derivatives and finite difference operators.

In a series of papers [1–4], Astengo et al have studied the problem of characterizing the image of $\mathcal{S}(\mathbb{H}^n)$ under the Fourier transform. By using multiple Fourier series, they have reduced the problem to the characterization of the image of polyradial Schwartz functions under the Gelfand transform (spherical Fourier transform). See the survey [14] for a readable account of these works with connections to spectral multipliers. In particular, for the class $\mathcal{S}_K(\mathbb{H}^n)$ of radial functions (recall that $K = U(n)$), they have proved the following elegant result.

Theorem 7.7 ([1, Astengo et al.]) *Let $\mathcal{S}(\Omega)$ stand for the space of restrictions of Schwartz functions on $\mathbb{R}^2$ to Ω equipped with the quotient topology $\mathcal{S}(\mathbb{R}^2)/\{f : f|_\Omega = 0\}$. The Gelfand transform $\mathcal{G}_n$ is a topological isomorphism between $\mathcal{S}_K(\mathbb{H}^n)$ and $\mathcal{S}(\Omega)$.*

Recall that the components $g_{\delta,j}$ of f_δ are defined by $r^{-p-q} f_{\delta,j}(r,t)$ where

$$f_{\delta,j}(r,t) = \int_{S^{2n-1}} f(r\omega,t) S_{\delta,j}(\omega)\, d\omega.$$

It is clear that when $f \in \mathcal{S}(\mathbb{H}^n)$, these functions $g_{\delta,j}$ considered as radial functions on $\mathbb{H}^{n+p+q}$ are Schwartz functions. For the sake of brevity, let us use the following notations: we write $\mathbb{H}^\delta$ in place of $\mathbb{H}^{n+p+q}$ and $U(\delta)$ in place of $U(n+p+q)$ so that $\mathbb{H}^n$ and $U(n)$ correspond to the trivial representation of $U(n)$. Following the previous authors, let $\mathcal{S}_{U(\delta)}(\mathbb{H}^\delta, \mathcal{S}_\delta)$ stand for the Schwartz space of $U(\delta)$ invariant functions on $\mathbb{H}^\delta$ taking values in $\mathcal{S}_\delta$. The Gelfand transform for the pair $(\mathbb{H}^\delta, U(\delta))$ will be denoted by $\mathcal{G}_\delta$. We then observe that for Schwartz class functions the decomposition (7.39) takes the form

$$\mathcal{S}(\mathbb{H}^n) = \bigoplus_{\delta \in \widehat{K}_0} \mathcal{S}_{U(\delta)}(\mathbb{H}^\delta, \mathcal{S}_\delta).$$

We also use the notation $\mathcal{S}(\mathbb{R}^2, \mathcal{S}_\delta)$ for the space of $\mathcal{S}_\delta$-valued Schwartz functions on $\mathbb{R}^2$. We define $\mathcal{S}(\Omega, \mathcal{S}_\delta)$ as in the scalar-valued case and take their orthogonal sum to define

$$\widehat{\mathcal{S}}(\Omega) = \bigoplus_{\delta \in \widehat{K}_0} \mathcal{S}(\Omega, \mathcal{S}_\delta).$$

With these notations, we can restate Theorem 7.7 in the following form.

Theorem 7.8 *The Strichartz Fourier transform is an isomorphism between $\mathcal{S}(\mathbb{H}^n)$ and $\widehat{\mathcal{S}}(\Omega)$.*

Even though this result is a consequence of Theorem 7.7, due to the various identifications we have made, the following explanations are in order. Given $f \in \mathcal{S}(\mathbb{H}^n)$ and f_δ defined as in (7.38), the function

$$\widetilde{f_\delta}(z,t) = \sum_{j=1}^{d(\delta)} g_{\delta,j}(z,t)\, S_{\delta,j}, \quad (z,t) \in \mathbb{H}^\delta,$$

is an element of the space $\mathcal{S}_{U(\delta)}(\mathbb{H}^\delta, \mathcal{S}_\delta)$. The Gelfand transform $\mathcal{G}_\delta \widetilde{f_\delta}$ is a function on the Heisenberg fan Ω_δ for the pair $(\mathbb{H}^\delta, U(\delta))$ which is a proper subset of Ω. By Theorem 7.7, there exists $m_\delta \in \mathcal{S}(\Omega, \mathcal{S}_\delta)$ such that $\mathcal{G}_\delta \widetilde{f_\delta}$ is the restriction of m_δ to Ω_δ. If $m_{\delta,j}$ are the components of m_δ, then we have the relation $\mathcal{G}_\delta \widetilde{f_\delta} = \sum_{j=1}^{d(\delta)} m_{\delta,j}\, S_{\delta,j}$ and the Strichartz Fourier transform of f_δ considered as a function on $\mathbb{H}^n$ is given by

$$\widehat{f_\delta}(a,z) = (2\pi)^{-p-q}|\lambda|^{p+q} \left(\sum_{j=1}^{d(\delta)} P_\delta^j(z)\, m_{\delta,j}(a(p,q)) \right) \varphi_{k-q,\lambda}^{n+p+q-1}(z)$$

for $\lambda > 0$ with a similar expression for $\lambda < 0$. Thus, the correspondence alluded to in the statement of the theorem is the one given by $\widehat{f_\delta} \to m_\delta$. Conversely, given $m_\delta \in \mathcal{S}(\Omega, \mathcal{S}_\delta)$, we can define $g_{\delta,j}$ by applying $\mathcal{G}_\delta^{-1}$ and the function f_δ is then defined as in (7.38). It is a routine matter to check that the resulting f is Schwartz.

7.4 Group Fourier Transform vs. Strichartz Fourier Transform

In this section, we investigate the relation between group Fourier transform and the Strichartz Fourier transform on $\mathbb{H}^n$. Though the former is defined in terms of the Schrödinger representations π_λ and the latter in terms of ρ_k^λ, there is a direct connection between $\widehat{f}(\lambda)$ and $\widehat{f}(a,z)$. We will show that this relation allows us to recast some theorems for the group Fourier transform in terms of the Strichartz Fourier transform.

Combining (7.18) and (7.15), we obtain the following relation between the Strichartz and the group Fourier transforms:

$$\int_{\mathbb{C}^n} |\widehat{f}(a,z)|^2 dz = (2\pi)^n\, |\lambda|^{-n}\, \|\widehat{f}(\lambda) P_k(\lambda)\|_{HS}^2.$$

As an application of this relation, let us rewrite theorems of Hardy and Ingham for the Heisenberg group in terms of the Strichartz Fourier transform.

Let p_a stand for the heat kernel associated with the sub-Laplacian $\mathcal{L}$ on $\mathbb{H}^n$. The Heisenberg analogue of the classical Hardy's uncertainty principle reads as follows (see [24, Theorem 2.9.2, p. 89]):

For a nontrivial function $f \in L^1(\mathbb{H}^n)$, the conditions

$$|f(z,t)| \le c\, p_\alpha(z,t), \quad \widehat{f}(\lambda)^*\widehat{f}(\lambda) \le c\, \widehat{p_{2\beta}}(\lambda)$$

cannot hold simultaneously unless $\alpha \ge \beta$. We can now restate this result in the following form.

Theorem 7.9 *Suppose* $f \in L^1(\mathbb{H}^n)$ *satisfies the conditions* $|f(z,t)| \le c\, p_\alpha(z,t)$ *and*

$$|\lambda|^n \int_{\mathbb{C}^n} |\widehat{f}(a,z)|^2 dz \le c\, \frac{(k+n-1)!}{k!(n-1)!}\, e^{-2\beta(2k+n)|\lambda|}$$

for every $(z,t) \in \mathbb{H}^n$ *and* $a \in R_k$. *Then,* $f = 0$ *whenever* $\alpha < \beta$.

The decay condition on $\widehat{f}(a,z)$ comes from the Hardy condition $\widehat{f}(\lambda)^*\widehat{f}(\lambda) \le c\, \widehat{p_{2\beta}}(\lambda)$. In fact, it is well known that $\widehat{p_\beta}(\lambda) = e^{-\beta H(\lambda)}$ is the semigroup generated by the Hermite operator and an easy calculation using the Hermite basis shows that

$$\|\widehat{f}(\lambda)P_k(\lambda)\|_{HS}^2 \le c\, \frac{(k+n-1)!}{k!(n-1)!}\, e^{-2\beta(2k+n)|\lambda|} \tag{7.42}$$

under the Hardy condition on $\widehat{f}(\lambda)$. An examination of the proof of Hardy's theorem presented in [24, Theorem 2.9.2] shows that the result holds under the weaker assumption (7.42). We remark that other more refined versions of Hardy's theorem can also be stated in terms of $\widehat{f}(a,z)$.

Ingham's uncertainty principle is another theorem that has received considerable attention in recent years. For functions on $\mathbb{R}$, Ingham proved in 1934 the following result:

Let θ *be a nonnegative even function on* $\mathbb{R}$ *which decreases to* 0 *at infinity. There exists a compactly supported function* f *on* $\mathbb{R}$ *whose Fourier transform satisfies the decay condition*

$$|\widehat{f}(y)| \le c\, e^{-|y|\,\theta(y)}$$

if and only if $\int_1^\infty \theta(t)t^{-1}dt < \infty$.

Recently, the following analogue of Ingham's theorem for the group Fourier transform has been proved in [5]:

Let Θ *be a nonnegative even function on* $\mathbb{R}$ *which decreases to* 0 *at infinity. There exists a compactly supported function* f *on* $\mathbb{H}^n$ *whose Fourier transform satisfies the decay condition*

$$|\widehat{f}(\lambda)| \le c\, e^{-\sqrt{H(\lambda)}\,\Theta(\sqrt{H(\lambda)})}$$

if and only if $\int_1^\infty \Theta(t)t^{-1}dt < \infty$.

As before using the relation between $\widehat{f}(\lambda)$ and $\widehat{f}(a,z)$, we can restate Ingham's theorem in the following form.

Theorem 7.10 *Let Θ be a nonnegative even function on $\mathbb{R}$ which decreases to 0 at infinity. There exists a compactly supported function f on $\mathbb{H}^n$ whose Fourier transform satisfies the decay condition*

$$|\lambda|^n \int_{\mathbb{C}^n} |\widehat{f}(a,z)|^2 dz \leq c \, \frac{(k+n-1)!}{k!(n-1)!} \, e^{-\sqrt{(2k+n)|\lambda|}} \, \Theta(\sqrt{(2k+n)|\lambda|})$$

if and only if $\int_1^\infty \Theta(t) t^{-1} dt < \infty$.

The proof of Ingham's theorem given in [5] can be easily modified to prove this version. We can also prove other refined versions stated and proved elsewhere. For more on Ingham's theorem and its close relatives, we refer to [9] and the references therein.

Acknowledgments This work is supported by J. C. Bose Fellowship from Department of Science and Technology, Government of India.

References

1. F. Astengo, Di Blasio and F. Ricci, Gelfand transforms of polyradial Schwartz functions on the Heisenberg group, *J. Funct. Anal.* **251** (2007), no. 2, 772–791.
2. F. Astengo, B. Di Blasio and F. Ricci, Gelfand pairs on the Heisenberg group and Schwartz functions, *J. Funct. Anal.* **256** (2009), no.5, 1565–1587.
3. F. Astengo, B. Di Blasio and F. Ricci, Fourier transform of Schwartz functions on the Heisenberg group, *Studia Math.*, **214** (2013), no.3, 201–222.
4. F. Astengo, B. Di Blasio and F. Ricci, Paley-Wiener theorems for the $U(n)$-spherical transform on the Heisenberg group, *Ann. Mat. Pura Appl. (4),* **194** (2015), no. 6, 1751–1774.
5. S. Bagchi, P. Ganguly, J. Sarkar and S. Thangavelu, An analogue of Ingham's theorem on the Heisenberg group, *Math. Ann.* DOI 10.1007/s00208-022-02479–5.
6. H. Bahouri, J-Y. Chemin and R. Danchin, A frequency space for the Heisenberg group *Ann. Inst. Fourier (Grenoble)*, **69** no. 1 (2019), 365–407.
7. C. Benson, J. Jenkins and G. Ratcliff, The spherical transform of a Schwartz function on the Heisenberg group, *J. Funct. Anal.* **154** (1998), no. 2, 379–423.
8. G. B. Folland, *Harmonic Analysis in Phase Space*, Ann. of Math. Stud., vol. 122, Princeton University Press, Princeton, N.J., 1989.
9. P. Ganguly and S. Thangavelu, Analogues of theorems of Chernoff and Ingham on the Heisenberg group, *J. d'Analyse Math.* **149** (2023), no. 1, 281–305
10. D. Geller, Fourier analysis on the Heisenberg group, I. Schwartz space, *J. Func. Anal.* **36** (1980), no. 2, 205–254.
11. S. Helgason, *Groups and Geometric Analysis,* Academic Press, (1984).
12. S. Helgason, *Geometric analysis on symmetric spaces,* Mathematical surveys and monographs, AMS, Vol 39, (1994).
13. R. Rawat, P. K. Ratnakumar and S. Thangavelu, A restriction theorem for the Heisenberg motion group, *Studia Math.* **126 (1)** (1997), 1–12.
14. F. Ricci, Schwartz functions on the Heisenberg group, spectral multipliers and Gelfand pairs, *Rev. Un. Mat. Argentina* **50** (2009), no. 2, 175–186.
15. E. M. Stein and G. Weiss, Interpolation of operators with change of measures, *Trans. Amer. Math. Soc.* **87** (1958), 159–172.

16. E. M. Stein and G. Weiss, *Introduction to Fourier Analysis in Euclidean spaces,* Princeton Univ. Press, Princeton, N.J., 1971. x+297 pp.
17. R. S. Strichartz, Harmonic analysis as spectral theory of Laplacians, *J. Funct. Anal.* **87** (1989), no. 1, 51–148.
18. R. S. Strichartz, L^p harmonic analysis and Radon transforms on the Heisenberg group, *J. Funct. Anal.* **96** (1991), no. 2, 350–406.
19. G. Szego, *Orthogonal polynomials*, Am. Math. Soc. Colloq. Pub. **23**, Providence, RI (1967).
20. M. Taylor, *Noncommutative harmonic analysis. Mathematical Surveys and Monographs,* **22** American Mathematical Society, Providence, RI, (1986).
21. S. Thangavelu, *Lectures on Hermite and Laguerre expansions*, Mathematical Notes **42**. Princeton University Press, Princeton, NJ, 1993.
22. S. Thangavelu, On Paley-Wiener theorems for the Heisenberg group, *J. Funct. Anal.* **115** (1993), no. 1, 24–44.
23. S. Thangavelu, *Harmonic analysis on the Heisenberg group,* Progr. Math. Vol. 159 Birkhäuser, Boston, MA, 1998.
24. S. Thangavelu, *An introduction to the uncertainty principle. Hardy's theorem on Lie groups. With a foreword by Gerald B. Folland*, Progress in Mathematics 217. Birkhäuser, Boston, MA, 2004.

Chapter 8
Asymptotic Behavior of the Heat Semigroup on Certain Riemannian Manifolds

Alexander Grigor'yan, Effie Papageorgiou, and Hong-Wei Zhang

8.1 Introduction

Let $\mathcal{M}$ be a Riemannian manifold of dimension $n \geq 2$ and Δ be the Laplace–Beltrami operator on $\mathcal{M}$. It is well understood that the long-time behavior of solutions to the heat equation

$$\partial_t u(t,x) = \Delta_x u(t,x), \qquad u(0,x) = u_0(x), \qquad t > 0,\ x \in \mathcal{M}, \tag{8.1}$$

is very much related to the global geometry of $\mathcal{M}$. This applies also to the heat kernel $h_t(x,y)$ that is the minimal positive fundamental solution of the heat equation or, equivalently, the integral kernel of the heat semigroup $\exp(t\Delta)$ (see for instance [7, 12, 17]).

If the initial function u_0 belongs to the space $L^p(\mathcal{M},\mu)$ with $p \in [1,\infty)$ (where μ is the Riemannian measure on $\mathcal{M}$), then the Cauchy problem (8.1) has a unique solution u such that $u(t,\cdot) \in L^p$ for any $t > 0$, and this solution is given by

$$u(t,x) = \int_{\mathcal{M}} h_t(x,y)\, u_0(y)\, \mathrm{d}\mu(y). \tag{8.2}$$

A. Grigor'yan (✉)
Fakultät für Mathematik, Universität Bielefeld, Bielefeld, Germany
e-mail: grigor@math.uni-bielefeld.de

E. Papageorgiou
Department of Mathematics and Applied Mathematics, University of Crete, Crete, Greece

H.-W. Zhang
Department of Mathematics, Analysis, Logic and Discrete Mathematics Ghent University, Ghent, Belgium
e-mail: hongwei.zhang@ugent.be

P. Alonso Ruiz et al. (eds.), *From Classical Analysis to Analysis on Fractals*, Applied and Numerical Harmonic Analysis, https://doi.org/10.1007/978-3-031-37800-3_8

The same is true for the case $p = \infty$, provided $\mathcal{M}$ is stochastically complete. Hence, by a solution of (8.1), we always mean the function (8.2).

The aim of this chapter is to investigate the connection between the long-time behavior of the solution $u\,(t, x)$ of (8.1) and that of the heat kernel $h_t\,(x, y)$. Let the initial function u_0 belong to $L^1(\mathcal{M})$, and denote by $M = \int_{\mathcal{M}} u_0(x)\,\mathrm{d}\mu(x)$ its mass. In the case when $\mathcal{M} = \mathbb{R}^n$ with the Euclidean metric, the heat kernel is given by

$$h_t(x, y) = (4\pi t)^{-\frac{n}{2}} e^{-\frac{|x-y|^2}{4t}},$$

and the solution to (8.1) satisfies as $t \to \infty$

$$\|u(t, .) - Mh_t(., x_0)\|_{L^1(\mathbb{R}^n)} \longrightarrow 0 \tag{8.3}$$

and

$$t^{\frac{n}{2}} \|u(t, .) - Mh_t(., x_0)\|_{L^\infty(\mathbb{R}^n)} \longrightarrow 0. \tag{8.4}$$

By interpolation, a similar convergence holds with respect to any L^p norm when $1 < p < \infty$:

$$t^{\frac{n}{2p'}} \|u(t, .) - Mh_t(., 0)\|_{L^p(\mathbb{R}^n)} \longrightarrow 0,$$

where p' is the Hölder conjugate of p.

Note that (8.3) and (8.4) hold for *any* choice of x_0, which means that in the long run the solution $u\,(t, x)$ and the heat kernel $h_t\,(x, x_0)$ "forget" about the initial function u_0 (respectively, initial point x_0). We refer to a recent survey [20] for more details about this property in the Euclidean setting.

The convergence properties (8.3) and (8.4) have an interesting probabilistic meaning. Let $\{X_t\}$ be the Brownian motion on $\mathcal{M}$ whose transition density is $h_t\,(x, y)$. Then, (8.4) means, in particular, that X_t eventually "forgets" about its starting point x_0, which corresponds to the fact that X_t escapes to ∞ rotating chaotically in angular direction.

The situation is drastically different in hyperbolic spaces. It was shown by Vázquez [21] that (8.3) fails for a general initial function $u_0 \in L^1\,(\mathbb{H}^n)$ but is still true if u_0 is spherically symmetric around x_0. Similar results were obtained in [2] in a more general setting of symmetric spaces of non-compact type by using tools of harmonic analysis. Note that these spaces have nonpositive sectional curvature. Recall that in hyperbolic spaces Brownian motion X_t tends to escape to ∞ along geodesics, which means that it "remembers" at least the direction of the starting point x_0.

Our main result is the following theorem that deals with manifolds of non-negative Ricci curvature. Denote by $B\,(x, r)$ the geodesic ball on $\mathcal{M}$ of radius r centered at $x \in \mathcal{M}$ and set $V\,(x, r) = \mu\,(B\,(x, r))$.

Theorem 8.1 *Let $\mathcal{M}$ be a complete, connected, and non-compact Riemannian manifold of non-negative Ricci curvature. Fix a base point $x_0 \in \mathcal{M}$ and suppose that $u_0 \in L^1(\mathcal{M})$.Then, the solution to the heat equation (8.1) satisfies as $t \to \infty$*

$$\|u(t,\,.) - Mh_t(\,.\,,x_0)\|_{L^1(\mathcal{M})} \to 0 \tag{8.5}$$

and

$$\|\,|u(t,\,.) - Mh_t(\,.\,,x_0)|\,V(\cdot,\sqrt{t})\|_{L^\infty(\mathcal{M})} \to 0. \tag{8.6}$$

Remark 8.1 By interpolation between (8.5) and (8.6), we obtain for any $p \in (1,\infty)$

$$\|\,|u(t,\,.) - Mh_t(\,.\,,x_0)|\,V(\cdot,\sqrt{t})^{1/p'}\|_{L^p(\mathcal{M})} \longrightarrow 0. \tag{8.7}$$

In Sect. 8.2, we will give a short review of different estimates of the heat kernel and reformulate Theorem 8.1 in more general ways. In Sect. 8.3, we prove Theorem 8.1. An essential idea of the proof is to describe the critical region where the heat kernel concentrates. In the last section, we show that an estimate as (8.6) fails on manifolds with two Euclidean ends.

8.2 Auxiliary Results

8.2.1 Heat Kernel Estimates

From now on, $\mathcal{M}$ denotes a complete, connected, and non-compact Riemannian manifold of dimension $n \geq 2$. Let μ be the Riemannian measure on $\mathcal{M}$. Let $d(x, y)$ be the geodesic distance between two points $x, y \in \mathcal{M}$ and $V(x,r) = \mu(B(x,r))$ be the Riemannian volume of the geodesic ball $B(x,r)$ of radius r centered at $x \in \mathcal{M}$.

Throughout the chapter, we follow the convention that $C, C_1, \ldots$ denote large positive constants, whereas $c, c_1, \ldots$ are small positive constants. These constants may depend on $\mathcal{M}$ but do not depend on the variables x, y, t. Moreover, the notation $A \lesssim B$ between two positive expressions means that $A \leq CB$, and $A \asymp B$ means $cB \leq A \leq CB$.

We say that $\mathcal{M}$ satisfies the volume doubling property if, for all $x \in \mathcal{M}$ and $r > 0$, we have

$$V(x,2r) \leq C\,V(x,r). \tag{8.8}$$

It follows from (8.8) that there exist some positive constants $\nu, \nu' > 0$ such that

$$c\left(\frac{R}{r}\right)^{\nu'} \leq \frac{V(x,R)}{V(x,r)} \leq C\left(\frac{R}{r}\right)^{\nu} \tag{8.9}$$

for all $x \in \mathcal{M}$ and $0 < r \leq R$ (see, for instance, [12, Section 15.6]). Moreover, (8.9) implies that, for all $x, y \in \mathcal{M}$ and $r > 0$,

$$\frac{V(x,r)}{V(y,r)} \leq C\left(1+\frac{d(x,y)}{r}\right)^{\nu}. \tag{8.10}$$

The integral kernel $h_t(x,y)$ of the heat semigroup $\exp(t\Delta)$ is the smallest positive fundamental solution to the heat equation (8.1). It is known that $h_t(x,y)$ is smooth in (t,x,y) and symmetric in x, y and satisfies the semigroup identity (see, for instance, [12, 19]). Besides, for all $y \in \mathcal{M}$ and $t > 0$,

$$\int_{\mathcal{M}} h_t(x,y)\,\mathrm{d}\mu(x) \leq 1.$$

The manifold $\mathcal{M}$ is called stochastically complete if for all $y \in \mathcal{M}$ and $t > 0$

$$\int_{\mathcal{M}} h_t(x,y)\,\mathrm{d}\mu(x) = 1.$$

It is known that if $\mathcal{M}$ is geodesically complete and, for some $x_0 \in \mathcal{M}$ and all large enough r,

$$V(x_0,r) \leq e^{Cr^2},$$

then $\mathcal{M}$ is stochastically complete. In particular, the volume doubling property (8.8) implies that $\mathcal{M}$ is stochastically complete.

When the Ricci curvature of $\mathcal{M}$ is non-negative, the following two-sided estimates of the heat kernel were proved by Li and Yau [15]:

$$\frac{c_1}{V(x,\sqrt{t})}\exp\left(-C_1\frac{d^2(x,y)}{t}\right) \leq h_t(x,y) \leq \frac{C_2}{V(x,\sqrt{t})}\exp\left(-c_2\frac{d^2(x,y)}{t}\right). \tag{8.11}$$

Besides, Li and Yau proved in [15] also the following gradient estimate for any positive solution $u(t,x)$ of the heat equation $\partial_t u = \Delta u$ on $\mathbb{R}_+ \times \mathcal{M}$:

$$\frac{|\nabla u|^2}{u^2} - \frac{\partial_t u}{u} \leq \frac{C}{t} \tag{8.12}$$

with $C = \frac{n}{2}$. By a result of [11], the upper bound of the heat kernel in (8.11), that is,

$$h_t(x, y) \le \frac{C}{V(x, \sqrt{t})} \exp\Big(- c\, \frac{d^2(x, y)}{t} \Big), \tag{8.13}$$

implies the following estimate of the time derivative:

$$\Big| \frac{\partial h_t}{\partial t}(x, y) \Big| \le \frac{C}{t\, V(x, \sqrt{t})} \exp\Big(- c \frac{d^2(x, y)}{t} \Big). \tag{8.14}$$

It follows from (8.12) that

$$|\nabla u|^2 \le u \partial_t u + \frac{C}{t} u^2$$

and, hence,

$$|\nabla u| \le \sqrt{u\, |\partial_t u|} + \frac{C}{\sqrt{t}} u.$$

Applying this for the function $u(t, y) = h_t(x, y)$ and combining with (8.13) and (8.14), we obtain that

$$|\nabla_y h_t(x, y)| \le \frac{C}{\sqrt{t}\, V(x, \sqrt{t})} \exp\Big(- c\, \frac{d^2(x, y)}{t} \Big). \tag{8.15}$$

It is known that the upper bound (8.13) of h_t (and, hence, its consequence (8.14)) holds on a larger class of Riemannian manifolds satisfying a so-called *relative Faber-Krahn inequality* (see, for example, [12]). But the estimate (8.15) of the gradient ∇h_t is much more subtle and requires more serious hypotheses, for example, non-negative Ricci curvature as we consider here.

Let us observe that the gradient estimate (8.15) is a particular (and limiting) case of the following Hölder estimate:

$$|h_t(x, y) - h_t(x, z)| \le \Big(\frac{d(y, z)}{\sqrt{t}} \Big)^\theta \frac{C}{V(x, \sqrt{t})} \exp\Big(- c \frac{d^2(x, y)}{t} \Big), \tag{8.16}$$

for some $0 < \theta \le 1$, and all $t > 0$, $x, y, z \in \mathcal{M}$ such that $d(y, z) \le \sqrt{t}$. It is important to mention that (8.16) is a consequence of the two-sided estimate (8.11) alone (for some $0 < \theta < 1$, see [17, Theorem 5.4.12]), and, hence, (8.16) holds on more general classes of manifolds than (8.15).

Finally, let us observe that on spaces of essentially negative curvature the above estimates of the heat kernel typically fail: for example, these are hyperbolic spaces [6], non-compact symmetric spaces [1], asymptotically hyperbolic manifolds [5], and fractal-like manifolds [3].

8.2.2 Alternative Statements of the Main Theorem

Now, let us reformulate Theorem 8.1 in a bit more general way.

Theorem 8.2 *Let $\mathcal{M}$ be a geodesically complete non-compact manifold that satisfies the following conditions:*

- *The volume doubling condition* (8.8)
- *The upper bound* (8.13) *of the heat kernel*
- *The Hölder regularity* (8.16) *of the heat kernel*

Then, the conclusions of Theorem 8.1 hold.

Apart from manifolds with non-negative Ricci curvature, the above-described manifolds cover many other examples. Let us recall that, on a complete Riemannian manifold, the following three properties are equivalent:

- The two-sided estimate (8.11) of the heat kernel
- The uniform parabolic Harnack inequality:

$$\sup_{(\frac{T}{4},\frac{T}{2})\times B(x,\frac{r}{2})} u(t,x) \le C \sup_{(\frac{3T}{4},T)\times B(x,\frac{r}{2})} u(t,x), \tag{8.17}$$

  where $u(t,x)$ is a non-negative solution of the heat equation $\partial_t u = \Delta u$ in a cylinder $(0,T) \times B(x,r)$ with $x \in \mathcal{M}$, $r > 0$ and $T = r^2$
- The conjunction of the volume doubling property (8.8) and the Poincaré inequality:

$$\int_{B(x,r)} |f - f_B|^2 \,\mathrm{d}\mu \le C r^2 \int_{B(x,r)} |\nabla f|^2 \,\mathrm{d}\mu, \tag{8.18}$$

  for all $x \in \mathcal{M}$, $t > 0$, and bounded Lipschitz functions f in $B(x,r)$. Here, f_B is the mean of f over $B(x,r)$.

See, for instance, [9, 10, 16, 17] for more details. Manifolds satisfying these equivalent conditions include complete manifolds with non-negative Ricci curvature, connected Lie groups with polynomial volume growth, co-compact covering manifolds whose deck transformation has polynomial growth, and many others. We refer to [18, pp. 417–418] for a list of examples. These equivalent conditions yield particularly the Hölder regularity (8.16) for some $0 < \theta < 1$, see [17, Theorem 5.4.12]. Hence, we obtain the following corollary.

Corollary 8.1 *Let $\mathcal{M}$ be a geodesically complete non-compact manifold that satisfies one of the following equivalent conditions:*

- *The two-sided estimate* (8.11) *of the heat kernel*
- *The uniform parabolic Harnack inequality* (8.17)

- *The conjunction of the volume doubling property* (8.8) *and the Poincaré inequality* (8.18)

Then, the conclusions of Theorem 8.1 hold.

In fact, we do not need a differential operator like the Laplacian (or sub-Laplacian). All we need is a function $h_t(x, y)$ satisfying certain estimates, and the result can be formulated as a property of integral operators. For example, our method can be applied on a metric space (X, d, μ) of homogeneous type, assuming that one has a semigroup of self-adjoint linear operators acting on $L^2(X, \mu)$ and admitting a good integral kernel, see, for instance, [8].

8.2.3 Two Preliminary Lemmas

We prove Theorem 8.2 in the next section. In the next two lemmas, we describe some consequences of the hypotheses of Theorem 8.2, in particular, the critical annulus where the heat kernel concentrates.

Lemma 8.1 *Under the volume doubling condition (8.8), we have, for any $c > 0$, $x_0 \in \mathcal{M}$, and $t > 0$,*

$$\int_{\mathcal{M}} \exp\left(-c\,\frac{d\,(x, x_0)^2}{t}\right) \frac{\mathrm{d}\mu(x)}{V(x_0, \sqrt{t})} \lesssim 1 \tag{8.19}$$

and, for any $N \in \mathbb{N}$ and all $r \geq \sqrt{t}$,

$$\int_{B(x_0,r)^c} \exp\left(-c\,\frac{d\,(x, x_0)^2}{t}\right) \frac{\mathrm{d}\mu(x)}{V(x_0, \sqrt{t})} \lesssim \left(\frac{r}{\sqrt{t}}\right)^{-N}. \tag{8.20}$$

Proof Let us prove first (8.20). Using (8.9), we have

$$\frac{1}{V(x_0, \sqrt{t})} \int_{d(x,x_0)\geq r} \exp\left(-c\,\frac{d\,(x, x_0)^2}{t}\right) \mathrm{d}\mu(x)$$

$$= \frac{1}{V(x_0, \sqrt{t})} \sum_{j=0}^{\infty} \int_{2^j r \leq d(x,x_0) < 2^{j+1} r} \exp\left(-c\,\frac{d\,(x, x_0)^2}{t}\right) \mathrm{d}\mu(x)$$

$$\lesssim \sum_{j=0}^{\infty} \frac{V(x_0, 2^{j+1} r)}{V(x_0, \sqrt{t})} \exp\left(-c\,\frac{\left(2^j r\right)^2}{t}\right)$$

$$\lesssim \sum_{j=0}^{\infty} \left(\frac{2^{j+1} r}{\sqrt{t}}\right)^{\nu} \exp\left(-c\,\frac{2^{2j} r^2}{t}\right).$$

Since $\exp(-s) \le (N!)\, s^{-N}$ for any $s > 0$ and $N \in \mathbb{N}$, we obtain that

$$\int_{B(x_0,r)^c} \exp\left(-c\,\frac{d(x,x_0)^2}{t}\right)\frac{d\mu(x)}{V(x_0,\sqrt{t})} \lesssim \sum_{j=0}^{\infty}\left(\frac{r}{\sqrt{t}}\right)^{\nu} 2^{\nu j-2Nj}\left(\frac{t}{r^2}\right)^{N}$$

$$\lesssim \left(\frac{r}{\sqrt{t}}\right)^{\nu-2N},$$

which proves (8.20).

In order to prove (8.19), we apply (8.20) with $r = \sqrt{t}$ and $N = 1$ and obtain

$$\int_{\mathcal{M}} \exp\left(-c\,\frac{d(x,x_0)^2}{t}\right)\frac{d\mu(x)}{V(x_0,\sqrt{t})}$$

$$= \left(\int_{B(x_0,\sqrt{t})^c} + \int_{B(x_0,\sqrt{t})}\right)\exp\left(-c\,\frac{d(x,x_0)^2}{t}\right)\frac{d\mu(x)}{V(x_0,\sqrt{t})}$$

$$\lesssim 1 + \int_{B(x_0,\sqrt{t})}\frac{d\mu(x)}{V(x_0,\sqrt{t})} = 2.$$

□

The next lemma describes the annulus where the heat kernel concentrates. Let us fix a point $x_0 \in \mathcal{M}$ and a positive function $\varphi(t)$ such that $\varphi(t) \to 0$ and $\varphi(t)\sqrt{t} \to \infty$ as $t \to \infty$. For any $t > 0$, define the following annulus in $\mathcal{M}$:

$$\Omega_t = \left\{x \in \mathcal{M} \mid \varphi(t)\sqrt{t} \le d(x,x_0) \le \frac{\sqrt{t}}{\varphi(t)}\right\}. \tag{8.21}$$

Lemma 8.2 *Under the hypotheses (8.8) and (8.13), we have for all large enough t*

$$\int_{\mathcal{M}\setminus\Omega_t} h_t(x,x_0)\, d\mu(x) \lesssim \varphi(t)^{\nu'}, \tag{8.22}$$

where ν' is the exponent from (8.9). Consequently,

$$\int_{\Omega_t} h_t(x,x_0)\, d\mu(x) \longrightarrow 1 \quad \text{as } t \longrightarrow \infty. \tag{8.23}$$

Proof Since $\mathcal{M}$ is stochastically complete, (8.23) follows from (8.22) and $\varphi(t) \to 0$. Since

$$\mathcal{M}\setminus\Omega_t = B\left(x_0, \varphi(t)\sqrt{t}\right) \cup B^c\left(x_0, \sqrt{t}/\varphi(t)\right),$$

we estimate the integrals over $B(x_0, \varphi(t)\sqrt{t})$ and $B^c(x_0, \sqrt{t}/\varphi(t))$ separately. Assume that t is large enough so that $\varphi(t) < 1$. Using (8.13) and (8.9), we obtain

$$\int_{d(x,x_0)<\varphi(t)\sqrt{t}} h_t(x, x_0)\, \mathrm{d}\mu(x) \lesssim \frac{V(x_0, \varphi(t)\sqrt{t})}{V(x_0, \sqrt{t})} \lesssim \varphi(t)^{\nu'}.$$

Using (8.13) and (8.20) with $r = \frac{\sqrt{t}}{\varphi(t)}$ and any $N \in \mathbb{N}$, we obtain

$$\begin{aligned}\int_{d(x,x_0)>\frac{\sqrt{t}}{\varphi(t)}} h_t(x, x_0)\, \mathrm{d}\mu(x) &\lesssim \int_{B(x_0,r)^c} \frac{\mathrm{d}\mu(x)}{V(x_0, \sqrt{t})} \exp\left(-c\,\frac{d\,(x, x_0)^2}{t}\right) \\ &\lesssim \left(\frac{r}{\sqrt{t}}\right)^{-N} = \varphi\,(t)^N,\end{aligned}$$

whence the claim follows. □

8.3 Proof of the Main Theorem

We start the proof of Theorem 8.2 with continuous compactly supported initial data and prove the asymptotic properties of the solution in the L^1 norm, working separately outside and inside the critical region Ω_t. Then, we show that these properties remain valid for all L^1 initial data by using a density argument. The L^∞ convergence is proved in the same spirit.

Proposition 8.1 *Let $\mathcal{M}$ be a manifold defined as in Theorem 8.2. Let $x_0 \in \mathcal{M}$ and $u_0 \in C_c(B(x_0, a))$ for some $a > 0$. Assume that $\varphi(t)$ is a positive function such that $\varphi(t) \to 0$ and $\varphi(t)\sqrt{t} \to \infty$ as $t \to \infty$. Then, the solution (8.2) satisfies*

$$\|u(t, .) - M h_t(\,.\,, x_0)\|_{L^1(\mathcal{M}\setminus\Omega_t)} \lesssim \varphi(t)^{\nu'} \tag{8.24}$$

and

$$\|u(t, .) - M h_t(\,.\,, x_0)\|_{L^1(\Omega_t)} \lesssim t^{-\frac{\theta}{2}}, \tag{8.25}$$

where Ω_t is defined by (8.21), ν' and θ are exponents from (8.9) and (8.16), *and*

$$M = \int_{\mathcal{M}} u_0(x)\, \mathrm{d}\mu(x).$$

Consequently,

$$\|u(t, .) - M h_t(\,.\,, x_0)\|_{L^1(\mathcal{M})} \lesssim t^{-\eta} \tag{8.26}$$

for any $\eta < \min(\nu', \theta)/2$.

Proof By the upper bound (8.13) of the heat kernel and Lemma 8.2, we have

$$\|h_t(\,.\,, x_0)\|_{L^1(\mathcal{M}\setminus\Omega_t)} \lesssim \varphi(t)^{\nu'}$$

so that (8.24) will follow if we prove that

$$\|u(t,\,.\,)\|_{L^1(\mathcal{M}\setminus\Omega_t)} \lesssim \varphi(t)^{\nu'}. \tag{8.27}$$

We write

$$\begin{aligned}\|u(t,\,.\,)\|_{L^1(\mathcal{M}\setminus\Omega_t)} &= \int_{\mathcal{M}\setminus\Omega_t} \left| \int_{B(x_0,a)} h_t(x,y)\, u_0(y)\, \mathrm{d}\mu(y) \right| \mathrm{d}\mu(x) \\ &\le \int_{B(x_0,a)} |u_0(y)| \left\{ \int_{\mathcal{M}\setminus\Omega_t} h_t(x,y)\, \mathrm{d}\mu(x) \right\} \mathrm{d}\mu(y).\end{aligned}$$

Notice that $x \in \mathcal{M}\setminus\Omega_t$ and $y \in B(x_0,a)$ imply $x \in \mathcal{M}\setminus\widetilde{\Omega}_{t,y}$, where

$$\widetilde{\Omega}_{t,y} = \left\{ x \in \mathcal{M} \,\middle|\, 2\,\varphi(t)\sqrt{t} \le d(x,y) \le \frac{1}{2}\frac{\sqrt{t}}{\varphi(t)} \right\},$$

provided t is large enough. Indeed, if $x \in \widetilde{\Omega}_{t,y}$, then

$$d(x,x_0) \le d(x,y) + d(y,x_0) \le \frac{1}{2}\frac{\sqrt{t}}{\varphi(t)} + a \le \frac{\sqrt{t}}{\varphi(t)}$$

and

$$d(x,x_0) \ge d(x,y) - d(y,x_0) \ge 2\,\varphi(t)\sqrt{t} - a \ge \varphi(t)\sqrt{t}$$

for t large enough, since $\varphi(t) \to 0$ and $\varphi(t)\sqrt{t} \to \infty$ as $t \to \infty$. It follows that $x \in \Omega_t$.

Applying (8.22) with $\widetilde{\Omega}_{t,y}$ instead of Ω_t, we obtain

$$\int_{\mathcal{M}\setminus\Omega_t} h_t(x,y)\, \mathrm{d}\mu(x) \le \int_{\mathcal{M}\setminus\widetilde{\Omega}_{t,y}} h_t(x,y)\, \mathrm{d}\mu(x) \lesssim \varphi(t)^{\nu'},$$

whence (8.27) follows.

Now, let us turn to (8.25). Observe that

$$u(t,x) - M h_t(x,x_0) = \int_{\mathcal{M}} u_0(y)\, \left(h_t(x,y) - h_t(x,x_0)\right) \mathrm{d}\mu(y)$$

$$= \int_{B(x_0,a)} u_0(y)\ (h_t(x,y) - h_t(x,x_0))\ \mathrm{d}\mu(y). \tag{8.28}$$

On the one hand, we deduce from the Hölder regularity (8.16) that

$$\begin{aligned}
&|u(t,x) - Mh_t(x,x_0)| \\
&\le \frac{C}{V(x,\sqrt{t})} \exp\Big(-c\,\frac{d^2(x,x_0)}{t}\Big) \int_{B(x_0,a)} \Big(\frac{d(x_0,y)}{\sqrt{t}}\Big)^{\theta} |u_0(y)|\ \mathrm{d}\mu(y) \\
&\lesssim t^{-\frac{\theta}{2}} \frac{1}{V(x,\sqrt{t})} \exp\Big(-c\,\frac{d^2(x,x_0)}{t}\Big),
\end{aligned} \tag{8.29}$$

for some $0 < \theta \le 1$ and for t large enough such that $d(x_0,y) \le a \le \sqrt{t}$. On the other hand, (8.10) implies that

$$\begin{aligned}
\frac{1}{V(x,\sqrt{t})} \exp\Big(-c\,\frac{d^2(x,x_0)}{t}\Big) &\lesssim \Big(\frac{d(x,x_0)}{\sqrt{t}} + 1\Big)^{\nu} \\
&\quad \times \frac{1}{V(x_0,\sqrt{t})} \exp\Big(-c\,\frac{d^2(x,x_0)}{t}\Big) \\
&\lesssim \frac{1}{V(x_0,\sqrt{t})} \exp\Big(-c\,\frac{d^2(x,x_0)}{2t}\Big).
\end{aligned} \tag{8.30}$$

Substituting (8.30) into the right-hand side of (8.29) and integrating in x over Ω_t, we obtain by (8.19)

$$\begin{aligned}
\int_{\Omega_t} |u(t,x) - Mh_t(x,x_0)|\,\mathrm{d}\mu(x) &\lesssim t^{-\frac{\theta}{2}} \int_{\Omega_t} \frac{\mathrm{d}\mu(x)}{V(x_0,\sqrt{t})} \\
&\quad \times \exp\Big(-c\,\frac{d^2(x,x_0)}{2t}\Big) \lesssim t^{-\frac{\theta}{2}}.
\end{aligned}$$

Finally, (8.26) follows by adding up (8.24) and (8.25) with $\varphi(t) = t^{\varepsilon-\frac{\theta}{2}}$ with small enough ε. □

Next, we prove our main theorem.

Proof of Theorem 8.2 Given $u_0 \in L^1(\mathcal{M})$, fix $\varepsilon > 0$ and choose $\widetilde{u}_0 \in C_c(\mathcal{M})$ such that

$$\|u_0 - \widetilde{u}_0\|_{L^1(\mathcal{M})} < \frac{\varepsilon}{3}.$$

Let us prove first (8.5). Setting $\widetilde{M} = \int_{M} \widetilde{u}_0(y)\, \mathrm{d}\mu(y)$, we have

$$|M - \widetilde{M}| \le \int_{M} \mathrm{d}\mu(y)\, |u_0(y) - \widetilde{u}_0(y)| = \|u_0 - \widetilde{u}_0\|_{L^1(M)} < \frac{\varepsilon}{3}. \tag{8.31}$$

It follows that

$$\|Mh_t(\,.\,, x_0) - \widetilde{M}h_t(\,.\,, x_0)\|_{L^1(M)} \le |M - \widetilde{M}|\, \|h_t(\,.\,, x_0)\|_{L^1(M)} < \frac{\varepsilon}{3}. \tag{8.32}$$

Let $\widetilde{u}(t, x)$ be the semigroup solution to the heat equation with the initial data $\widetilde{u}_0$. Then, we have

$$\begin{aligned} &\|u(t,\,.) - \widetilde{u}(t,\,.)\|_{L^1(M)} \\ &\le \int_{M} |u_0(y) - \widetilde{u}_0(y)| \left\{ \int_{M} \mathrm{d}\mu(x)\, h_t(x, y) \right\} \mathrm{d}\mu(y) < \frac{\varepsilon}{3}. \end{aligned} \tag{8.33}$$

By the estimate (8.26) of Proposition 8.1, we have, for sufficiently large t,

$$\|\widetilde{u}(t,\,.) - \widetilde{M}h_t(\,.\,, x_0)\|_{L^1(M)} < \frac{\varepsilon}{3}. \tag{8.34}$$

Combining (8.32), (8.33), and (8.34) together, we obtain

$$\|u(t,\,.) - Mh_t(\,.\,, x_0)\|_{L^1(M)} < \varepsilon,$$

which finishes the proof of (8.5).

Let us turn to the sup-norm convergence (8.6). Since $\widetilde{u}_0$ is compactly supported, we can apply (8.29) which gives for all x

$$|\widetilde{u}(t, x) - \widetilde{M}h_t(x, x_0)| \le \frac{C}{t^{\theta/2} V(x, \sqrt{t})} \le \frac{\varepsilon}{V(x, \sqrt{t})} \tag{8.35}$$

for sufficiently large t. Next, we have

$$\begin{aligned} |u(t, x) - \widetilde{u}(t, x)| &\le \int_{M} h_t(x, y)\, |u_0(y) - \widetilde{u}_0(y)|\, \mathrm{d}\mu(y) \\ &\le \frac{C\, \|u_0 - \widetilde{u}_0\|_{L^1}}{V(x, \sqrt{t})} \lesssim \frac{\varepsilon}{V(x, \sqrt{t})} \end{aligned} \tag{8.36}$$

and

$$|Mh_t(x, x_0) - \widetilde{M}h_t(x, x_0)| = |M - \widetilde{M}|\, h_t(x, x_0) \lesssim \frac{\varepsilon}{V(x, \sqrt{t})}. \tag{8.37}$$

Putting together (8.35), (8.36), and (8.37), we conclude that

$$|u(t, x) - Mh_t(x, x_0)| \lesssim \frac{\varepsilon}{V(x, \sqrt{t})},$$

whence (8.6) follows. □

8.4 A Counterexample

Let $n \geq 3$. Consider a manifold $\mathcal{M} = \mathbb{R}^n \# \mathbb{R}^n$ that is a connected sum of two copies of $\mathbb{R}^n$. This means that $\mathcal{M}$ is a union of a compact part K and two Euclidean ends $E^{\pm} = \mathbb{R}^n \setminus B(0, R)$ (see Fig. 8.1). This is a classical setting where the two-sided heat kernel estimate (8.11) fails, see, for instance, [13].

Note that $V(x, R) \asymp R^n$ for all $x \in \mathcal{M}$, and the heat kernel on $\mathcal{M}$ satisfies the upper bound (8.13) (see, for instance, [13, Corollary 4.6, p.1946]). However, the essential hypotheses, Hölder regularity (8.16) or pointwise gradient estimate (8.15), fail in the present setting (and so does the Poincaré inequality [14]). Indeed, it was shown in [4, Proposition 6.1] that, for large t,

$$\sup_{x,y\in\mathcal{M}} \left|\nabla_y h_t(x, y)\right| \geq c\, t^{-n/2},$$

which is incompatible with (8.15).

Let us verify that also the conclusion (8.6) of Theorem 8.1 fails in this setting. For that, consider on E^+ for any $t > 1$ a point x_t with $|x_t| = \sqrt{t}$ and with a fixed direction $\omega = x_t/\sqrt{t}$. It was proved in [4, Proposition 6.1] that, for any $x \in K$,

$$\lim_{t\to\infty} t^{\frac{n}{2}} h_t(x_t, x) = \Phi(x), \tag{8.38}$$

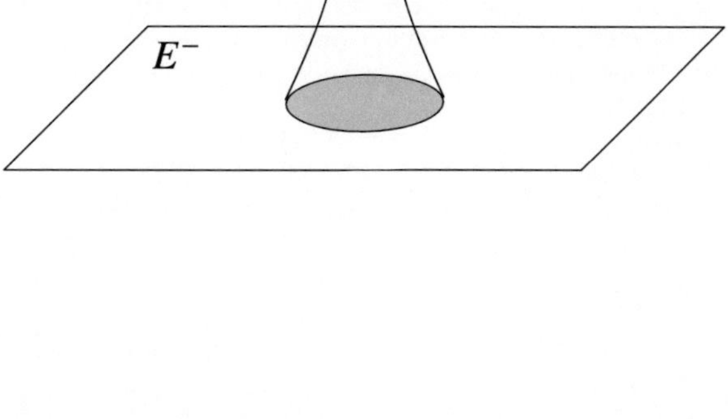

Fig. 8.1 Manifold $\mathcal{M} = \mathbb{R}^n \# \mathbb{R}^n$

where $\Phi(x)$ is a positive harmonic function on $\mathcal{M}$ that tends to a positive constant as $x \to \infty$ at the end E^+ and tends to 0 as $x \to \infty$ at the end E^-. Let us fix two points x_1 and x_2 in K such that $\Phi(x_1) \neq \Phi(x_2)$. Then, (8.38) implies that

$$t^{\frac{n}{2}} \left(h_t(x_t, x_1) - h_t(x_t, x_2)\right) \longrightarrow \left(\Phi(x_1) - \Phi(x_2)\right) \quad \text{as} \quad t \longrightarrow \infty, \tag{8.39}$$

while (8.6) would imply that

$$t^{\frac{n}{2}} \|h_t(\cdot, x_1) - h_t(\cdot, x_2)\|_{L^\infty(\mathcal{M})} \longrightarrow 0 \quad \text{as} \quad t \longrightarrow \infty,$$

and hence,

$$t^{\frac{n}{2}} (h_t(x_t, x_1) - h_t(x_t, x_2)) \longrightarrow 0,$$

which contradicts to (8.39).

Acknowledgments The first author is funded by the Deutsche Forschungsgemeinschaft (DFG, German Research Foundation)—Project-ID 317210226—SFB 1283. The second author is supported by the Hellenic Foundation for Research and Innovation, Project HFRI-FM17-1733. The last author acknowledges financial support from the Methusalem Programme *Analysis and Partial Differential Equations (Grant number 01M01021)* during his postdoc stay at Ghent University.

References

1. Anker, J.-Ph., Ostellari, P.: The heat kernel on noncompact symmetric spaces. In: Lie Groups And Symmetric Spaces, pp. 27–46. Amer. Math. Soc., Providence, RI (2003)
2. Anker, J.-Ph., Papageorgiou, E., Zhang, H.-W.: Asymptotic behavior of solutions to the heat equation on noncompact symmetric spaces. J. Funct. Anal. **284**, Paper No. 109828 (2023)
3. Barlow, M.: Diffusions on fractals. In: Lectures On Probability Theory And Statistics (Saint-Flour, 1995), pp. 1–121. Springer, Berlin (1998)
4. Carron, G., Coulhon, T., Hassell, A.: Riesz transform and L^p-cohomology for manifolds with Euclidean ends. Duke Math. J. **133**, 59–93 (2006)
5. Chen, X., Hassell, A.: The heat kernel on asymptotically hyperbolic manifolds. Comm. Partial Differential Equations **45**, 1031–1071 (2020)
6. Davies, E., Mandouvalos, N.: Heat kernel bounds on hyperbolic space and Kleinian groups. Proc. London Math. Soc. (3) **57**, 182–208 (1988)
7. Davies, E.: Heat kernels and spectral theory. Cambridge University Press (1990)
8. Dziubański, J., Preisner, M.: Hardy spaces for semigroups with Gaussian bounds. Ann. Mat. Pura Appl. (4). **197**, 965–987 (2018)
9. Fabes, E., Stroock, D.: A new proof of Moser's parabolic Harnack inequality using the old ideas of Nash. Arch. Ration. Mech. Anal. **96**, 327–338 (1986)
10. Grigor'yan, A.: The heat equation on noncompact Riemannian manifolds. Mat. Sb. **182**, 55–87 (1991)
11. Grigor'yan, A.: Upper bounds of derivatives of the heat kernel on an arbitrary complete manifold. J. Funct. Anal. **127**, 363–389 (1995)
12. Grigor'yan, A.: Heat kernel and analysis on manifolds. AMS International Press (2009)

13. Grigor'yan, A., Saloff-Coste, L.: Heat kernel on manifolds with ends. Ann. Inst. Fourier (Grenoble) **59**, 1917–1997 (2009)
14. Grigor'yan, A., Ishiwata, S., Saloff-Coste, L.: Geometric analysis on manifolds with ends. In: Analysis And Partial Differential Equations On Manifolds, Fractals And Graphs, pp. 325–343, Adv. Anal. Geom. **3**, De Gruyter, Berlin (2021)
15. Li, P., Yau, S.: On the parabolic kernel of the Schrödinger operator. Acta Math. **156**, 153–201 (1986)
16. Saloff-Coste, L.: A note on Poincaré, Sobolev, and Harnack inequalities. Internat. Math. Res. Notices, 27–38 (1992)
17. Saloff-Coste, L.: Aspects of Sobolev-type inequalities. Cambridge University Press (2002)
18. Saloff-Coste, L.: The heat kernel and its estimates. In: Probabilistic Approach To Geometry, pp. 405–436. Adv. Stud. Pure Math. **57**, Math. Soc. Japan, Tokyo (2010)
19. Strichartz, R.: Analysis of the Laplacian on the complete Riemannian manifold. J. Functional Analysis **52**, 48–79 (1983)
20. Vázquez, J.: Asymptotic behaviour methods for the heat equation. Convergence to the Gaussian. preprint, arXiv:1706.10034
21. Vázquez, J. Asymptotic behaviour for the heat equation in hyperbolic space. To appear in Comm. Analysis and Geometry, arXiv:1811.09034

Part IV
Intrinsic Analysis on Fractals

Chapter 9
Fourier Series for Fractals in Two Dimensions

John E. Herr, Palle E. T. Jorgensen, and Eric S. Weber

Dedicated to the memory of Robert Strichartz.[1]

9.1 Introduction

The past two decades have seen a flurry of interest in exploring Fourier expansions for $L^2(\mu)$ in the case that μ is in a class of fractal measures, with particular emphasis on the case when μ is singular with respect to Lebesgue measure. These studies have fallen into two groups: (i) the case of orthogonal expansions (analyzed for Cantor measures and related IFS measures) and (ii) the case of frame-like expansions. In both cases, there now exist explicit algorithms for analysis and synthesis for the corresponding Hilbert spaces $L^2(\mu)$ [22, 28]. By this we mean that there exists a sequence of frequencies $\{\lambda_n\}$ such that for every $f \in L^2(\mu)$,

$$f(x) = \sum_{n=0}^{\infty} c_n e^{2\pi i \lambda_n \cdot x}. \tag{9.1}$$

[1] The present chapter expands on a talk by the second named author at the 7th Cornell Conference on Analysis, Probability, and Mathematical Physics on Fractals, June 4–8, 2022, in the memory of R. S. Strichartz.

J. E. Herr (✉)
Department of Mathematical Sciences, Butler University, Indianapolis, IN, USA
e-mail: jeherr@butler.edu

P. E. T. Jorgensen
Department of Mathematics, University of Iowa, Iowa City, IA, USA
e-mail: palle-jorgensen@uiowa.edu

E. S. Weber
Department of Mathematics, Iowa State University, Ames, IA, USA
e-mail: esweber@iastate.edu

P. Alonso Ruiz et al. (eds.), *From Classical Analysis to Analysis on Fractals*, Applied and Numerical Harmonic Analysis, https://doi.org/10.1007/978-3-031-37800-3_9

The coefficients are given by inner products, which in many circumstances are not taken against the exponentials themselves. Moreover, the convergence may be conditional only.

Orthogonal expansions of the form in Eq. (9.1) for fractal measures were initiated by Pedersen and the second named author in [28] and have been built upon by numerous authors; we highlight just a few subsequent papers [14, 15, 32], particularly those by Strichartz [51, 52]. We note that the initial constructions of [28] were in one spatial dimension but are generalizable to higher dimensions. However, only certain fractal measures (namely spectral measures) admit an orthogonal expansion. Non-orthogonal expansions as in Eq. (9.1) for arbitrary singular measures μ were introduced by the first and third named authors in [22] using so-called effective sequences obtained via the Kaczmarz algorithm. Recent work by all three co-authors [19–21] has focused on connections between these Fourier expansions and model spaces in the form of de Branges Hilbert spaces of analytic functions. However, up to now these methods have been largely limited to the cases when the measure μ at hand has its support in one spatial dimension. Our chapter will extend the Kaczmarz Fourier expansions to measures in two spatial dimensions that satisfy a "slice-singular" condition. We will do so by developing a theory of an operator-valued Kaczmarz algorithm.

In this chapter, we present a generalized Fourier expansion for a specific family of singular measures μ admitting realizations in $\mathbb{R}^2$. The family we consider, called slice-singular measures, includes IFS measures μ, and it covers in particular a variety of Sierpinski IFS measure constructions. In this multi-variable setting, we present explicit non-orthogonal Fourier series expansion in $L^2(\mu)$. These Fourier series expansions go far beyond the related and better-known orthogonal expansions, first introduced by Jorgensen and Pedersen [28] and explored later in multiple papers by Strichartz [49–52]. Still, the Sierpinski-type class comprises only a small class of Cantor-type measures. In the present chapter, we cover $L^2(\mu)$ Fourier expansions for all slice-singular multivariable measures μ. We also stress why the orthogonality requirement for the earlier $L^2(\mu)$ Fourier expansions in the Jorgensen-Pedersen/Strichartz family is extremely restrictive.

Our main result can be loosely stated thusly:

Theorem *Suppose that μ is a slice-singular Borel probability measure on* $[0, 1]^2$. *For any $f \in L^2(\mu)$, there exists a sequence of coefficients $\{c_{nm}\}$ such that*

$$f(x, y) = \sum_{n,m=0}^{\infty} c_{nm} e^{2\pi i(nx+my)},$$

where the sum converges in norm, possibly conditionally. □

The order of the double sum depends on the slice-singular nature of the measure μ (see Theorem 9.1 and the discussion following).

The chapter is organized as follows: The remainder of this section discusses the requisite measure theory (e.g., the Rokhlin Decomposition Theorem and the

Kakutani Dichotomy Theorem) and discusses the Kaczmarz construction of Fourier series expansions for singular measures in one dimension. Section 9.2 defines slice-singular measures and derives several properties of such measures. We show that the Sierpinski gasket and carpet are both slice-singular. Section 9.3 lays out the proof of our main results concerning the construction of Fourier series expansions for slice-singular measures. Our proof utilizes the Kaczmarz algorithm for operators, described in Sect. 9.4. The proof of our main results uses the crucial result of Sect. 9.4, which is that the operators we define in Sect. 9.3 are effective.

Remark 9.1 While the present work is focused on a particular Fourier-based harmonic analysis on affine IFS-fractal structures, we stress that there are many other approaches. The following list gives a glimpse into these diverse approaches: [6, 7, 12, 13, 17, 33, 40].

9.1.1 Self-similar Fractals and Measures

Self-similar fractals are the invariant set of an iterated function system (IFS) consisting of transformations acting on Euclidean space. Well-known examples of such fractals are Cantor sets and Mandelbrot sets. Fractals are commonly studied objects in many contexts. Cantor sets, in particular, appear in the context of analysis [5, 50], number theory [8, 48], probability [4, 27, 36], geometry [41, 42], physics [1, 53, 54], and harmonic analysis [28, 44, 51].

Our main results concerning Fourier series expansions will apply to a large class of measures, those that we refer to as "slice-singular" (Definition 9.6). Notably, a common set of measures that satisfy our definition are fractal measures generated by IFSs.

Definition 9.1 Suppose $\Psi = \{\psi_1, \ldots, \psi_N\}$ is a set of maps acting on a metric space (X, d). We say that $A \subset X$ is invariant for Ψ if $A = \cup_{i=1}^{N} \psi_i(A)$.

In [25], Hutchinson laid out the main relationship between fractals and iterated function systems (IFS); this relationship is the foundation of our results. Recall that $\psi : X \to X$ is Lipschitz if for all $x, y \in X$, there exists $C > 0$ such that

$$d(\psi(x), \psi(y)) \leq Cd(x, y).$$

The Lipschitz constant of ψ is the infimum of all such C. We say that ψ is a contraction if it has Lipschitz constant less than 1. Hutchinson's Theorem is as follows:

Theorem A *Let $X = (X, d)$ be a complete metric space and $\Psi = \{\psi_1, \ldots, \psi_N\}$ a finite set of contraction maps on X. Then there exists a unique closed bounded set A such that*

$$A = \bigcup_{i=1}^{N} \psi_i(A).$$

Furthermore, A is compact and is the closure of the set of fixed points of finite compositions of members of Ψ. Moreover, for a closed and bounded set K, $\Psi^p(K) \to A$ in the Hausdorff metric.

Moreover, given a set of weights $w_1, \ldots, w_N > 0$ such that $\sum w_j = 1$, there exists a unique Borel probability measure μ supported on A that satisfies the invariance equation

$$\int f d\mu = \sum_{j=1}^{N} w_j \int f \circ \psi_j d\mu \tag{9.2}$$

for every continuous function f. □

The measure μ is called the invariant measure for the IFS $\{\psi_1, \ldots, \psi_N\}$ with weights $\{w_1, \ldots, w_N\}$.

Note that in Eq. (9.2), some of the ψ_j's can be repeated, in which case we can sum the corresponding weights together to obtain the same invariance equation but with the ψ_j's all distinct. We shall use this observation in several of our examples, and unless otherwise stated, we assume this distinctiveness condition on our IFS.

Our IFSs often take the form of affine transformations acting on $\mathbb{R}^d$. We do this by specifying a scaling parameter $R > 1$ and a set of digits $B = \{b_j\}_{n=1}^N \subset \mathbb{R}^d$ and defining

$$\psi_j(x) = \frac{x + b_j}{R}.$$

As we will discuss in the examples in Sect. 9.2.3, the Sierpinski gasket and Sierpinski carpet both arise in this fashion.

An application of the Kakutani Dichotomy Theorem [23, 30] says that given two invariant measures of an IFS acting on the same metric space with corresponding weights, the measures are either identical or mutually singular. Consequently, if their weights are different, then the two measures are mutually singular. We formalize this statement for future reference; though Kakutani's Theorem did not concern invariant measures directly, we present the result as such:

Theorem B (Kakutani Dichotomy Theorem) *Suppose $\{\psi_1, \ldots, \psi_N\}$ are distinct contractions acting on the metric space (X, d). Let μ and $\tilde{\mu}$ denote the invariant measures arising from these contractions using weights $\{w_1, \ldots, w_N\}$ and $\{\tilde{w}_1, \ldots, \tilde{w}_N\}$, respectively. Then:*

1. *The measures $\mu = \tilde{\mu}$ if and only if $w_j = \tilde{w}_j$ for all j.*
2. *The measures μ and $\tilde{\mu}$ are mutually singular whenever $w_j \neq \tilde{w}_j$ for some j.* □

9.1.2 Review of Fourier Series for 1D Fractals

Let $d \in \mathbb{N}$ be a number of dimensions, let $X \subseteq \mathbb{R}^d$, and let μ be a probability measure on X. By a Fourier series for $f \in L^2(\mu)$, we mean an ordered countable set of frequencies $\Lambda \subset \mathbb{R}^d$ and sequence of coefficients $\{c_\lambda\}$, where $c_\lambda \in \mathbb{C}$ for all $\lambda \in \Lambda$, such that

$$f = \sum_{\lambda \in \Lambda} c_\lambda e^{2\pi i \langle \lambda, x \rangle},$$

where the convergence occurs in the norm of $L^2(\mu)$. For concision, we will adopt the notation $e_\lambda := e^{2\pi i \langle \lambda, x \rangle}$, and we will refer to such functions as "(complex) exponential functions."

We begin by reviewing the method used in [22] to establish Fourier series for functions in $L^2(\mu)$, where μ is a one-dimensional singular probability measure on $[0, 1)$. The present work generalizes this method.

Remark 9.2 The expansions here are done with a choice of Hilbert space and inner products in the context of $L^2(\mu)$ for some Borel measure μ on $\mathbb{R}^d$ or, in our discussion of general frame expansions, an abstract Hilbert space (Definitions 9.2, 9.3, 9.4, and 9.5). In each case, the inner products occurring inside the expansions are written $\langle \cdot, \cdot \rangle$, and they will correspond to these Hilbert spaces. In Theorem D we return to the case of $L^2(\mu)$.

Such a new method was necessary due to the general insufficiency of orthonormal bases and frames to yield such series. For example, if there exists a countable set $\Lambda \subseteq \mathbb{R}$ such that $\left\{e^{2\pi i \lambda x}\right\}_{\lambda \in \Lambda}$ is an orthonormal basis of $L^2(\mu)$, then clearly

$$f = \sum_{\lambda \in \Lambda} \langle f, e_\lambda \rangle \, e^{2\pi i \lambda x}$$

for any $f \in L^2(\mu)$. The measure μ is then said to be spectral with spectrum Λ, and we have Fourier series representations.

Sometimes $L^2(\mu)$ is spectral, and sometimes it is not [28]. For example, in the case $d = 1$, $R = 4$, and $B = \{0, 2\}$, we get the IFS $\left\{\psi_0(x) = \frac{x}{4}, \psi_2(x) = \frac{x+2}{4}\right\}$, set invariant set A from Hutchinson's Theorem is the quaternary Cantor set, and the corresponding invariant measure is the quaternary Cantor measure, denoted ν_4, which is spectral. On the other hand, in the case that $d = 1$, $R = 3$, and $B = \{0, 2\}$, the set A is the famous ternary or "middle-thirds" Cantor set, and μ is the ternary Cantor measure, denoted ν_3. Jorgensen and Pedersen showed that there cannot exist three mutually orthogonal complex exponential functions in this situation, and so there cannot be an orthonormal basis of complex exponential functions. Therefore, we cannot rely on orthonormal bases of complex exponential functions for the construction of Fourier series.

Frames would be another option [16, 34]:

Definition 9.2 A sequence $\{f_n\}_{n=0}^{\infty}$ in a Hilbert space $\mathbb{H}$ is said to be *Bessel* if there exists a constant $B > 0$ such that for any $x \in \mathbb{H}$,

$$\sum_{n=0}^{\infty} |\langle x, f_n \rangle|^2 \leq B\|x\|^2. \tag{9.3}$$

This is equivalent to

$$\left\| \sum_{n=0}^{K} c_n f_n \right\| \leq \sqrt{B \sum_{n=0}^{K} |c_n|^2}$$

for any finite sequence $\{c_0, c_1, \ldots, c_K\}$ of complex numbers. The sequence is called a *frame* if in addition there exists a constant $A > 0$ such that for any $x \in \mathbb{H}$,

$$A\|x\|^2 \leq \sum_{n=0}^{\infty} |\langle x, f_n \rangle|^2 \leq B\|x\|^2. \tag{9.4}$$

If $A = B$, then the frame is said to be *tight*. If $A = B = 1$, then $\{f_n\}_{n=0}^{\infty}$ is a *Parseval frame*. The constant A is called the *lower frame bound* and the constant B is called the *upper frame bound* or *Bessel bound*.

If $\{f_n\}$ is a frame, then there exists a dual frame $\{g_n\}$ such that for all $x \in \mathbb{H}$,

$$x = \sum \langle x, g_n \rangle f_n = \sum \langle x, f_n \rangle g_n.$$

Therefore, if there were a collection of exponential functions $\{e_{\ell_n}\}$ that is a frame in $L^2(\mu)$, then taking a dual frame $\{g_n\}$, we would have

$$f = \sum_{n=0}^{\infty} \langle f, g_n \rangle e_{\ell_n}.$$

Unfortunately, for non-spectral singular probability measures μ, it is generally unknown whether $L^2(\mu)$ possesses a complex exponential frame. In fact, it is still an open question of whether the middle-third Cantor set possesses an exponential frame, first posed by Strichartz [49].

Moreover, the full sequence $\{e_n\}_{n=1}^{\infty}$ is not Bessel if μ is singular, and so it cannot be a frame. The fact that orthogonal bases and frames do not readily work to yield Fourier series inspired the authors in [22] to turn to effective sequences:

Definition 9.3 (Effective Sequences) Let $\{\varphi_n\}_{n=0}^{\infty}$ be a linearly dense sequence of unit vectors in a Hilbert space $\mathbb{H}$. Given any element $x \in \mathbb{H}$, we may define a sequence $\{x_n\}_{n=0}^{\infty}$ in the following manner:

$$x_0 = \langle x, \varphi_0 \rangle \varphi_0$$
$$x_n = x_{n-1} + \langle x - x_{n-1}, \varphi_n \rangle \varphi_n.$$

If $\lim_{n\to\infty} \|x - x_n\| = 0$ regardless of the choice of x, then the sequence $\{\varphi_n\}_{n=0}^{\infty}$ is said to be effective.

The above formula for producing $\{x_n\}$ is known as the Kaczmarz algorithm. The Kaczmarz sequence begins with the projection of x onto φ_0. Each next x_n is then the result of moving from x_{n-1} in the direction of φ_n to the extent that it takes us closer to x. In other words, x_n is the projection of x_{n-1} onto the affine subspace of $\mathbb{H}$ containing x parallel to the subspace $\varphi_n^{\perp}$.

In 1937, Stefan Kaczmarz [29] proved the effectivity of linearly dense periodic sequences in the finite-dimensional case. In 2001, these results were extended to infinite-dimensional Banach spaces under certain conditions by Kwapień and Mycielski [31]. These two also gave the following formula for the sequence $\{x_n\}_{n=0}^{\infty}$, which we state here for the Hilbert space setting: Define

$$g_0 = \varphi_0$$
$$g_n = \varphi_n - \sum_{i=0}^{n-1} \langle \varphi_n, \varphi_i \rangle g_i. \tag{9.5}$$

Then

$$x_n = \sum_{i=0}^{n} \langle x, g_i \rangle \varphi_i. \tag{9.6}$$

As shown by [31], and also more clearly for the Hilbert space setting by [18], we have

$$\|x\|^2 - \lim_{n\to\infty} \|x - x_n\|^2 = \sum_{n=0}^{\infty} |\langle x, g_n \rangle|^2,$$

from which it follows that $\{\varphi_n\}_{n=0}^{\infty}$ is effective if and only if

$$\sum_{n=0}^{\infty} |\langle x, g_n \rangle|^2 = \|x\|^2. \tag{9.7}$$

That is to say, $\{\varphi_n\}_{n=0}^{\infty}$ is effective if and only if the associated sequence $\{g_n\}_{n=0}^{\infty}$ is a Parseval frame. We call $\{g_n\}_{n=0}^{\infty}$ the auxiliary sequence of $\{\varphi_n\}_{n=0}^{\infty}$.

If $\{\varphi_n\}_{n=0}^{\infty}$ is effective, then (9.6) implies that for any $x \in \mathbb{H}$, $\sum_{i=0}^{\infty}\langle x, g_i\rangle\varphi_i$ converges to x in norm, and as noted, $\{g_n\}_{n=0}^{\infty}$ is a Parseval frame. This does not mean that $\{g_n\}_{n=0}^{\infty}$ and $\{\varphi_n\}_{n=0}^{\infty}$ are dual frames, since $\{\varphi_n\}_{n=0}^{\infty}$ need not even be a frame. However, $\{\varphi_n\}_{n=0}^{\infty}$ and $\{g_n\}_{n=0}^{\infty}$ are pseudo-dual in the following sense, first given by Li and Ogawa in [35]:

Definition 9.4 Let $\mathbb{H}$ be a separable Hilbert space. Two sequences $\{\varphi_n\}$ and $\{\varphi_n^{\star}\}$ in $\mathbb{H}$ form a pair of *pseudoframes* for $\mathbb{H}$ if for all $x, y \in \mathbb{H}$, $\langle x, y\rangle = \sum_n \langle x, \varphi_n^{\star}\rangle\langle\varphi_n, y\rangle$.

All frames are pseudoframes, but not the converse. Observe that if $x, y \in \mathbb{H}$ and $\{\varphi_n\}_{n=0}^{\infty}$ is effective, then

$$\begin{aligned}\langle x, y\rangle &= \left\langle \sum_{m=0}^{\infty}\langle x, g_m\rangle\varphi_m, y\right\rangle \\ &= \sum_{m=0}^{\infty}\langle x, g_m\rangle\,\langle\varphi_m, y\rangle\end{aligned}$$

and so $\{\varphi_n\}_{n=0}^{\infty}$ and $\{g_n\}_{n=0}^{\infty}$ are pseudo-dual.

Of course, since $\{g_n\}_{n=0}^{\infty}$ is a Parseval frame, it is a true dual frame for itself. We also employ the following definition:

Definition 9.5 (Dextroduality) Let $\mathcal{H}$ be a separable Hilbert space. Let $\{f_n\}$ and $\{g_n\}$ be two sequences in $\mathcal{H}$. We say that $\{g_n\}$ is dextrodual to $\{f_n\}$ if

$$\sum_{n=0}^{\infty}\langle x, g_n\rangle\, f_n = x$$

for all $x \in \mathbb{H}$.

In other words, if T_g is the analysis operator of $\{g_n\}$ and T_f^* is the synthesis operator of $\{f_n\}$, then $T_f^*T_g = I_{\mathcal{H}}$. The appearance of the synthesis operator T_g on the right side of the product is the reason for using the prefix "dextro-."

Thus, if $\{\varphi_n\}$ is effective, then $\{g_n\}$ is dextrodual to $\{\varphi_n\}$.

The aforementioned theorem of Kwapień and Mycielski in [31] demonstrates a condition under which $\{\varphi_n\}$ may be effective in an infinite-dimensional separable Hilbert space:

Theorem C (Kwapień and Mycielski) *A stationary sequence of unit vectors that is linearly dense in a Hilbert space is effective if and only if its spectral measure either coincides with the normalized Lebesgue measure or is singular with respect to Lebesgue measure.* □

By a stationary sequence, it is meant a sequence $\{\varphi_n\}$ such that $\langle\varphi_m, \varphi_n\rangle = \langle\varphi_{m+k}, \varphi_{n+k}\rangle$ for all $m, n, k \in \mathbb{N}_0$. Thus, the expression $\langle\varphi_m, \varphi_{m+k}\rangle$ depends on

k alone. Then the spectral measure of $\{\varphi\}$ is the unique positive Borel measure σ satisfying

$$\langle \varphi_m, \varphi_{m+k} \rangle = \int_0^1 e^{-2\pi i k x} \, d\sigma(x).$$

It was observed in [22] that in the case that μ is a Borel probability measure on $[0, 1)$, then the sequence $\{e_n\}_{n=0}^{\infty}$ of complex exponential functions is stationary and linearly dense in $L^2(\mu)$, and μ is itself the spectral measure of $\{e_n\}$. It follows by the theorem of Kwapień and Mycielski that $\{e_n\}$ is effective in $L^2(\mu)$ if and only if μ is Lebesgue measure or singular with respect to Lebesgue measure. Thus, the following result was obtained:

Theorem D (Herr and Weber) *If μ is a singular Borel probability measure on $[0, 1)$, then the sequence $\{e_n\}_{n=0}^{\infty}$ is effective in $L^2(\mu)$. As a consequence, any element $f \in L^2(\mu)$ possesses a Fourier series*

$$f(x) = \sum_{n=0}^{\infty} c_n e^{2\pi i n x},$$

where

$$c_n = \int_0^1 f(x) \overline{g_n(x)} \, d\mu(x)$$

and $\{g_n\}_{n=0}^{\infty}$ is the auxiliary sequence associated to $\{e_n\}_{n=0}^{\infty}$ via Equation (9.5). *The sum converges in norm, and Parseval's identity $\|f\|^2 = \sum_{n=0}^{\infty} |c_n|^2$ holds.* □

Thus, singular probability measures on $[0, 1)$ do yield Fourier series, and they come from performing the Kaczmarz algorithm with the sequence $\{e_n\}_{n=0}^{\infty}$. In this chapter, we turn to the problem of obtaining the same result in higher dimensions. The main obstacle to applying the Kwapień-Mycielski Theorem in the same way as before is the condition of stationarity. In $[0, 1)^d$ with μ a Borel probability measure on $[0, 1)^d$, the complex exponential functions are now of the form $e_\lambda := e^{2\pi i \langle \lambda, x \rangle}$, where $\lambda, x \in \mathbb{R}^d$. The set $\{e_n : n \in \mathbb{N}_0^d\}$ can be shown to be linearly dense in $L^2(\mu)$, but when ordered into a sequence will not be stationary. It is this issue that we address in this chapter.

The construction of Fourier Series for singular measures in 1 dimension via the Kaczmarz algorithm is enriched by its connection to the de Branges-Rovnyak subspaces of the classical Hardy space. By the Herglotz Representation Theorem, there is a 1-to-1 correspondence between the singular nonnegative Borel measures μ on $[0, 1)$ and the nonconstant inner functions b in the Hardy space H^2 given by the Poisson integral:

$$\mathrm{Re}\left(\frac{1+b(z)}{1-b(z)}\right)=\int_0^1\frac{1-|z|^2}{\left|e^{2\pi ix}-z\right|^2}\,d\mu(x).$$

Since b is inner, the de Branges-Rovnyak subspace $\mathcal{H}(b)$ of H^2 is equal to $H^2\ominus bH^2$ with the same norm as H^2. The normalized Cauchy transform V_μ gives a mapping of $L^2(\mu)$ onto $\mathcal{H}(b)$ via

$$V_\mu f(z)=\frac{\int_0^1\frac{f(x)}{1-ze^{-2\pi ix}}\,d\mu(x)}{\int_0^1\frac{1}{1-ze^{-2\pi ix}}\,d\mu(x)}.$$

In the case that μ is a probability measure (i.e., $\|\mu\|=1$, or equivalently $b(0)=0$), which is the case to which we restrict ourselves in this chapter, V_μ is a unitary transformation.

Suppose μ is a singular Borel probability measure on $[0,1)$, and $f\in L^2(\mu)$. Let $V_\mu f(z)=\sum_{n=0}^\infty a_nz^n$. Then, Theorem 1.1 of [43] implies that

$$\sum_{n=0}^\infty a_ne^{2\pi inx}=f(x)\tag{9.8}$$

in the $L^2(\mu)$ norm. In fact, however, this series turns out to be the same series as constructed from the Kaczmarz algorithm in Theorem D via the Kwapień-Mycielski Theorem.

9.1.3 Rokhlin Disintegration Theorem

The notion of direct integrals in the setting of analysis of operators in Hilbert spaces arises in such diverse applications as the theory of unitary representations of groups, decompositions used in the study of von Neumann algebras, mathematical physics, ergodic theory, machine learning models, statistics, and probability theory [24, 26, 38, 39, 46, 47].

Perhaps closest to our present analysis is the use of "slice decompositions" arising in Bayesian probability theory. There one studies joint distributions of systems of random variables, and then one introduces associated marginal measures and the corresponding conditional distributions. "Bayesian" derives from Thomas Bayes, who offered the first mathematical treatment of statistical data analysis, introducing what is now known as Bayesian inference.

For our purpose at hand, the readers might find it useful to think of our present direct integrals as extensions of orthogonal direct sums (the case of counting measure) to direct integrals, or alternatively, extending the notion of discrete frame expansions to their wider measurable counterparts.

Here, we stress the use of direct integrals in a very specific context: that of reducing a particular harmonic analysis in two variables to the study of one-dimensional slices, which then comprise the particular direct integral decompositions at hand. In our present application, we make precise our particular direct integrals, both for the Hilbert spaces at hand and for the resulting direct integral operators.

The Rokhlin Disintegration Theorem [46] generalizes the Fubini-Tonelli theorems for product measures. It allows for measures on product spaces to be decomposed in such a way that integration can be accomplished using iterated integrals.

Theorem E (Rokhlin Disintegration Theorem) *If μ is a Borel probability measure on $A \times B$ (with A, B metric spaces), then there exist a Borel probability measure σ on B and a parametrized family of Borel probability measures $\{\gamma^b\}_{b\in B}$ on A such that*

1. *For a Borel set $E \subset A \times B$, $\mu(E) = \int_B \gamma^b(E \cap (A \times \{b\}))\, \sigma(db)$.*
2. *For $f \in L^1(\mu)$, for σ-a.e. b, $f(\cdot, b) \in L^1(\gamma^b)$.*
3. *For $f \in L^1(\mu)$, the mapping $b \mapsto \int_B f(a, b)\gamma^b(da)$ is measurable and integrable w.r.t σ.* □

The measure σ is called the B-marginal of μ, and we refer to the $\{\gamma^b\}$ as the *slice measures*.

Note that the disintegration of μ can also be done using the A-marginal and the slices $\{\gamma_a\}_{a\in A}$ in the obvious manner. Further discussion of the Rokhlin theory can be found in [2, 3, 9, 45].

9.2 Slice-Singular Measures

In this section we present the analytic details needed for our direct integral decomposition of slice-singular measures μ in two dimensions. We shall refer to these direct integral decompositions as slice decompositions, and the measures μ considered in two dimensions are assumed to be slice-singular. Our aim is multi-variable Fourier expansions for $L^2(\mu)$, generally non-orthogonal.

For a given measure μ in two dimensions, the notion "slice-singular" is made precise below; it refers to assumptions regarding both the marginal measures defined from μ and the corresponding conditional measures. The resulting decompositions for μ may be viewed as Bayes rules, but our setting is more general, and we shall refer to the general decompositions as Rokhlin-disintegration-decompositions.

If μ is supported in a subset of $\mathbb{R}^2$, then there are two marginal measures, each one a one-dimensional measure. To each of these marginal measure there is associated a one-dimensional conditional measure. Our singularity assumptions will pertain to these measures that are obtained after disintegration.

9.2.1 Projections of Invariant Measures

We let π_1 and π_2 be the projections of $\mathbb{R}^2$ onto $\mathbb{R}$ given by $(x, y) \mapsto x$ and $(x, y) \mapsto y$, respectively.

We consider a set $\{\psi_1, \ldots, \psi_N\}$ of strict contractions on $\mathbb{R}^2$. Fix nonnegative weights $\{c_1, \ldots, c_N\}$ such that $\sum c_j = 1$, and let μ be the invariant measure given by Hutchinson's Theorem.

We say that the mapping $\psi : \mathbb{R}^2 \to \mathbb{R}^2$ is *Cartesian* if there exist mappings $\eta_1, \eta_2 : \mathbb{R} \to \mathbb{R}$ such that $\psi(x, y) = (\eta_1(x), \eta_2(y))$. This is equivalent to the condition $\eta_1 \circ \pi_1 = \pi_1 \circ \psi$ and $\eta_2 \circ \pi_2 = \pi_2 \circ \psi$.

If $\{\psi_1, \ldots, \psi_N\}$ are Cartesian, then the maps $\{\phi_1, \ldots, \phi_N\}$ given by

$$\phi_j \circ \pi_1 = \pi_1 \circ \psi_j \tag{9.9}$$

are a set of strict contractions on $\mathbb{R}$. Therefore, again by Hutchinson's Theorem, there exists a unique probability measure ρ such that for every continuous function $g : \mathbb{R} \to \mathbb{R}$,

$$\int g \, \rho(dx) = \sum_{j=1}^{N} c_j \int g \circ \phi_j \, \rho(dx).$$

Lemma 9.1 *Let $\{\psi_1, \ldots, \psi_N\}$ be a set of Cartesian strict contractions on $\mathbb{R}^2$ with invariant measure μ for the weights $\{c_1, \ldots, c_N\}$. Let μ_1 be the marginal of μ in the x-direction. Let $\{\phi_1, \ldots, \phi_N\}$ be given by Eq.* (9.9). *Then, for every continuous function $g : \mathbb{R} \to \mathbb{R}$,*

$$\int g \, \mu_1(dx) = \sum_{j=1}^{N} c_j \int g \circ \phi_j \, \mu_1(dx).$$

In other words, the x-marginal for μ is the (unique) invariant probability measure for the contractions $\{\phi_1, \ldots, \phi_N\}$ with weights $\{c_1, \ldots, c_N\}$.

Proof Let g be a continuous function on $\mathbb{R}$. We have that

$$\int g \circ \pi_1 \, \mu(dx\, dy) = \sum_{j=1}^{N} c_j \int g \circ \pi_1 \circ \psi_j \, \mu(dx\, dy) \tag{9.10}$$

$$= \sum_{j=1}^{N} c_j \int g \circ \phi_j \circ \pi_1 \, \mu(dx\, dy) \tag{9.11}$$

and

$$\int g \circ \pi_1 \, \mu(dx\,dy) = \int g \, \mu_1(dx).$$

Consequently, we have

$$\int g \, \mu_1(dx) = \sum_{j=1}^{N} c_j \int g \circ \phi_j \, \mu_1(dx).$$

□

Remark 9.3 Even though the contractions $\{\psi_1, \ldots, \psi_N\}$ in Lemma 9.1 may be distinct, there may be repetition in the projected contractions $\{\phi_1, \ldots, \phi_N\}$.

9.2.2 Slice-Singular Measures

Let μ be a Borel measure on $[0, 1] \times [0, 1]$. The Rokhlin decomposition states that there exists a measure μ_1 on $[0, 1]$, and for every $x \in [0, 1]$, there exists a measure ρ_x on $[0, 1]$ such that

$$\mu(dx\,dy) = \rho_x(dy)\mu_1(dx).$$

The measure μ_1 is called the *marginal* of μ (in the x-direction). It is given by

$$\mu_1(A) = \mu(\pi_1^{-1}(A))$$

for Borel sets $A \subset \mathbb{R}$. In short, $\mu_1 = \mu \circ \pi_1^{-1}$.

Likewise, the decomposition can be obtained in the other order:

$$\mu(dx\,dy) = \rho^y(dx)\mu_2(dy),$$

where $\mu_2 = \mu \circ \pi_2^{-1}$.

Definition 9.6 We say that a Borel measure on $[0, 1] \times [0, 1]$ is x-slice-singular if in the Rokhlin decomposition $\mu(dx\,dy) = \rho_x(dy)\mu_1(dx)$:

1. μ_1 is singular.
2. For μ_1 a.e. x, ρ_x is singular.

Similarly, we say that a Borel measure on $[0, 1] \times [0, 1]$ is y-slice-singular if in the Rokhlin decomposition $\mu(dx\,dy) = \rho^y(dx)\mu_2(dy)$:

1. μ_2 is singular.
2. For μ_2 a.e. y, ρ^y is singular.

We say that μ is slice-singular if it is so in either direction, and we say that μ is bi-slice-singular if it is slice-singular in both directions.

Our canonical examples of slice-singular measures correspond to affine fractals, such as the Sierpinski gasket and the Sierpinski carpet. These are generated by affine iterated function systems and generally are bi-slice-singular, as we show in Sect. 9.2.3. However, we also give an example of a measure that is slice-singular in one direction but not both in Example 3.

The following is a useful lemma for determining when a measure is slice-singular. We let λ denote Lebesgue measure on $\mathbb{R}$.

Lemma 9.2 *Suppose μ is a Borel probability measure on $[0, 1]^2$ such that both the x- and y-marginal measures are singular. Then, μ is bi-slice-singular.*

Proof Since the x-marginal measure μ_1 is singular, there exists a Borel set $A \subseteq [0, 1]$ such that $\lambda(A^C) = 0$ and $\mu_1(A) = 0$. It follows that

$$\mu(A \times [0, 1]) = \int_A \int_0^1 \rho_x(dy)\, \mu_1(dx) = 0.$$

Assume, for the sake of contradiction, that there exists a Borel set $B \subseteq [0, 1]$ of positive μ_2 measure such that for each $y \in B$, the y-marginal measure ρ^y is not purely singular, say $\rho^y = \nu^y_{\text{sing}} + \nu^y_{\text{cont}}$, where ν^y_{sing} is singular, ν^y_{cont} is absolutely continuous, and $\nu^y_{\text{cont}} \not\equiv 0$. It follows that $\nu^y_{\text{cont}}(A^C) = 0$, and so $\nu^y_{\text{cont}}(A) > 0$. Thus, $\rho^y(A) > 0$ for each $y \in B$. Therefore,

$$\mu(A \times [0, 1]) \geq \mu(A \times B) = \int_B \rho^y(A)\, \mu_2(dy) > 0,$$

which is a contradiction. By a symmetric argument, μ is also x-slice-singular, and so μ is bi-slice-singular. □

Lemma 9.3 *Suppose that μ is slice-singular. Then the set of exponential functions*

$$\{e^{2\pi i(nx+my)} : n, m \in \mathbb{N}_0\}$$

is dense in $L^2(\mu)$.

Proof Suppose that μ is y-slice-singular. Let $f \in L^2(\mu)$ be such that

$$\langle f(x, y), e^{2\pi i(nx+my)}\rangle_\mu = 0, \ \forall n, m \in \mathbb{N}_0.$$

For a fixed n, we have that

$$\int \left(\int f(x, y) e^{2\pi inx} \rho^y(dx) \right) e^{2\pi imy} \mu_2(dy) = 0.$$

Since μ_2 is singular, $\{e^{2\pi imy} : m \in \mathbb{N}_0\}$ is dense in $L^2(\mu_2)$, so we must have that

$$\int f(x, y)e^{2\pi inx} \rho^y(dx) = 0$$

for μ_2-a.e. y. It follows that for μ_2-a.e. y,

$$\int f(x, y)e^{2\pi inx} \rho^y(dx) = 0$$

for every $n \in \mathbb{N}_0$. Since ρ^y is also singular for μ_2-a.e. y, for ρ^y-a.e. x, we have $f(x, y) = 0$. It follows that $f(x, y) = 0$ for μ-a.e. (x, y). □

9.2.3 Examples

We will now give some examples. Of special interest are choices of planar measures μ that have the structure of IFS measures defined from the family of systems of 2D affine maps. We concentrate on Sierpinski constructions from the more familiar Sierpinski geometries in the plane, the Sierpinski gasket and carpet. As we will show, they are slice-singular IFS measures. In each case, using our results above applied to the particular $L^2(\mu)$-settings at hand, we can obtain corresponding non-orthogonal Fourier expansions.

Lemma 9.4 *Suppose μ is a Borel probability measure on $[0, 1]^2$ that is invariant under the reflection about the line $y = x$. Then μ is x-slice-singular if and only if it is y-slice-singular.*

Proof The proof is easy and is left as an exercise to the reader. □

Lemma 9.5 *Suppose $\Psi = \{\psi_1, \ldots, \psi_N\}$ are affine contractions of the form*

$$\psi_j(x) = \frac{x + b_j}{R}$$

with $\{b_j\} \subset \mathbb{Z}^2 \cap [0, R)^2$ and $R \nmid N$. If μ is the invariant measure for Ψ with equal weights $\frac{1}{N}$, then the marginal measures μ_1 and μ_2 of μ are singular with respect to Lebesgue measure.

Proof By Lemma 9.1, the marginal measure μ_1 is obtained by projecting Ψ onto the x-axis in $\mathbb{R}^2$. Doing so yields the IFS acting on $\mathbb{R}$ with generators

$$\phi_j(x) = \frac{x + \pi_1(b_j)}{R} \tag{9.12}$$

with weights $\frac{1}{N}$. If there are no repetitions among the $\{\phi_j\}$ in Eq. (9.12), we must have that $\{\pi_1(b_j)\} \subsetneq \{0, \ldots, R-1\}$, so μ_1 is singular. If there are repetitions, then we combine the generators by summing the corresponding weights to obtain μ_1 as the invariant measure for an IFS with generators

$$\tilde{\phi}_k(x) = \frac{x + c_k}{R}, \quad k = 0, \ldots, K-1 \tag{9.13}$$

with weights $\{\tilde{w}_0, \ldots, \tilde{w}_{K-1}\}$. Since $R \nmid N$, we must have that these weights are not uniformly $\frac{1}{K}$, and so, by the Kakutani Dichotomy Theorem, μ_1 is singular. □

Example 1 Our first example is the Sierpinski gasket, which for convenience we align with the coordinate axes. The generators for the gasket are given by

$$\varphi_0(x, y) = \left(\frac{x}{2}, \frac{y}{2}\right) \quad \varphi_1(x, y) = \left(\frac{x+1}{2}, \frac{y}{2}\right) \quad \varphi_2(x, y) = \left(\frac{x}{2}, \frac{y+1}{2}\right). \tag{9.14}$$

The measure for the gasket is obtained by using equal weights. The projection onto the x-axis yields the IFS with generators

$$\psi_0(x) = \frac{x}{2} \quad \psi_1(x) = \frac{x+1}{2} \quad \psi_2(x) = \frac{x}{2} \tag{9.15}$$

with equal weights, which is equivalent to the IFS $\{\psi_0, \psi_1\}$ with weights $(\frac{2}{3}, \frac{1}{3})$. Thus, by the Kakutani Dichotomy Theorem, the x-marginal measure for the Sierpinski gasket is singular.

It follows by symmetry that the y-marginal measure is also singular. Hence, by Lemmas 9.2 and 9.4, the Sierpinksi gasket is bi-slice-singular.

Example 2 Our second example is the Sierpinski carpet. The IFS generators are given by

$$\begin{aligned}
\varphi_0(x, y) &= \left(\frac{x}{3}, \frac{y}{3}\right) & \varphi_1(x, y) &= \left(\frac{x+1}{3}, \frac{y}{3}\right) \\
\varphi_2(x, y) &= \left(\frac{x+2}{3}, \frac{y}{3}\right) & \varphi_3(x, y) &= \left(\frac{x}{3}, \frac{y+1}{3}\right) \\
\varphi_4(x, y) &= \left(\frac{x+2}{3}, \frac{y+1}{3}\right) & \varphi_5(x, y) &= \left(\frac{x}{3}, \frac{y+2}{3}\right) \\
\varphi_6(x, y) &= \left(\frac{x+1}{3}, \frac{y+2}{3}\right) & \varphi_7(x, y) &= \left(\frac{x+2}{3}, \frac{y+2}{3}\right).
\end{aligned} \tag{9.16}$$

The projection onto the x-axis is equivalent to

$$\psi_0(x) = \frac{x}{3} \quad \psi_1(x) = \frac{x+1}{3} \quad \psi_2(x) = \frac{x+2}{3} \tag{9.17}$$

with weights $(\frac{3}{8}, \frac{2}{8}, \frac{3}{8})$. Thus, by the Kakutani Dichotomy Theorem, the x-marginal is singular. By symmetry, it now follows by Lemmas 9.2 and 9.4 that the Sierpinksi carpet is bi-slice-singular.

Remark 9.4 Observe that Examples 1 and 2 also follow immediately from Lemma 9.5.

The next example illustrates that slice-singular measures can be so in one direction but not the other.

Example 3 Consider the IFS generated by the functions

$$\begin{aligned} \varphi_0(x, y) &= \left(\frac{x}{4}, \frac{y}{4}\right) & \varphi_1(x, y) &= \left(\frac{x+1}{4}, \frac{y}{4}\right) \\ \varphi_2(x, y) &= \left(\frac{x+2}{4}, \frac{y+2}{4}\right) & \varphi_3(x, y) &= \left(\frac{x+3}{4}, \frac{y+2}{4}\right), \end{aligned} \tag{9.18}$$

and let μ be the invariant measure corresponding to equal weights. We immediately see that the projection of this IFS onto the x-axis yields the following generators:

$$\psi_0(x) = \frac{x}{4} \quad \psi_1(x) = \frac{x+1}{4} \quad \psi_2(x) = \frac{x+2}{4} \quad \psi_3(x) = \frac{x+3}{4} \tag{9.19}$$

together with equal weights. Consequently, the x-marginal of the invariant measure μ is Lebesgue measure on $[0, 1]$, so μ is not x-slice-singular.

On the other hand, the projection onto the y-axis yields the generators

$$\gamma_0(y) = \frac{y}{4} \quad \gamma_1(y) = \frac{y}{4} \quad \gamma_2(y) = \frac{y+2}{4} \quad \gamma_3(y) = \frac{y+2}{4} \tag{9.20}$$

with equal weights (of $\frac{1}{4}$), whose invariant measure is identical to that with generators $\{\gamma_0, \gamma_2\}$ with equal weights (of $\frac{1}{2}$). Therefore, the y-marginal of μ is the (singular) spectral measure ν_4 of Jorgensen and Pedersen [28] supported on the Cantor set C_4.

We claim that for $y \in C_4$, the slice measure ρ^y is a translation of the invariant measure ν for the IFS $\{\lambda_0, \lambda_1\}$, with equal weights, where

$$\lambda_0(x) = \frac{x}{4}, \qquad \lambda_1(x) = \frac{x+1}{4}. \tag{9.21}$$

Indeed, we claim that ρ^y is ν shifted by y:

$$\int f(x)\rho^y(dx) = \int f(x-y)\nu(dx). \tag{9.22}$$

This can be seen by showing that the measure $\tilde{\mu}(dxdy) = \rho^y(dx)\nu_4(dy)$ is invariant under the IFS in Eq. (9.18). For a continuous function f, we calculate

$$\begin{aligned}
&\int f\left(\frac{x}{4}, \frac{y}{4}\right)\tilde{\mu}(dxdy) + \int f\left(\frac{x+1}{4}, \frac{y}{4}\right)\tilde{\mu}(dxdy) \\
&\quad = \int f\left(\frac{x}{4}, \frac{y}{4}\right)\rho^y(dx)\nu_4(dy) + \int f\left(\frac{x+1}{4}, \frac{y}{4}\right)\rho^y(dx)\nu_4(dy) \\
&\quad = \int f\left(\frac{x-y}{4}, \frac{y}{4}\right)\nu(dx)\nu_4(dy) \\
&\qquad + \int f\left(\frac{x-y+1}{4}, \frac{y}{4}\right)\nu(dx)\nu_4(dy) \\
&\quad = 2\int f\left(x-y, \frac{y}{4}\right)\nu(dx)\nu_4(dy) \\
&\quad = 2\int f\left(x, \frac{y}{4}\right)\rho^y(dx)\nu_4(dy).
\end{aligned}$$

Therefore, we have

$$\begin{aligned}
&\frac{1}{4}\left[\int f\left(\frac{x}{4}, \frac{y}{4}\right)\tilde{\mu}(dxdy) + \int f\left(\frac{x+1}{4}, \frac{y}{4}\right)\tilde{\mu}(dxdy)\right. \\
&\qquad \left. + \int f\left(\frac{x+2}{4}, \frac{y+2}{4}\right)\tilde{\mu}(dxdy) + \int f\left(\frac{x+3}{4}, \frac{y+2}{4}\right)\tilde{\mu}(dxdy)\right] \\
&\quad = \frac{1}{4}\left[2\int f\left(x, \frac{y}{4}\right)\rho^y(dx)\nu_4(dy) + 2\int f\left(x, \frac{y+2}{4}\right)\rho^y(dx)\nu_4(dy)\right] \\
&\quad = \int f(x, y)\rho^y(dx)\nu_4(dy).
\end{aligned}$$

Therefore, the y-slice measures are singular, and so μ is y-slice-singular.

9.3 Fourier Series for Slice-Singular Measures

In this section, we present the analytic details of the Kaczmarz algorithm that yield the generalized Fourier expansions for $L^2(\mu)$, when μ is defined in two dimensions. However, we shall present the Kaczmarz algorithm in the framework of (infinite-dimensional) Hilbert spaces and operator theory. The main features of our algorithm will be reviewed below, and we shall refer to our earlier papers, especially [19–22], for additional details. Our recent application is to slice-singular measures μ and associated multi-variable Fourier expansions for $L^2(\mu)$, generally non-orthogonal.

Our main result is the following:

Theorem 9.1 *Suppose μ is a y-slice-singular Borel probability measure on* $[0, 1)^2$. *For any $f \in L^2(\mu)$, f possesses a Fourier series expansion of the form*

$$f(x, y) = \sum_{n=0}^{\infty} \sum_{m=0}^{\infty} d_{nm} e^{2\pi i(nx+my)}. \tag{9.23}$$

The series converges conditionally in norm, and the coefficients are as expressed in (9.46).

The notation presented in (9.23) is of an iterated sum and must be interpreted as such. For each fixed n, the series on m converges conditionally in norm. The series on n also converges conditionally in norm.

Note that the order of the double sum is dependent on the decomposition of the measure μ. If the measure is bi-slice-singular, then the Fourier series expansions can be obtained in either order. However, the *coefficients* still depend on the decomposition:

Corollary 9.1 *Suppose μ is a bi-slice-singular Borel probability measure on* $[0, 1)^2$. *For any $f \in L^2(\mu)$, f possesses Fourier series expansions of the form*

$$f(x, y) = \sum_{m=0}^{\infty} \sum_{n=0}^{\infty} c_{mn} e^{2\pi i(nx+my)} \tag{9.24}$$

$$= \sum_{n=0}^{\infty} \sum_{m=0}^{\infty} d_{nm} e^{2\pi i(nx+my)}. \tag{9.25}$$

The series converge conditionally in norm.

For the remainder of this section, we will assume that μ is y-slice-singular. Our construction is oriented in the y direction; for a measure that is x-slice-singular, our construction can be modified in the obvious manner.

Our proof of the existence of the Fourier series expansions will utilize the for operators. We will construct a sequence of operators corresponding to projections onto subspaces of $L^2(\mu)$ that are effective in the following operator-theoretic sense, which is explained in full detail in Sect. 9.4:

Definition (Effective Sequences of Operators) Let $\mathcal{H}$ be a Hilbert space, and for each $j \in \mathbb{N}_0$, let $\mathcal{H}_j$ be a Hilbert space and let $R_j : \mathcal{H} \to \mathcal{H}_j$ be a bounded surjective operator. For $x \in \mathcal{H}$, define $x_0 = R_0^* R_0 x$, and for each $j \in \mathbb{N}$, let x_j be the orthogonal projection of x_{j-1} onto the affine subspace of $\mathcal{H}$ containing x parallel to $\ker R_j$. If for any $x \in \mathcal{H}$, $\{x_j\} \to x$, then we say the sequence of operators $\{R_j\}$ is effective. If the $\mathcal{H}_j$ are closed subspaces of $\mathcal{H}$ and the R_j are the orthogonal projections onto the $\mathcal{H}_j$, then we may also say that $\{\mathcal{H}_j\}$ is a sequence of effective subspaces of $\mathcal{H}$. □

The subspaces will have the property that functions within the subspaces have Fourier series expansions arising from the Kaczmarz algorithm. Then, we will obtain that for $f \in L^2(\mu)$, the sequence of projections can be used to reconstruct f in the form of a Fourier series, yielding a doubly indexed Fourier series expansion for f as in Eq. (9.23). The coefficients of the expansion are obtained via a careful analysis of the operator auxiliary sequence, which is defined by matrix inversion as described in Sect. 9.4.

Specifically, for μ a y-slice-singular measure, we will apply the Kaczmarz algorithm to the following sequences (indexed by $\mathbb{N}_0$) of subspaces and operators:

$$\mathcal{M}_n = \overline{span}\{e^{2\pi i(nx+my)} : m \in \mathbb{N}_0\};$$

$$P_n = \text{orthogonal projection onto } \mathcal{M}_n;$$

$$R_n : L^2(\mu) \to L^2(\mu_2) : f(x,y) \to \left(y \mapsto \int f(x,y)e^{-2\pi i n x}\rho^y(dx)\right); \tag{9.26}$$

$$S : L^2(\mu) \to L^2(\mu) : f(x,y) \mapsto e^{2\pi i x} f(x,y). \tag{9.27}$$

By the Rokhlin decomposition theorem, R_n is well-defined. Simple calculations show that

$$\mathcal{M}_n = \{e^{2\pi i n x} g(y) : g \in L^2(\mu_2)\}; \tag{9.28}$$

$$R_n^* : L^2(\mu_2) \to L^2(\mu) : g(y) \mapsto e^{2\pi i n x} g(y);$$

$$P_n = R_n^* R_n;$$

$$R_{n+j} = R_n S^{-j} \text{ for } n, j \geq 0. \tag{9.29}$$

Our aim is to show that the sequence of operators $\{R_n\}_{n=0}^{\infty}$ as defined above is effective (Def 9.8), which will then be used to show that every $f \in L^2(\mu)$ can be written as a doubly indexed Fourier series. To that end, we make the following definition:

Definition 9.7 For a sequence of operators $\{B_n\}_{n=0}^{\infty} \subset \mathcal{B}(H, K)$, we say the sequence is *stationary* if there exists a unitary $S \in \mathcal{B}(H)$ such that $B_{n+j} = B_n S^{-j}$ for $n, j \geq 0$.

We immediately see that our operators $\{R_n\}$ form a stationary sequence. Section 9.4 will be devoted to proving that the sequence $\{R_n\}$ is an effective sequence. Utilizing this fact, as articulated in Theorem 9.3, we can present the proof of Theorem 9.1.

The following lemma is an immediate consequence of Theorem D:

Lemma 9.6 *If μ is y-slice-singular, with Rokhlin decomposition $\mu(dx\,dy) = \rho^y(dx)\mu_2(dy)$, then $\{e^{2\pi imy}\}_{m=0}^{\infty}$ is an effective sequence in $L^2(\mu_2)$, and consequently for every $f \in L^2(\mu_2)$, there exists a sequence of coefficients $\{a_m\}_{m=0}^{\infty} \subset \mathbb{C}$ such that*

$$f(y) = \sum_{m=0}^{\infty} a_m e^{2\pi imy},$$

with convergence of the series in norm, given by $a_m = \langle f, g_m\rangle_{\mu_2}$, where $\{g_m\}$ is the auxiliary sequence of $\{e^{2\pi imy}\}$ in $L^2(\mu_2)$.

Proposition 9.1 *Suppose μ is y-slice-singular, with Rokhlin decomposition given by $\mu(dx\,dy) = \rho^y(dx)\mu_2(dy)$. For $f \in L^2(\mu)$, the function $P_n f$ possesses a Fourier series expansion of the form*

$$P_n f(x, y) = \sum_{m=0}^{\infty} a_{nm} e^{2\pi i(nx+my)}. \tag{9.30}$$

Proof The sequence $\{e^{2\pi i(nx+my)}\}_{m\in\mathbb{N}_0}$ is a stationary sequence with μ_2 as its spectral measure, by virtue of Lemma 9.6 and the fact that R_n^* is an isometry. The result now follows from Theorem D. □

We can say more about the nature of the Fourier series expansion in Eq. (9.30). For $f_n \in \mathcal{M}_n$, we have

$$f_n(x, y) = e^{2\pi inx}[R_n f_n](y),$$

since $P_n = R_n^* R_n$. Therefore, if $\{g_m(y)\}_{m\in\mathbb{N}_0}$ is the auxiliary sequence of $\{e^{2\pi imy}\}_{m\in\mathbb{N}_0}$ in $L^2(\mu_2)$, then by Lemma 9.6,

$$[P_n f](x, y) = e^{2\pi inx}[R_n f_n](y) \tag{9.31}$$

$$= e^{2\pi inx}\left(\sum_{m=0}^{\infty} \langle [R_n f_n](y), g_m(y)\rangle_{\mu_2} e^{2\pi imy}\right). \tag{9.32}$$

Thus far, we have used only the auxiliary sequence $\{g_m\}$ of $\{e^{2\pi imy}\}$ in $L^2(\mu_2)$ coming from the vector Kaczmarz algorithm used in Theorem D, but the operator Kaczmarz algorithm applied to sequence of operators also induces an auxiliary sequence of operators $\{G_n\}$. The general construction of $\{G_n\}$ is described in Sect. 9.4.1. The sequence $\{G_n\}$ is given explicitly by Eq. (9.47).

In our present setting, the auxiliary sequence of $\{R_n\}$ is concretized as follows: For each y such that ρ^y is singular, we have that $\{e^{2\pi inx}\}_{n=0}^{\infty}$ is effective in $L^2(\rho^y)$. We let $\{g_n^{(y)}\}_{n=0}^{\infty}$ denote the corresponding auxiliary sequence. When μ is y-slice-singular, this holds for μ_2 a.e. y. We first claim that for each n, $g_n^{(y)}(x)$ is a measurable function of x and y:

Lemma 9.7 *Let μ be a y-slice-singular measure on $[0, 1)^2$, and $\left\{g_n^{(y)}\right\}$ the auxiliary sequence of $L^2(\rho^y)$. Then, for each n, $g_n^{(y)}(x) : [0, 1)^2 \to \mathbb{C}$ is measurable as a function of two variables.*

Proof By [22, Prop 1], we have that $g_n^{(y)}(x) = \sum_{j=0}^{n} \alpha_{n-j}(y) e_n(x)$, where for each fixed y, the sequence $\{\alpha_k(y)\}$ is defined by

$$\frac{1}{\sum_{j=0}^{\infty} \widehat{\rho^y}(j) z^j} = \sum_{j=0}^{\infty} \alpha_j(y) z^j.$$

Now, $\widehat{\rho^y}(j) = \int_0^1 e^{2\pi i j x} \rho^y(dx)$ is a measurable function of y by the Rokhlin Disintegration Theorem. It follows that each $\alpha_j(y)$ is a measurable function of y (cf. [22, Lemma 2]). Therefore, each $g_n^{(y)}(x)$ is measurable in the variable y and continuous in the variable x and hence measurable. □

We now explicitly identify $\{G_n\}$ in our present setting:

Proposition 9.2 *We have that*

$$[G_n f](y) = \int f(x, y) \overline{g_n^{(y)}(x)} \rho^y(dx). \tag{9.33}$$

Proof In Sect. 9.4.1, Eq. (9.48), we observe that the auxiliary sequence $\{G_n\}$ uniquely solves the system

$$R_n = \sum_{j=0}^{n} R_n R_j^* G_j \tag{9.34}$$

for all $n \in \mathbb{N}_0$. Thus, we need only to show that G_j as defined in Eq. (9.33) satisfies Eq. (9.34).

Now a simple calculation shows that

$$[R_n R_j^* h](y) = \int e^{2\pi i j x} h(y) e^{-2\pi i n x} \rho^y(dx) \tag{9.35}$$

$$= \widehat{\rho^y}(n - j) h(y). \tag{9.36}$$

Thus, from [18], we have that for any fixed y,

$$\sum_{j=0}^{n} \overline{\widehat{\rho^y}(n - j) g_j^{(y)}(x)} = e^{2\pi i n x}. \tag{9.37}$$

We therefore calculate

$$\sum_{j=0}^{n}[R_n R_j^* G_j f](y) = \sum_{j=0}^{n} \widehat{\rho^y}(n-j) \int f(x,y)\overline{g_j^{(y)}(x)}\rho^y(dx) \tag{9.38}$$

$$= \int f(x,y)\overline{\sum_{j=0}^{n} \overline{\widehat{\rho^y}(n-j)} g_j^{(y)}(x)}\rho^y(dx) \tag{9.39}$$

$$= [R_n f](y). \tag{9.40}$$

Equation (9.33) now follows immediately. □

We will prove in Theorem 9.3 that the sequence $\{R_n\}$ is effective. As a consequence of this theorem, we have the following:

Proposition 9.3 *Suppose μ is y-slice-singular, with Rokhlin decomposition given by $\mu(dx\,dy) = \rho^y(dx)\mu_2(dy)$. For every $f \in L^2(\mu)$, we can express f as*

$$f(x,y) = \sum_{n=0}^{\infty}[G_n f](y)e^{2\pi i n x} \tag{9.41}$$

$$= \sum_{n=0}^{\infty}\left[\sum_{m=0}^{\infty} a_{nm}e^{2\pi i m y}\right]e^{2\pi i n x}, \tag{9.42}$$

where the sequence $\{G_n\}_{n=0}^{\infty}$ is defined by Eq. (9.33). The convergence is in norm, conditional, and order-dependent.

Proof Since the sequence of operators $\{R_n\}$ is effective, Theorem F implies that the sum $\sum_{n=0}^{\infty} R_n^* G_n$ converges in the SOT. Consequently,

$$f = \sum_{n=0}^{\infty} R_n^* G_n f \tag{9.43}$$

$$= \sum_{n=0}^{\infty}\left(\sum_{m=0}^{\infty}\langle [G_n f](y), g_m(y)\rangle_{\mu_2} e^{2\pi i m y}\right)e^{2\pi i n x} \tag{9.44}$$

by applying Lemma 9.6 to $G_n f$. □

Combining these results, we are ready to prove our main result:

Proof of Theorem 9.1 By combining (9.44) and (9.33), for any $f \in L^2(\mu)$, we have

$$f = \sum_{n=0}^{\infty}\left(\sum_{m=0}^{\infty}\left(\int f(x,y)\overline{g_n^{(y)}(x) g_m(y)}\,\mu(dx\,dy)\right)e^{2\pi i m y}\right)e^{2\pi i n x} \tag{9.45}$$

$$= \sum_{n=0}^{\infty}\sum_{m=0}^{\infty}\left\langle f(x,y), g_n^{(y)}(x)g_m(y)\right\rangle e^{2\pi i(nx+my)}. \tag{9.46}$$

Thus, setting $d_{nm} = \left\langle f(x,y), g_n^{(y)}(x)g_m(y)\right\rangle_{\mu}$, Theorem 1 is established. □

The convergence in (9.46) is in norm and conditional. Therefore, we have a pseudo-duality between the sequence $\{e^{2\pi i(nx+my)}\}_{m,n=0}^{\infty}$ and $\{g_n^{(y)}(x)g_m(y)\}_{m,n=0}^{\infty}$. We will expand on this in Appendix 9.5.1.

9.4 The Kaczmarz Algorithm for Bounded Operators

The purpose of the present chapter is to establish explicit multi-variable Fourier expansions for $L^2(\mu)$ in the case of planar measures μ that are assumed to satisfy the slice-singular property (see Sect. 9.2.2). In principle (as noted in Sect. 9.1), these multi-variable Fourier series results parallel the corresponding results for the case when μ is instead assumed to be singular and supported on an interval.

In both cases, for the proofs, we rely on our (infinite-dimensional) Hilbert space framework for the Kaczmarz algorithm (see Sect. 9.1.2). However, the extension to the case of planar measures μ is non-trivial. It entails additional considerations involving operator theory, in particular a new analysis of infinite-by-infinite block-matrices, i.e., referring to the case of operator-valued entries (see, e.g., Theorem 9.2 below). It also entails a new framework involving general Hardy spaces (Sect. 9.4.3) relying on ideas from de Branges, especially from [11]. Since this material is not readily available, or widely known, it will be presented in the present section.

Let $\mathcal{H}$ be a Hilbert space, and for each $j \in \mathbb{N}_0$, let $\mathcal{H}_j$ be a Hilbert space (which in some situations may be closed subspaces of $\mathcal{H}$). Let $R_j : \mathcal{H} \to \mathcal{H}_j$ be bounded surjective operators. In the classical case, we typically have a sequence of vectors $\{\phi_n\} \subset \mathcal{H}$ (ultimately identified as an "effective" sequence), $\mathcal{H}_j = \mathbb{C}$ for all j, and the operators R_j are the analysis operators of the ϕ_j, that is to say, $R_j f = \langle f, \phi_j\rangle_{\mathcal{H}}$. Thus, in the classical case, $\ker R_j = \{f \in \mathcal{H} : f \perp \phi_j\}$.

Let $b_j \in \mathcal{H}_j$ for all j. The objective of the Kaczmarz algorithm is to find an $f \in \mathcal{H}$ that solves the system $R_j f = b_j$ for all j.

With the b_j's fixed, for each $j \in \mathbb{N}_0$, let $h_j \in \mathcal{H}$ be the unique solution to $R_j h_j = b_j$ lying in $(\ker R_j)^{\perp}$, that is, $h_j = R_j^*(R_j R_j^*)^{-1} b_j$. Let Q_j be the orthogonal projection onto $\ker R_j$.

Suppose we create such a situation as described above by fixing an $x \in \mathcal{H}$ and then letting $b_j = R_j x$. Define P_j by

$$\begin{aligned} P_j &= Q_j + h_j \\ &= Q_j + R_j^*(R_j R_j^*)^{-1} b_j \end{aligned}$$

$$= Q_j + R_j^*(R_j R_j^*)^{-1} R_j x.$$

Observe that $P_j f$ is the orthogonal projection of f onto the affine subspace of $\mathcal{H}$ containing x parallel to $\ker R_j$.

In a typical situation, we will know the b_j measurements, but we will not know the vector x by which the measurements were made. We attempt to recover x by carrying out the Kaczmarz algorithm: Define the Kaczmarz sequence $\{x_j\}$ by $x_0 = R_0^* b_0 = R_0^* R_0 x$, and for $j > 0$,

$$\begin{aligned}
x_j &= P_j x_{j-1} \\
&= Q_j x_{j-1} + R_j^*(R_j R_j^*)^{-1} R_j x \\
&= Q_j x_{j-1} + R_j^*(R_j R_j^*)^{-1} R_j x_{j-1} - R_j^*(R_j R_j^*)^{-1} R_j x_{j-1} + R_j^*(R_j R_j^*)^{-1} R_j x \\
&= x_{j-1} + R_j^*(R_j R_j^*)^{-1} R_j (x - x_{j-1}) \\
&= x_{j-1} + R_j^*(R_j R_j^*)^{-1} (b_j - R_j x_{j-1}).
\end{aligned}$$

Definition 9.8 If $\{x_j\} \to x$ regardless of which x was chosen, then we say the sequence of operators $\{R_j\}$ is effective. If the $\mathcal{H}_j$ are closed subspaces of $\mathcal{H}$ and the R_j are the orthogonal projections onto the $\mathcal{H}_j$, then we may also say that $\{\mathcal{H}_j\}$ is a sequence of effective subspaces of $\mathcal{H}$.

Natterer [37] introduces the Kaczmarz algorithm for bounded operators in the finite regime–there exist finitely many $\mathcal{H}_j$'s and R_j's, and he proves, in analogy to Kaczmarz's original paper [29], that the periodized sequence of R_j's is effective. We will prove an operator analogue of the Kwapień-Mycielski result [31] for stationary sequences.

Definition 9.9 Given a sequence of Hilbert spaces $\{\mathcal{H}_j\}$, we will need the following spaces:

1. $\oplus\mathcal{H}_j$ denotes the vector space of sequences of vectors whose j-th component is a vector in $\mathcal{H}_j$.
2. $\mathcal{F}(\oplus\mathcal{H}_j)$ denotes the inner-product space consisting of sequences in $\oplus\mathcal{H}_j$ with all but finitely many components equal to 0.
3. $\ell^2(\oplus\mathcal{H}_j)$ denotes the Hilbert space consisting of sequences in $\oplus\mathcal{H}_j$ that are square-summable in norm.

We will consider sequences in these spaces as column vectors.

9.4.1 The Auxiliary Sequence for the Operator Kaczmarz Algorithm

We define the auxiliary sequence $\{G_j\}$ of $\{R_j\}$ as follows: Let $M : \oplus_{k=0}^{\infty}\mathcal{H}_k \to \oplus_{j=0}^{\infty}\mathcal{H}_j$ be defined by the infinite strictly lower-triangular matrix of operators whose j, k-th entry is $R_j R_k^*$ if $j > k$ and the zero operator $0 : \mathcal{H}_k \to \mathcal{H}_j$ otherwise. (Here, the indices j and k start at 0.) Let I be the identity operator on $\oplus_{k=0}^{\infty}\mathcal{H}_k$, which can be defined by the infinite diagonal matrix whose j, j-th entry is the identity operator on $\mathcal{H}_j$.

Define $R : \mathcal{H} \to \oplus_{j=0}^{\infty}\mathcal{H}_j$ to be the 1-column matrix consisting of the R_j's. Since $(I + M)$ is lower-triangular, it is invertible. Therefore, we may define $G : \mathcal{H} \to \oplus_{j=0}^{\infty}\mathcal{H}_j$ to be the unique 1-column matrix satisfying

$$(I + M)G = R. \tag{9.47}$$

The m-th entry of G, $G_m : \mathcal{H} \to \mathcal{H}_m$, is the m-th auxiliary sequence operator. Because $(I + M)G = R$, we have

$$R_n = \sum_{j=0}^{n} R_n R_j^* G_j \tag{9.48}$$

for each $j \in \mathbb{N}_0$.

Note that there exists an operator $U : \oplus_{j=0}^{\infty}\mathcal{H}_j \to \oplus_{k=0}^{\infty}\mathcal{H}_k$ such that

$$(I + M)^{-1} = (I + U),$$

where U is realizable as an infinite strictly lower-triangular matrix whose j, k-th entry is an operator $C_{jk} : \mathcal{H}_k \to \mathcal{H}_j$ if $j > k$ and the zero operator $0 : \mathcal{H}_k \to \mathcal{H}_j$ otherwise. Then,

$$G := (I + U)R. \tag{9.49}$$

Theorem F (Natterer) $\sum_{k=0}^{n} R_k^* G_k x = x_n$.

The proof can be found in [37, Page 133].

Proposition 9.4 *For $n \geq 1$,*

$$\|x - x_{n-1}\|_{\mathcal{H}}^2 = \|x - x_n\|_{\mathcal{H}}^2 + \left\| R_n^* G_n x \right\|_{\mathcal{H}}^2.$$

Proof Recall that

$$x_n = x_{n-1} + R_n^*(R_n R_n^*)^{-1}(R_n x - R_n x_{n-1}).$$

Applying R_n to both sides,

$$R_n x_n = R_n x_{n-1} + R_n x - R_n x_{n-1} = R_n x.$$

It follows that $(x - x_n) \in \ker R_n$. Since by the Fredholm Alternative the range of R_n^* is perpendicular to $\ker R_n$, it follows that

$$\begin{aligned}\|x - x_n\|_{\mathcal{H}}^2 + \left\|R_n^* G_n x\right\|_{\mathcal{H}}^2 &= \left\|x - x_n + R_n^* G_n x\right\|_{\mathcal{H}}^2 \\ &= \|x - x_{n-1}\|_{\mathcal{H}}^2\end{aligned}$$

by Theorem F. □

Corollary 9.2 *If R_n^* is an isometry (as is the case in the classical situation when the ϕ_j are unit vectors), then $\left\|R_n^* G_n x\right\|_{\mathcal{H}}^2 = \|G_n x\|_{\mathcal{H}_n}^2$, and consequently it is also true that*

$$\|x - x_{n-1}\|_{\mathcal{H}}^2 = \|x - x_n\|_{\mathcal{H}}^2 + \|G_n x\|_{\mathcal{H}_n}^2 .$$

Corollary 9.3 *The sequence $\{R_n\}_{n=0}^{\infty}$ is effective if and only if $\sum_{n=0}^{\infty} \left\|R_n^* G_n x\right\|_{\mathcal{H}}^2 = \|x\|_{\mathcal{H}}^2$.*

Moreover, under the assumption that each R_n^* is an isometry, $\{R_n\}_{n=0}^{\infty}$ is effective if and only if $\sum_{n=0}^{\infty} \|G_n x\|_{\mathcal{H}_n}^2 = \|x\|_{\mathcal{H}}^2$.

Proof Note that $G_0 = R_0$. Since $x_0 = R_0^* R_0 x$, x_0 is in the range of R_0^* and is therefore perpendicular to $x - x_0$. So by Proposition 9.4,

$$\begin{aligned}\sum_{n=0}^{k} \left\|R_n^* G_n x\right\|_{\mathcal{H}}^2 &= \left\|R_0^* R_0 x\right\|_{\mathcal{H}}^2 + \sum_{n=1}^{k} (\|x - x_{n-1}\|_{\mathcal{H}}^2 - \|x - x_n\|_{\mathcal{H}}^2) \\ &= \|x_0\|_{\mathcal{H}}^2 + \|x - x_0\|_{\mathcal{H}}^2 - \|x - x_k\|_{\mathcal{H}}^2 \qquad (9.50) \\ &= \|x\|_{\mathcal{H}}^2 - \|x - x_k\|_{\mathcal{H}}^2 .\end{aligned}$$

Taking a limit of both sides as $k \to \infty$, the result follows. □

9.4.2 A Matrix Characterization of Effectivity

Haller and Szwarc [18] prove a general statement about when a sequence of vectors $\{\phi_j\}_{j=0}^{\infty} \subset \mathcal{H}$ is effective. Specifically, they prove that the sequence is effective if and only if the matrix $\mathscr{U}$ is a partial isometry on $\ell^2(\mathbb{N}_0)$, where

$$(I+\mathscr{U})(I+\mathscr{M})=I, \qquad \mathscr{M}_{j,k}=\begin{cases}\langle\phi_k,\phi_j\rangle, & \text{for } j>k,\\ 0, & \text{otherwise.}\end{cases}$$

We prove the analogue of the Haller-Szwarc Theorem for the Kaczmarz Algorithm for Operators. The proof proceeds nearly identically to the proof presented in [18]; we include it here for completeness.

Theorem 9.2 *Suppose every R_n^* is an isometry. Then, $\{R_j\}_{j=0}^{\infty}$ is effective if and only $\cup(\ker(R_j)^{\perp})$ is linearly dense in $\mathcal{H}$ and the matrix*

$$U=\begin{bmatrix} 0 & & & & 0\\ C_{10} & 0 & & & \\ C_{20} & C_{21} & 0 & & \\ C_{30} & C_{31} & C_{32} & 0 & \\ \vdots & & & & \ddots \end{bmatrix}$$

is a partial isometry on $\ell^2(\oplus_{j=0}^{\infty}\mathcal{H}_j)$.

To prove this theorem, we first establish some lemmas and definitions.

For a matrix A with operator entries, let $\hat{A}_n$ denote the nth principal submatrix of A, that is, the matrix A with entries changed to the zero operator in rows or columns beyond the nth.

If $A:\oplus\mathcal{H}_j\to\oplus\mathcal{H}_j$, we say that A is positive definite on $\mathcal{F}(\oplus\mathcal{H}_j)$ if $\left\langle\hat{A}_n x, x\right\rangle_{\mathcal{F}(\oplus\mathcal{H}_j)}>0$ for all $x\in\mathcal{F}(\oplus\mathcal{H}_j)$ and all $n\in\mathbb{N}_0$.

Proposition 9.5 *U is a contraction on $\ell^2(\oplus_{j=0}^{\infty}\mathcal{H}_j)$ if and only if $M+M^*+I$ is positive definite on $\mathcal{F}(\oplus_{j=0}^{\infty}\mathcal{H}_j)$.*

Proof $\hat{M}_n$ and $\hat{U}_n$ are bounded on $\ell^2(\oplus_{j=0}^{\infty}\mathcal{H}_j)$, and by assumption, $\hat{M}_n\hat{U}_n=\hat{U}_n\hat{M}_n=-\hat{U}_n-\hat{M}_n$. Assume the matrix $M+M^*+I$ is positive definite on $\mathcal{F}(\oplus_{j=0}^{\infty}\mathcal{H}_j)$. Then, the matrix $\hat{M}_n+\hat{M}_n^*+I$ corresponds to a positive bounded operator on $\ell^2(\oplus_{j=0}^{\infty}\mathcal{H}_j)$. Thus,

$$\begin{aligned}0&\le(\hat{U}_n^*+I)(\hat{M}_n+\hat{M}_n^*+I)(\hat{U}_n+I)\\&=(\hat{U}_n^*+I)((\hat{M}_n+I)(\hat{U}_n+I)+\hat{M}_n^*(\hat{U}_n+I))\\&=(\hat{U}_n^*+I)(I+\hat{M}_n^*\hat{U}_n+\hat{M}_n^*)\\&=(\hat{U}_n^*+I)(\hat{M}_n^*+I)+(\hat{U}_n^*+I)\hat{M}_n^*\hat{U}_n\\&=I+\hat{U}_n^*\hat{M}_n^*\hat{U}_n+\hat{M}_n^*\hat{U}_n\\&=I+(-\hat{U}_n^*-\hat{M}_n^*)\hat{U}_n+\hat{M}_n^*\hat{U}_n^*\\&=I-\hat{U}_n^*\hat{U}_n.\end{aligned}$$

Hence, $\left\|\hat{U}_n\right\| \leq 1$, where this denotes the operator norm of $\mathcal{B}(\ell^2(\oplus\mathcal{H}_j))$. Consequently, we obtain $\|U\| \leq 1$.

Conversely, suppose $\|U\| \leq 1$, where this is the norm of $\mathcal{B}(\ell^2(\oplus\mathcal{H}_j))$. Then, $\left\|\hat{U}_n\right\| \leq 1$ for all n, and so $I - \hat{U}_n^*\hat{U}_n \geq 0$ on $\ell^2(\oplus\mathcal{H}_j)$. Therefore,

$$
\begin{aligned}
0 &\leq (\hat{M}_n^* + I)(I - \hat{U}_n^*\hat{U}_n)(\hat{M}_n + I) \\
&= (\hat{M}_n^* + I)(\hat{M}_n + I - \hat{U}_n^*\hat{U}_n\hat{M}_n - \hat{U}_n^*\hat{U}_n) \\
&= (\hat{M}_n^* + I)(\hat{M}_n + I - \hat{U}_n^*(-\hat{U}_n - \hat{M}_n) - \hat{U}_n^*\hat{U}_n) \\
&= (\hat{M}_n^* + I)(\hat{M}_n + I + \hat{U}_n^*\hat{M}_n) \\
&= \hat{M}_n^*\hat{M}_n + \hat{M}_n^* + \hat{M}_n^*\hat{U}_n^*\hat{M}_n + \hat{M}_n + I + \hat{U}_n^*\hat{M}_n \\
&= \hat{M}_n^*\hat{M}_n + \hat{M}_n^* + (-\hat{U}_n^* - \hat{M}_n^*)\hat{M}_n + \hat{M}_n + I + \hat{U}_n^*\hat{M}_n \\
&= \hat{M}_n + \hat{M}_n^* + I.
\end{aligned}
$$

It follows that $M + M^* + I$ is positive definite.

□

Lemma 9.8 *Suppose every R_n^* is an isometry. Then, $\sum_{n=0}^{\infty} G_n^*G_n = I_{\mathcal{H}}$ in the weak operator topology, if and only if $\{R_j\}_{j=0}^{\infty}$ is effective.*

Proof Suppose $\sum_{n=0}^{\infty} G_n^*G_n = I_{\mathcal{H}}$ in the weak operator topology. Let $x \in \mathcal{H}$. Observe that

$$
\begin{aligned}
\sum_{n=0}^{\infty} \|R_n^*G_n x\|_{\mathcal{H}}^2 &= \sum_{n=0}^{\infty} \|G_n x\|_{\mathcal{H}_n}^2 \\
&= \sum_{n=0}^{\infty} \langle G_n x, G_n x\rangle_{\mathcal{H}_n} \\
&= \sum_{n=0}^{\infty} \langle G_n^*G_n x, x\rangle_{\mathcal{H}} \\
&= \lim_{k\to\infty} \left\langle \sum_{n=0}^{k} G_n^*G_n x, x \right\rangle_{\mathcal{H}} \\
&= \langle Ix, x\rangle_{\mathcal{H}} \\
&= \|x\|_{\mathcal{H}}^2.
\end{aligned}
$$

Therefore, by Corollary 9.3, the Kaczmarz sequence $x_n \to x$. Since x was arbitrary, it follows that $\{R_j\}$ is effective.

Conversely, suppose $\{R_j\}_{j=0}^{\infty}$ is effective. Let $x \in \mathcal{H}$. Then, the Kaczmarz sequence $x_n \to x$. Because of (9.50), we may define $T : \mathcal{H} \to \ell^2(\oplus \mathcal{H}_n)$ by

$$Tx = \begin{bmatrix} G_0 x \\ G_1 x \\ G_2 x \\ \vdots \end{bmatrix}. \tag{9.51}$$

(Indeed, by (9.50), it follows that $\|T\|_{\ell^2(\oplus \mathcal{H}_j)} \leq 1$ regardless of whether $\{R_j\}_{j=0}^{\infty}$ is effective.) By Corollary 9.3, since $\{R_j\}_{j=0}^{\infty}$ is effective, we have

$$\begin{aligned} \|Tx\|_{\ell^2(\oplus \mathcal{H}_j)} &= \sqrt{\sum_{n=0}^{\infty} \|G_n x\|_{\mathcal{H}_n}^2} \\ &= \|x\|_{\mathcal{H}}. \end{aligned}$$

Thus, T is an isometry. Let $y \in \mathcal{H}$. Then,

$$\begin{aligned} \langle x, y \rangle_{\mathcal{H}} &= \langle Tx, Ty \rangle_{\ell^2(\oplus \mathcal{H}_j)} \\ &= \sum_{n=0}^{\infty} \langle G_n x, G_n y \rangle_{\mathcal{H}_n} \\ &= \sum_{n=0}^{\infty} \langle G_n^* G_n x, y \rangle_{\mathcal{H}} \\ &= \lim_{k \to \infty} \left\langle \sum_{n=0}^{k} G_n^* G_n x, y \right\rangle_{\mathcal{H}}. \end{aligned} \tag{9.52}$$

Since x and y were arbitrary, it follows that $\sum_{n=0}^{\infty} G_n^* G_n = I_{\mathcal{H}}$ in the weak operator topology. □

Suppose $\{R_j\}$ is effective. Then, by Lemma 9.8, $\sum_{n=0}^{\infty} G_n^* G_n = I_{\mathcal{H}}$ in the weak operator topology. It follows that for any $j, k \in \mathbb{N}_0$,

$$R_j R_k^* = \sum_{n=0}^{\infty} R_j G_n^* G_n R_k^* \tag{9.53}$$

in the weak operator topology.

For each j, define $D_j : \mathcal{H}_j \to \oplus_{k=0}^{\infty}\mathcal{H}_k$ to be the 1-column matrix whose nth entry is the zero operator $0 : \mathcal{H}_j \to \mathcal{H}_n$ if $n \neq j$ and the identity operator $I_{\mathcal{H}_j}$ if $n = j$.

Lemma 9.9

$$R_j G_n^* = D_j^*(UM^* + M^* + I)^* D_n.$$

Proof Set $C_{nn} = M_{nn} = I_{\mathcal{H}_n}$. Recall that since $G = (I + U)R$, we have

$$G_n = \sum_{k=0}^{n} C_{nk} R_k.$$

Therefore,

$$\begin{aligned}
R_j G_n^* &= R_j \left(\sum_{k=0}^{n} R_k^* C_{nk}^* \right) \\
&= \sum_{k=0}^{n} R_j R_k^* C_{nk}^* \\
&= \begin{cases} \sum_{k=0}^{n} M_{jk} C_{nk}^* & \text{if } j > n \\ \sum_{k=0}^{j-1} M_{jk} C_{nk}^* + \sum_{k=j}^{n} R_j R_k^* C_{nk}^* & \text{if } j \leq n \end{cases} \\
&= \begin{cases} \sum_{k=0}^{n} M_{jk} C_{nk}^* & \text{if } j > n \\ \sum_{k=0}^{j-1} M_{jk} C_{nk}^* + \sum_{k=j}^{n} \left(R_k R_j^* \right)^* C_{nk}^* & \text{if } j \leq n \end{cases} \\
&= \begin{cases} \sum_{k=0}^{n} M_{jk} C_{nk}^* & \text{if } j > n \\ \sum_{k=0}^{j-1} M_{jk} C_{nk}^* + \sum_{k=j}^{n} (M_{kj})^* C_{nk}^* & \text{if } j \leq n \end{cases} \\
&= \begin{cases} \sum_{k=0}^{n} M_{jk} C_{nk}^* & \text{if } j > n \\ \sum_{k=0}^{j-1} M_{jk} C_{nk}^* + \left(\sum_{k=j}^{n} C_{nk} M_{kj} \right)^* & \text{if } j \leq n. \end{cases}
\end{aligned}$$

Since $(I + U)(I + M) = I$, for $j \leq n$, we get

$$\sum_{k=j}^{n} C_{nk} M_{kj} = D_{jn}.$$

Therefore,

$$R_j G_n^* = \begin{cases} \sum_{k=0}^{n} M_{jk} C_{nk}^* & \text{if } j > n \\ \sum_{k=0}^{j-1} M_{jk} C_{nk}^* + D_{jn}^* & \text{if } j \leq n \end{cases}$$

$$= \begin{cases} D_j^* M(U^* + I) D_n & \text{if } j > n \\ D_j^* M U^* D_n + D_j^* D_n & \text{if } j \leq n \end{cases}$$

$$= \begin{cases} D_j^* M(U^* + I) D_n + D_j^* D_n & \text{if } j > n \\ D_j^* M U^* D_n + D_j^* D_n + D_j^* M D_n & \text{if } j \leq n \end{cases}$$

because $D_j^* D_n : \mathcal{H}_n \to \mathcal{H}_j$ is the zero operator when $j \neq n$ and because $D_j^* M D_n : \mathcal{H}_n \to \mathcal{H}_j$ is the zero operator when $j \leq n$. Therefore,

$$\begin{aligned} R_j G_n^* &= D_j^* \left(M U^* + M + I\right) D_n \\ &= D_j^* \left(U M^* + M^* + I\right)^* D_n. \end{aligned}$$

□

Lemma 9.10 *$\sum_{n=0}^{\infty} D_n D_n^* = I$ in the weak operator topology.*

Proof Let $x, y \in \ell^2(\oplus\mathcal{H}_j)$. We have

$$\begin{aligned} \lim_{k\to\infty} \left\langle \sum_{n=0}^{k} D_n D_n^* x, y \right\rangle_{\ell^2(\oplus\mathcal{H}_j)} &= \lim_{k\to\infty} \sum_{n=0}^{k} \langle D_n D_n^* x, y\rangle_{\ell^2(\oplus\mathcal{H}_j)} \\ &= \lim_{k\to\infty} \sum_{n=0}^{k} \langle D_n^* x, D_n^* y\rangle_{\mathcal{H}_n} \\ &=: \langle x, y\rangle_{\ell^2(\oplus\mathcal{H}_j)} . \end{aligned}$$

□

Proof of Theorem 9.2 Let $A = UM^* + M^* + I$. Let $j, k \in \mathbb{N}_0$, and let $x \in \mathcal{H}_k$ and $y \in \mathcal{H}_j$. By applying Lemmas 9.9 and 9.10 and Eq. (9.53), we get

$$\begin{aligned} \left\langle (R_j R_k^* - D_j^* A^* A D_k) x, y \right\rangle_{\mathcal{H}_j} &= \lim_{N\to\infty} \left\langle \left(R_j R_k^* - \sum_{n=0}^{N} D_j^* A^* D_n D_n^* A D_k \right) x, y \right\rangle_{\mathcal{H}_j} \\ &= \lim_{N\to\infty} \left\langle \left(R_j R_k^* - \sum_{n=0}^{N} R_j G_n^* G_n R_k^* \right) x, y \right\rangle_{\mathcal{H}_j} \\ &= \left\langle \left(R_j R_k^* - R_j R_k^*\right) x, y \right\rangle_{\mathcal{H}_j} \\ &= 0. \end{aligned}$$

Since y was arbitrary, it follows that

$$R_j R_k^* = D_j^* A^* A D_k.$$

We claim that

$$D_j^* M U^* U M^* D_k = D_j^* M M^* D_k.$$

By the relation $MU = -U - M$, we have

$$\begin{aligned} A^* A &= (UM^* + M^* + I)^*(UM^* + M^* + I) \\ &= (MU^* + M + I)(UM^* + M^* + I) \\ &= MU^*UM^* + MU^*M^* + MU^* + MUM^* + MM^* \\ &\quad + M + UM^* + M^* + I \\ &= MU^*UM^* + MU^*M^* + MU^* - UM^* - MM^* + MM^* \\ &\quad + M + UM^* + M^* + I \\ &= MU^*UM^* + MU^*M^* + MU^* + M + M^* + I \end{aligned}$$

If $j > k$, then $D_j^* D_k = 0$ and $D_j^* M^* D_k = 0$, so $D_j^* M D_k + D_j^* M^* D_k + D_j D_k = D_j^* M D_k = R_j R_k^*$.

If $j < k$, then $D_j^* D_k = 0$ and $D_j^* M D_k = 0$, so $D_j^* M D_k + D_j^* M^* D_k + D_j D_k = D_j^* M^* D_k = (R_k R_j^*)^* = R_j R_k^*$.

If $j = k$, then $D_j^* M D_k = 0$ and $D_j^* M^* D_k = 0$, so $D_j^* M D_k + D_j^* M^* D_k + D_j D_k = D_j^* D_j = I = R_j R_j^*$.

Hence,

$$\begin{aligned} & R_j R_k^* = D_j^* A^* A D_k = \\ & D_j^* M U^* U M^* D_k + D_j^* M U^* M^* D_k + D_j^* M U^* D_k \\ & \quad + D_j^* M D_k + D_j^* M^* D_k + D_j^* D_k \\ & = D_j^* M U^* U M^* D_k + D_j^* M U^* M^* D_k + D_j^* M U^* D_k + R_j R_k^*. \end{aligned}$$

Therefore,

$$\begin{aligned} D_j^* M U^* U M^* D_k &= D_j^*(-MU^* M^* - MU^*) D_k \\ &= D_j^*(M(-U^* M^* - U^*)) D_k \\ &= D_j^* M M^* D_k. \end{aligned} \tag{9.54}$$

Recall that $\mathcal{F}(\oplus\mathcal{H}_j)$ is the subspace of $\oplus\mathcal{H}_j$ consisting of elements whose entries are zero in all but finitely many components, i.e., $x \in \oplus\mathcal{H}_j$ such that $D_n^* D_n x \neq 0$ for only finitely many $n \in \mathbb{N}_0$. Now, define

$$\mathbb{H} = \overline{M^*\mathcal{F}(\oplus\mathcal{H}_j)}.$$

Let $x, y \in \mathcal{F}(\oplus\mathcal{H}_j)$. This means there exist $x_0, x_1, \ldots, x_N$ and $y_0, y_1, \ldots, y_N$, where $x_j, y_j \in \mathcal{H}_j$, such that $x = \sum_{n=0}^N D_n x_n$ and $y = \sum_{n=0}^N D_n y_n$. By Eq. (9.54), we then have

$$\begin{aligned}
\langle UM^*x, UM^*y\rangle_{\ell^2(\oplus\mathcal{H}_j)} &= \left\langle UM^* \sum_{n=0}^N D_n x_n, UM^* \sum_{m=0}^N D_m y_m \right\rangle_{\ell^2(\oplus\mathcal{H}_j)} \\
&= \sum_{n=0}^N \sum_{m=0}^N \langle D_m^* MU^*UM^* D_n x_n, y_m\rangle_{\mathcal{H}_m} \\
&= \sum_{n=0}^N \sum_{m=0}^N \langle D_m^* MM^* D_n x_n, y_m\rangle_{\mathcal{H}_m} \\
&= \langle M^*x, M^*y\rangle_{\ell^2(\oplus\mathcal{H}_j)}.
\end{aligned}$$

This establishes that U is isometric on $M^*\mathcal{F}(\oplus\mathcal{H}_j)$. Because U is represented by a matrix, it must be the unique bounded extension of its restriction to $M^*\mathcal{F}(\oplus\mathcal{H}_j)$, and hence U is isometric on $\mathbb{H} = \overline{M^*\mathcal{F}(\oplus\mathcal{H}_j)}$.

It suffices to show that U vanishes on $\mathbb{H}^\perp$. To this end, observe that the matrices U^* and M^* leave the subspace $\mathcal{F}(\oplus\mathcal{H}_j)$ invariant. We have

$$\begin{aligned}
M^*(U^* + I) &= M^*U^* + M^* \\
&= -U^*.
\end{aligned}$$

Therefore,

$$\begin{aligned}
U^*(\mathcal{F}(\oplus\mathcal{H}_j)) &= -M^*(U^* + I)(\mathcal{F}(\oplus\mathcal{H}_j)) \\
&\subseteq -M^*(\mathcal{F}(\oplus\mathcal{H}_j)) \\
&\subseteq \mathbb{H}.
\end{aligned}$$

Thus, by the Fredholm Alternative,

$$\begin{aligned}
\mathbb{H}^\perp &\subseteq (U^*(\mathcal{F}(\oplus\mathcal{H}_j)))^\perp \\
&\subseteq (U^*M^*\mathcal{F}(\oplus\mathcal{H}_j))^\perp
\end{aligned}$$

$$
\begin{aligned}
&= (\mathrm{ran}_{M^*\mathcal{F}(\oplus\mathcal{H}_j)} U^*)^{\perp} \\
&= \ker_{M^*\mathcal{F}(\oplus\mathcal{H}_j)} U \\
&\subseteq \ker_{\mathbb{H}} U.
\end{aligned}
$$

Thus, U is a partial isometry.

Conversely, let U be a partial isometry on $\ell^2(\oplus\mathcal{H}_j)$. Hence, U is isometric on

$$
\begin{aligned}
(\ker_{\ell^2(\oplus\mathcal{H}_j)} U)^{\perp} &= \overline{\mathrm{ran}_{\ell^2(\oplus\mathcal{H}_j)} U^*} \\
&= \overline{\mathrm{ran}_{\mathcal{F}(\oplus\mathcal{H}_j)} U^*}.
\end{aligned}
$$

The formula $U^*(M^* + I) = -M^*$ implies that U is isometric on $M^*(\mathcal{F}(\oplus\mathcal{H}_j))$. This is equivalent to

$$
\langle UM^*x, UM^*y\rangle_{\ell^2(\oplus\mathcal{H}_j)} = \langle M^*x, M^*y\rangle_{\ell^2(\oplus\mathcal{H}_j)}
$$

for $x, y \in \mathcal{F}(\oplus\mathcal{H}_j)$. Tracking backward, the proof of the first part implies the formula (9.53). That is to say, we obtain

$$
R_j R_k^* = \sum_{n=0}^{\infty} R_j G_n^* G_n R_k^*
$$

in the weak operator topology of $B(\mathcal{H}_k, \mathcal{H}_j)$.

Let $T : \mathcal{H} \to \ell^2(\oplus\mathcal{H}_j)$ be as in (9.51), and note that $\|T\| \leq 1$. By (9.52), $\sum_{n=0}^{\infty} G_n^* G_n = T^*T$ in the weak operator topology of $B(\mathcal{H})$.

Let $\hat{x}_0, \hat{x}_1, \ldots, \hat{x}_N$ and $\hat{y}_0, \hat{y}_1, \ldots, \hat{y}_N$ be such that $\hat{x}_j, \hat{y}_j \in \mathcal{H}_j$. Then,

$$
\begin{aligned}
\left\langle T^*T \sum_{k=0}^{N} R_k^* \hat{x}_k, \sum_{j=0}^{N} R_k^* \hat{y}_k \right\rangle_{\mathcal{H}} &= \lim_{m\to\infty} \left\langle \sum_{n=0}^{m} G_n^* G_n \sum_{k=0}^{N} R_k^* \hat{x}_k, \sum_{j=0}^{N} R_j^* \hat{y}_k \right\rangle_{\mathcal{H}} \\
&= \lim_{m\to\infty} \sum_{n=0}^{m} \sum_{j=0}^{N} \sum_{k=0}^{N} \left\langle G_n^* G_n R_k^* \hat{x}_k, R_j^* \hat{y}_j \right\rangle_{\mathcal{H}} \\
&= \lim_{m\to\infty} \sum_{n=0}^{m} \sum_{j=0}^{N} \sum_{k=0}^{N} \langle R_j G_n^* G_n R_k^* \hat{x}_k, \hat{y}_j \rangle_{\mathcal{H}_j} \\
&= \sum_{j=0}^{N} \sum_{k=0}^{N} \lim_{m\to\infty} \sum_{n=0}^{m} \langle R_j G_n^* G_n R_k^* \hat{x}_k, \hat{y}_j \rangle_{\mathcal{H}_j}
\end{aligned}
$$

$$= \sum_{j=0}^{N} \sum_{k=0}^{N} \langle R_j R_k^* \hat{x}_k, \hat{y}_j \rangle_{\mathcal{H}_j}$$

$$= \left\langle \sum_{k=0}^{N} R_k^* \hat{x}_k, \sum_{j=0}^{N} R_j^* \hat{y}_j \right\rangle_{\mathcal{H}}.$$

Then, by the assumption that $\cup(\ker R_j)^{\perp}$ is linearly dense in $\mathcal{H}$ and the fact that T^*T is a bounded operator, we have that $T^*T = I$. It follows that $\sum_{n=0}^{\infty} G_n^* G_n = I$ in the weak operator topology. Therefore, $\{R_j\}_{j=0}^{\infty}$ is effective by Lemma 9.8. □

9.4.3 Stationary Sequences of Operators

Recall from Eq. (9.29) that $R_{n+j} S^{-j} = R_n$. This provides an operator analogue of a stationary sequence of vectors. We want to show that for a slice-singular measure, the stationary sequence of operators $\{R_n\}$ is effective. We need to show that the operators satisfy Theorem 9.2. First, consider $(I + M)$, which becomes

$$I + M = \begin{bmatrix} I & & & 0 \\ R_0 S^{-1} R_0^* & I & & \\ R_0 S^{-2} R_0^* & R_0 S^{-1} R_0^* & I & \\ R_0 S^{-3} R_0^* & R_0 S^{-2} R_0^* & R_0 S^{-1} R_0^* & \ddots \\ \vdots & & \vdots & \ddots \end{bmatrix}. \tag{9.55}$$

Note that in our notation, $\mathcal{H}_j = \mathcal{H}_0 = L^2(\mu_2)$ (or $L^2(\mu_1)$, depending on the direction of the slice-singularity of μ). We will think of $\mathcal{H}_0$ as a subspace of $L^2(\mu)$ via its image under R_0^*.

We write the formal inverse of $I + M$ as

$$I + U = \begin{bmatrix} I & & & 0 \\ A_1 & I & & \\ A_2 & A_1 & I & \\ A_3 & A_2 & A_1 & \ddots \\ \vdots & & \vdots & \ddots \end{bmatrix}. \tag{9.56}$$

By Theorem 9.2, the stationary sequence of operators $\{R_n\}_{n=0}^{\infty}$ is effective if and only if U is a partial isometry on $\ell^2(\oplus \mathcal{H}_0)$. We will show that if the unitary S has a spectral representation that corresponds to a slice-singular measure, then the matrix is an isometry.

Inspired by [11, Lemma 2], we will define $B(z) \in \mathcal{B}(\mathcal{H}_0)$ by

$$(I - B(z))^{-1} = \sum_{n=0}^{\infty} R_0 S^{-n} R_0^* z^n \tag{9.57}$$

so that $B(z) = \sum_{n=1}^{\infty} A_n z^n$. In this form, $B(z)$ also acts on the space

$$\mathcal{H}_0(z) = \left\{ \sum_{n=0}^{\infty} f_n z^n \,\middle|\, f_n \in \mathcal{H}_0,\ \| \sum_{n=0}^{\infty} f_n z^n \|^2 = \sum_n \| f_n \|^2 \right\}.$$

From [11, Lemma 2], we have that there exists an operator-valued function $B(z)$ that satisfies the equation

$$\left\langle \frac{I + B(z)}{I - B(z)} a, c \right\rangle_{\mathcal{H}_0} = \left\langle \frac{I + zS^*}{I - zS^*} a, c \right\rangle_{L^2(\mu)} \tag{9.58}$$

for $a, c \in \mathcal{H}_0$. We claim that the $B(z)$ as defined in Eqs. (9.57) and (9.58) coincide.

As calculated in de Branges's proof of [11, Lemma 2], the power series expansion of the RHS of Eq. (9.58) can be written as

$$\langle I + 2zS^{-1} + 2z^2 S^{-2} \ldots a, c \rangle,$$

which we write as

$$\langle a, c \rangle + 2z \langle R_0 S^{-1} R_0^* a, c \rangle + 2z^2 \langle R_0 S^{-2} R_0^* a, c \rangle + \cdots$$

or

$$2 \langle S_+(z) a, c \rangle - \langle a, c \rangle = \langle (2S_+(z) - I) a, c \rangle,$$

with $S_+(z) = \sum_{n=0}^{\infty} R_0 S^{-n} R_0^* z^n$.

Formally, then, we have that

$$\frac{I + B(z)}{I - B(z)} = 2S_+(z) - I.$$

Solving for $B(z)$ (as in the scalar case), we obtain

$$B(z) = I - [S_+(z)]^{-1},$$

from which our claim follows.

Now, we want to calculate explicitly the action of $B(z)$ on $\mathcal{H}_0(z)$. We suppose that μ is y-slice-singular and write $\mu(dx\,dy) = \rho^y(dx)\mu_2(dy)$ as before. We let

$b_y(z)$ be the inner function associated with the singular measure ρ^y according to the Herglotz representation:

$$\frac{1+b_y(z)}{1-b_y(z)} = \int \frac{e^{2\pi ix}+z}{e^{2\pi ix}-z}\rho^y(dx).$$

Note that for a fixed z, by the Rokhlin Disintegration Theorem, $b_y(z)$ is a measurable function in y.

Consider $a = a(y)$ and $c = c(y)$ as elements of $\mathcal{H}_0 = L^2(\mu_2)$. By the functional calculus, the action of $(I+zS^*)(I-zS^*)^{-1}$ on $L^2(\mu)$ corresponds to multiplication by $(1+ze^{-2\pi ix})(1-ze^{-2\pi ix})^{-1}$. Therefore, we have

$$\begin{aligned}
\left\langle \frac{1+zS^*}{1-zS^*}a, c\right\rangle_{L^2(\mu)} &= \int \frac{1+ze^{-2\pi ix}}{1-ze^{-2\pi ix}}a(y)\overline{c(y)}\mu(dx\,dy)\\
&= \int\int \frac{1+ze^{-2\pi ix}}{1-ze^{-2\pi ix}}\rho^y(dx)\; a(y)\overline{c(y)}\;\mu_2(dy)\\
&= \int \frac{1+b_y(z)}{1-b_y(z)}a(y)\overline{c(y)}\mu_2(dy)\\
&= \left\langle \frac{1+B(z)}{1-B(z)}a, c\right\rangle_{\mathcal{H}_0}. \qquad (9.59)
\end{aligned}$$

Again, by the functional calculus, we have that the action of $B(z)$ on $\mathcal{H}_0$ is given by

$$[B(z)a](y) = b_y(z)a(y).$$

Consider $F \in \mathcal{H}_0(z)$, which we write as

$$F(z) = \sum_{n=0}^{\infty} f_n z^n$$

$$F_y(z) = \sum_{n=0}^{\infty} f_n(y)z^n.$$

The action of $B(z)$ on $\mathcal{H}_0(z)$ can then be written as

$$\begin{aligned}
[BF]_y(z) &= B(z)F_y(z)\\
&= b_y(z)F_y(z).
\end{aligned}$$

For μ_2 a.e. y, $F_y \in H^2(\mathbb{D})$, and $\|b_yF_y\|_{H^2} = \|F_y\|_{H^2}$. Moreover,

$$\|F\|^2_{\mathcal{H}_0(z)} = \int \|F_y\|^2_{H^2}\mu_2(dy),$$

so we obtain that

$$\|BF\|_{\mathcal{H}_0(z)} = \|F\|_{\mathcal{H}_0(z)}. \tag{9.60}$$

We now have that B, whose matrix representation is given by U in Eq. (9.56), is an isometry on $\ell^2(\oplus_{n=0}^{\infty}\mathcal{H}_0)$. We therefore obtain the following theorem.

Theorem 9.3 *Suppose μ is a Borel probability measure on $[0, 1]^2$ and is y-slice-singular. The operators $\{R_n\}$ as defined in Eq. (9.26) are effective in $L^2(\mu)$.*

9.5 Appendix

9.5.1 Duality

In Sect. 9.3, we derived our multi-variable $L^2(\mu)$ Fourier expansions with the use of some lemmas from the theory of Parseval frames in Hilbert spaces (see especially Sects. 9.1.2 and 9.4.1). The purpose of the present section is to present certain needed parts of frame theory.

Proposition 9.6 *If $\{g_n\}$ is a Parseval frame in $L^2(\mu)$, then $\|g_n\|_\mu \leq 1$.*

Proof For any $k \in \mathbb{N}_0$,

$$\|g_k\|_\mu^2 = \sum_n \left|\langle g_k, g_n\rangle_\mu\right|^2 = \|g_k\|_\mu^4 + \sum_{n\neq k} \left|\langle g_k, g_n\rangle_\mu\right|^2.$$

Therefore,

$$\|g_k\|_\mu^2 \,(1 - \|g_k\|_\mu^2) = \sum_{n\neq k} \left|\langle g_k, g_n\rangle_\mu\right|^2.$$

It follows that $\|g_k\|_\mu \leq 1$, or else the left side above would be negative, contrary to the right side being nonnegative. □

Proposition 9.7 *If μ disintegrates as $\rho^y(dx)\,\mu_2(dy)$, then $g_n^{(y)}(x)g_m(y) \in L^2(\mu)$.*

Proof

$$\begin{aligned}\int\int \left|g_n^{(y)}(x)g_m(y)\right|^2 d\rho_x(y)\,\mu_2(dy) &= \int |g_m(y)|^2 \int \left|g_n^{(y)}(x)\right|^2 \rho^y(dx)\,\mu_2(dy)\\ &= \int |g_m(y)|^2 \left\|g_n^{(x)}(\cdot)\right\|_{\rho_x}^2 \mu_2(dy)\\ &\leq \int |g_m(y)|^2\, \mu_2(dy)\end{aligned}$$

$$= \|g_m(\cdot)\|^2_{\mu_1}$$
$$< \infty.$$

□

Proposition 9.8 *If* $F(x, y) \in L^2(\mu)$, *then* $\int F(\cdot, y)\overline{g_n^{(\cdot)}(y)}\, \rho^y(dx) \in L^2(\mu_2)$.

Proof

$$\begin{aligned}
\int \left| \int F(x, y)\overline{g_n^{(y)}(x)}\, \rho^y(dx) \right|^2 \mu_2(dy) &\le \int \left(\int \left| F(x, y)\overline{g_n^{(y)}(x)} \right| \rho^y(dx) \right)^2 \mu_2(dy) \\
&\le \int \left(\|F(x, \cdot)\|_{\rho_x} \left\| g_n^{(x)}(\cdot) \right\|_{\rho_x} \right)^2 \mu_2(dy) \\
&\le \int \|F(x, \cdot)\|^2_{\rho_x}\, \mu_2(dy) \\
&= \int \int |F(x, y)|^2\, \rho^y(dx)\, \mu_2(dy) \\
&= \|F(x, y)\|^2_\mu < \infty.
\end{aligned}$$

□

Let μ be a y-slice-singular measure on $[0, 1)^2$. Let $\mathbf{n} = (m, n)$. Define $g_{\mathbf{n}}(x, y) = g_m(y)g_n^{(y)}(x)$, where $\{g_m\}$ is the auxiliary sequence of the marginal measure μ_2, and for μ_2-almost-every y, $g_n^{(y)}$ is the auxiliary sequence of the slice measure ρ^y. Also, define $e_{\mathbf{n}}(x, y) := e^{2\pi i m y}e^{2\pi i n x}$. (Note: for an x-slice-singular measure, we would define $g_{\mathbf{n}}(x, y) = g_m(x)g_n^{(x)}(y)$ and $e_{\mathbf{n}}(x, y) := e^{2\pi i m x}e^{2\pi i n y}$.)

Theorem 9.4 *If* μ *is* y*-slice-singular, then* $\{g_{\mathbf{n}}\}_{\mathbf{n}\in\mathbb{N}_0^2}$ *is a Parseval frame in* $L^2(\mu)$ *in the* y-x *order.*

Let $F \in L^2(\mu)$. Then,

$$\begin{aligned}
&\sum_n \sum_m \left| \left\langle F, g_m(y)g_n^{(y)}(x) \right\rangle_\mu \right|^2 \\
&= \sum_n \sum_m \left| \int \int F(x, y)\overline{g_m(y)}\overline{g_n^{(y)}(x)}\, \rho^y(dx)\, \mu_2(dy) \right|^2 \\
&= \sum_n \sum_m \left| \int \overline{g_m(y)} \int F(x, y)\overline{g_n^{(y)}(x)}\, \rho^y(dx)\, \mu_2(dy) \right|^2
\end{aligned}$$

$$= \sum_n \sum_m \left| \left\langle \int F(x,y)\overline{g_n^{(y)}(x)}\,\rho^y(dx), g_m(y) \right\rangle_{\mu_2} \right|^2 \qquad \text{[By Proposition 9.8]}$$

$$= \sum_n \left\| \int F(x,y)\overline{g_n^{(y)}(x)}\,\rho^y(dx) \right\|_{\mu_2}^2 \quad \text{[Because } \{g_m\} \text{ is a Parseval frame in } L^2(\mu_2)]$$

$$= \sum_n \int \left| \int F(x,y)\overline{g_n^{(y)}(x)}\,\rho^y(dx) \right|^2 d\mu_2(y)$$

$$= \int \sum_n \left| \int F(x,y)\overline{g_n^{(y)}(x)}\,\rho^y(dx) \right|^2 d\mu_2(y)$$

$$= \int \sum_n \left| \left\langle F(x,y), g_n^{(y)}(x) \right\rangle_{\rho^y} \right|^2 d\mu_2(y)$$

$$= \int \|F(x,y)\|_{\rho^y}^2\, d\mu_2(y) \quad \text{[Because } \left\{g_n^{(y)}\right\} \text{ is a Parseval frame in } L^2(\rho^y)]$$

$$= \int \int |F(x,y)|^2\, \rho^y(dx)\, d\mu_2(dy)$$

$$= \|F\|_\mu^2 .$$

Theorem 9.5 *If μ is y-slice-singular, then $\{g_{\mathbf{n}}\}_{\mathbf{n}\in\mathbb{N}_0^2}$ is dextrodual to $\{e_{\mathbf{n}}\}_{\mathbf{n}\in\mathbb{N}_0^2}$ in $L^2(\mu)$ in the y-x order.*

Proof

$$\sum_n \sum_m \langle F, g_{\mathbf{n}}\rangle_\mu\, e_{\mathbf{n}}(\tilde{x},\tilde{y})$$

$$= \sum_n \sum_m \left(\int \int F(x,y)\overline{g_m(y)}\,\overline{g_n^{(y)}(x)}\,\rho^y(dx)\,\mu_2(dy) \right) e^{2\pi i m\tilde{y}} e^{2\pi i n\tilde{x}}$$

$$= \sum_n \sum_m \left(\int \int F(x,y)\overline{g_n^{(y)}(x)}\,\rho^y(dx)\, e^{2\pi i n\tilde{x}}\, \overline{g_m(y)}\,\mu_2(dy) \right) e^{2\pi i m\tilde{y}}$$

$$= \sum_m \left(\int \left(\sum_n \int F(x,y)\overline{g_n^{(y)}(x)}\,\rho^y(dx)\, e^{2\pi i n\tilde{x}} \right) \overline{g_m(y)}\,\mu_2(dy) \right) e^{2\pi i m\tilde{y}}$$

$$= \sum_m \left(\int \left(\sum_n \left\langle F(x,y), g_n^{(y)}(x) \right\rangle_{\rho^y} e_n(\tilde{x}) \right) \overline{g_m(y)}\,\mu_2(dy) \right) e^{2\pi i m\tilde{y}}$$

$$= \sum_m \left(\int F(\tilde{x},y)\overline{g_m(y)}\,\mu_2(dy) \right) e^{2\pi i m\tilde{y}}$$

$$= \sum_m \langle F(\tilde{x}, y), g_m(y)\rangle_{\mu_2} e_m(\tilde{y})$$

$$= F(\tilde{x}, \tilde{y}).$$

□

9.5.2 Direct Integrals

The proof of Theorem 9.3 can be simplified by using the formalism of direct-integral theory. Doing so also reveals additional structure of the Kaczmarz algorithm in terms of the action of $B(z)$ on $\mathcal{H}_0$ in analogy to the one-dimensional version as we presented in Eq. (9.8), namely in terms of the Clark theory [22]. We avoided the direct-integral theory in our initial proof for the benefit of the reader who is not familiar with the subject.

For the y-slice-singular measure μ, we can obtain a direct-integral decomposition of our space as

$$L^2(\mu) = \int^{\oplus} L^2(\rho^y)\ \mu_2(dy).$$

In the context of the Kaczmarz algorithm, the subspace $\mathcal{H}_0 = L^2(\mu_2)$ then yields a similar direct integral decomposition

$$\mathcal{H}_0(z) = \int^{\oplus} H^2_y(\mathbb{D})\ \mu_2(dy).$$

Here, $H^2_y(\mathbb{D}) = H^2(\mathbb{D})$; we simply use the subscript y to indicate the fibers of the decomposition are indexed by y.

Now, the action of the operator $B(z)$ has a particularly simple form: for $F = \int^{\oplus} f_y\ \mu_2(dy) \in \mathcal{H}_0(z)$,

$$BF = \int^{\oplus} b_y f_y\ \mu_2(dy).$$

From here, we immediately see that B is an isometry. However, we see even more, since we obtain that the range of B is given by

$$B\left(\mathcal{H}_0(z)\right) = \int^{\oplus} \mathcal{H}(b_y)\ \mu_2(dy),$$

where $\mathcal{H}(b_y)$ is the de Branges-Rovnyak space $H^2 \ominus b_y H^2$.

Moreover, we see the dilation theory of de Branges [11] as well as a significant part of the Clark theory on model subspaces and the Normalized Cauchy Transform [10].

Indeed, recall that in the one-variable case, the Normalized Cauchy Transform acts as a unitary operator from $L^2(\nu)$ to $\mathcal{H}(b)$, where ν is a singular measure on $\mathbb{T}$ and $\mathcal{H}(b) \subset H^2(\mathbb{D})$ is the model subspace corresponding to the inner function b. Via the direct integral theory, we can identify $L^2(\mu) \simeq \int^{\oplus} \mathcal{H}(b_y)\,\mu_2(dy)$ via the operator $\int^{\oplus} V_{\rho^y}\,\mu_2(dy)$ acting on $\int^{\oplus} L^2(\rho^y)\,\mu_2(dy)$.

Regarding the dilation theory of de Branges, the direct integral decomposition also provides a concrete representation of the dilation spaces that de Branges constructs, e.g., [11, Lemmas 4 and 11]. In fact, in de Branges' notation, we have $\mathscr{C} = L^2(\mu_2)$ and $\mathscr{C}(z) = \int^{\oplus} H^2_y(\mathbb{D})\mu_2(dy)$. Then, the operator $B(z)$ arising from the Herglotz representation decomposes the dilation space $\int H^2_y(\mathbb{D})\mu_2(dy)$ into

$$B(\mathscr{C}(z)) = \int^{\oplus} \mathcal{H}(b_y)\,\mu_2(dy) \simeq \int^{\oplus} L^2(\rho^y)\,\mu_2(dy) = L^2(\mu).$$

This space de Branges refers to as $\mathcal{H}(B)$, the operator-valued analogy of the model space $\mathcal{H}(b)$. In this view, we can extend the Normalized Cauchy Transform to act on $L^2(\mu)$ by the mapping

$$f \mapsto \sum_n \sum_m \langle f(x,y), g_n^{(y)}(x) g_m(y)\rangle z_1^n z_2^m, \tag{9.61}$$

which is a function in $H^2(\mathbb{D}^2)$. This mapping is an isometry (as a consequence of Theorem 9.4), and its image is a vector-valued analog of the model space $\mathcal{H}(b)$. Moreover, this mapping can be represented as an iteration of one-variable Normalized Cauchy Transforms, as follows:

Let $V_{\mu_2} : L^2(\mu_2) \to H^2$ and $V_{\rho^y} : L^2(\rho^y) \to H^2$ be the normalized Cauchy transforms of μ_2 and ρ^y, respectively. Then, for an $f(x,y) \in L^2(\mu)$, $f(\cdot, y) \in L^2(\rho^y)$ for μ_2-almost-every $y \in [0,1]$.

Then, for $f(x,y) \in L^2(\mu)$, consider $V_{\mu_2}[V_{\rho^y}[f(\cdot,y)](z_2)](z_1)$, where first V_{ρ^y} acts on $f(x,y)$ as a function of x, returning for μ_2-almost-every $y \in [0,1]$ a function $F_y(z_2) := V_{\rho^y}[f(\cdot,y)](z_2) \in H^2$. Then, regarding z_2 as fixed, V_{μ_2} acts on $F_y(z_2)$ as a function of y, returning for each fixed $z_2 \in \mathbb{D}$ a function $G(z_1,z_2) := V_{\mu_2}[F_y(z_2)](z_1) \in H^2$ as a function of z_1. We will now verify that $G(z_1,z_2)$ is the function returned by the mapping (9.61).

Recall that for a measure ν on $[0,1]$, the normalized Cauchy transform $V_\nu f : L^2(\nu) \to H^2(\mathbb{D})$ is given by

$$V_\nu f(z) = \frac{\int_0^1 \frac{f(x)}{1-ze^{-2\pi i x}}\,\nu(dx)}{\int_0^1 \frac{1}{1-ze^{-2\pi i x}}\,\nu(dx)}.$$

In [22, Proposition 1], it was proved that if ν is singular, then $V_\nu f(z) = \sum_{n=0}^{\infty} \langle f, g_n \rangle z^n$, where $\{g_n\}$ is the auxiliary sequence of $\{e_n\}$ in $L^2(\nu)$. Therefore, we have

$$
\begin{aligned}
&V_{\mu_2}[V_{\rho^y}[f(\cdot, y)](z_2)](z_1)\\
&=V_{\mu_2}[\sum_{n=0}^{\infty} \left\langle f(x,y), g_n^{(y)}(x)\right\rangle_{\rho^y(x)} z_2^n](z_1)\\
&=\sum_{m=0}^{\infty}\left\langle \sum_{n=0}^{\infty}\left\langle f(x,y), g_n^{(y)}(x)\right\rangle_{\rho^y(x)} z_2^n, g_m(y)\right\rangle_{\mu_2(y)} z_1^m\\
&=\sum_{m=0}^{\infty}\sum_{n=0}^{\infty}\left\langle \int f(x,y)\overline{g_n^{(y)}(x)}\,\rho^y(dx) z_2^n, g_m(y)\right\rangle_{\mu_2(y)} z_1^m\\
&=\sum_{m=0}^{\infty}\sum_{n=0}^{\infty}\int\left(\int f(x,y)\overline{g_n^{(y)}(x)}\,\rho^y(dx) z_2^n\right)\overline{g_m(y)}\,\mu_2(dy) z_1^m\\
&=\sum_{m=0}^{\infty}\sum_{n=0}^{\infty}\int\int f(x,y)\overline{g_n^{(y)}(x) g_m(y)}\,\rho^{(y)}(dx)\,\mu_2(dy) z_1^m z_2^n\\
&=\sum_{m=0}^{\infty}\sum_{n=0}^{\infty}\int f(x,y)\overline{g_n^{(y)}(x) g_m(y)}\,\mu(dx\,dy) z_1^m z_2^n\\
&=\sum_{m=0}^{\infty}\sum_{n=0}^{\infty}\left\langle f(x,y), g_n^{(y)}(x) g_m(y)\right\rangle_{\mu} z_1^m z_2^n.
\end{aligned}
$$

We will expound on these connections more in a subsequent paper. In addition, this dilated view also presents the opportunity for analyzing the boundary representations of subspaces of $H^2(\mathbb{D}^2)$ as we did in [20, 21]; this too will be expounded upon in a subsequent paper.

9.5.3 Higher Dimensions

Our results have concerned two-dimensional slice-singular measures. Much of our work can be extended easily to higher dimensions once we have a definition of slice-singular.

We define a d-dimensional slice-singular measure by the $d-1$-dimensional case as follows. A positive Borel measure μ in $\mathbb{R}^d$ is said to be x_j-slice-singular if its marginal measure $\mu \circ \pi_j^{-1}$ is singular and the corresponding conditional measures are $d-1$ slice-singular. We then say that μ is slice-singular if it is x_j-slice-singular for some $j \in \{1, \ldots, d\}$. Recall that when μ is a measure on $\mathbb{R}^d$, then each of the indexed systems of conditional measures $\mu(\cdot|\pi_j = a)$, $a \in \mathbb{R}$, is supported in a "hyperplane," or rather subspace, and so is a measure in $d - 1$ dimensions. The conditional measures may be viewed as Borel measures on $\mathbb{R}^{d-1}$, allowing us to define slice-singular by induction.

Our results in Sects. 9.2.1 and 9.2.2 extend naturally in the higher dimensional case. The results in Sect. 9.4 are dimension independent. The one result that is not immediate in higher dimensions is Theorem 9.3, particularly the question of whether the operator $B(z)$ is an isometry on $\mathcal{H}_0(z)$. We will address this issue in a subsequent paper.

Acknowledgments Eric S. Weber was supported in part by the National Science Foundation and the National Geospatial Intelligence Agency under award #1830254.

References

1. C. Allain and M. Cloitre, *Characterizing the lacunarity of random and deterministic fractal sets*, Phys. Rev. A **44** (1991), 3552–3558.
2. Sergey Bezuglyi and Palle E. T. Jorgensen, *Graph Laplace and Markov operators on a measure space*, Linear systems, signal processing and hypercomplex analysis, Oper. Theory Adv. Appl., vol. 275, Birkhäuser/Springer, Cham, 2019, pp. 67–138. MR 3968033
3. ——, *Symmetric measures, continuous networks, and dynamics*, New directions in function theory: from complex to hypercomplex to non-commutative, Oper. Theory Adv. Appl., vol. 286, Birkhäuser/Springer, Cham, [2021] ©2021, pp. 139–197. MR 4368036
4. Allison Byars, Evan Camrud, Steven N. Harding, Sarah McCarty, Keith Sullivan, and Eric S. Weber, *Sampling and interpolation of cumulative distribution functions of Cantor sets in* [0, 1], Dem. Math. **54** (2021), 85–109.
5. Georg Cantor, *De la puissance des ensembles parfaits de points*, Acta Mathematica **4** (1884), no. 1, 381–392.
6. Shiping Cao, Malte S. Haß ler, Hua Qiu, Ely Sandine, and Robert S. Strichartz, *Existence and uniqueness of diffusions on the Julia sets of Misiurewicz-Sierpinski maps*, Adv. Math. **389** (2021), Paper No. 107922, 41. MR 4289048
7. Shiping Cao, Yiqi Huang, Hua Qiu, Robert S. Strichartz, and Xiaohan Zhu, *Spectral analysis beyond* ℓ^2 *on Sierpinski lattices*, J. Fourier Anal. Appl. **27** (2021), no. 3, Paper No. 55, 19. MR 4269450
8. J. W. S. Cassels, *On a problem of Steinhaus about normal numbers*, Colloq. Math. **7** (1959), 95–101. MR 0113863
9. Joseph T Chang and David Pollard, *Conditioning as disintegration*, Statistica Neerlandica **51** (1997), no. 3, 287–317.
10. Douglas N. Clark, *One dimensional perturbations of restricted shifts*, J. Analyse Math. **25** (1972), 169–191. MR 0301534 (46 #692)
11. Louis de Branges, *Some Hilbert spaces of entire functions*, Trans. Amer. Math. Soc. **96** (1960), 259–295. MR 133455
12. Jessica L. DeGrado, Luke G. Rogers, and Robert S. Strichartz, *Gradients of Laplacian eigenfunctions on the Sierpinski gasket*, Proc. Amer. Math. Soc. **137** (2009), no. 2, 531–540. MR 2448573
13. Qi-Rong Deng and Ka-Sing Lau, *Structure of the class of iterated function systems that generate the same self-similar set*, J. Fractal Geom. **4** (2017), no. 1, 43–71. MR 3631376

14. Dorin Ervin Dutkay, Deguang Han, and Qiyu Sun, *On the spectra of a Cantor measure*, Adv. Math. **221** (2009), no. 1, 251–276. MR MR2509326
15. Dorin Ervin Dutkay, John Haussermann, and Chun-Kit Lai, *Hadamard triples generate self-affine spectral measures*, Trans. Amer. Math. Soc. **371** (2019), no. 2, 1439–1481. MR 3885185
16. Dorin Ervin Dutkay and Chun-Kit Lai, *Uniformity of measures with Fourier frames*, Adv. Math. **252** (2014), 684–707. MR 3144246
17. Qingsong Gu and Ka-Sing Lau, *Resistance estimates and critical exponents of Dirichlet forms on fractals*, Analysis and partial differential equations on manifolds, fractals and graphs, Adv. Anal. Geom., vol. 3, De Gruyter, Berlin, [2021] ©2021, pp. 71–102. MR 4320086
18. Rainis Haller and Ryszard Szwarc, *Kaczmarz algorithm in Hilbert space*, Studia Math. **169** (2005), no. 2, 123–132. MR 2140451 (2006b:41049)
19. John E. Herr, Palle E. T. Jorgensen, and Eric S. Weber, *A matrix characterization of boundary representations of positive matrices in the Hardy space*, Frames and harmonic analysis, Contemp. Math., vol. 706, Amer. Math. Soc., Providence, RI, 2018, pp. 255–270. MR 3796641
20. ——, *Positive matrices in the Hardy space with prescribed boundary representations via the Kaczmarz algorithm*, J. Anal. Math. **138** (2019), no. 1, 209–234. MR 3996038
21. ——, *Harmonic analysis of fractal measures: Basis and frame algorithms*, Analysis, Probability and Mathematical Physics on Fractals (P. Ruiz, J. Chen, L. Rogers, R. Strichartz, and A. Teplyaev, eds.), Fractals and Dynamics in Mathematics, Science, and the Arts, vol. 5, World Scientific, 2020, pp. 163–221.
22. John E. Herr and Eric S. Weber, *Fourier series for singular measures*, Axioms **6** (2017), no. 2:7, 13 ppg., http://dx.doi.org/10.3390/axioms6020007.
23. Takeyuki Hida, *Brownian motion*, Applications of Mathematics, vol. 11, Springer-Verlag, New York-Berlin, 1980, Translated from the Japanese by the author and T. P. Speed. MR 562914
24. Alexander S. Holevo and Juha-Pekka Pellonpää, *Extreme covariant observables for type I symmetry groups*, Found. Phys. **39** (2009), no. 6, 625–641. MR 2548411
25. John E. Hutchinson, *Fractals and self similarity*, Indiana University Mathematics Journal **30** (1981), no. 5, 713–747.
26. David Jekel, *Conditional expectation, entropy, and transport for convex Gibbs laws in free probability*, Int. Math. Res. Not. IMRN (2022), no. 6, 4514–4619. MR 4391896
27. Palle E. T. Jorgensen, *Analysis and probability: wavelets, signals, fractals*, Graduate Texts in Mathematics, vol. 234, Springer, New York, 2006. MR MR2254502 (2008a:42030)
28. Palle E. T. Jorgensen and Steen Pedersen, *Dense analytic subspaces in fractal L^2-spaces*, J. Anal. Math. **75** (1998), 185–228. MR 1655831 (2000a:46045)
29. Stefan Kaczmarz, *Angenäherte Auflöösung von Systemen linearer Gleichungen*, Bulletin International de l'Académie Polonaise des Sciences et des Lettres. Classe des Sciences Mathématiques et Naturelles. Série A, Sciences Mathématiques (1937), 355–357.
30. Shizuo Kakutani, *On equivalence of infinite product measures*, Ann. of Math. (2) **49** (1948), 214–224. MR 23331
31. Stanisław Kwapień and Jan Mycielski, *On the Kaczmarz algorithm of approximation in infinite-dimensional spaces*, Studia Math. **148** (2001), no. 1, 75–86. MR 1881441 (2003a:60102)
32. Izabella Łaba and Yang Wang, *On spectral Cantor measures*, J. Funct. Anal. **193** (2002), no. 2, 409–420. MR 1929508 (2003g:28017)
33. Chun-Kit Lai, Ka-Sing Lau, and Hui Rao, *Classification of tile digit sets as product-forms*, Trans. Amer. Math. Soc. **369** (2017), no. 1, 623–644. MR 3557788
34. Chun-Kit Lai and Yang Wang, *Non-spectral fractal measures with Fourier frames*, J. Fractal Geom. **4** (2017), no. 3, 305–327. MR 3702855
35. Shidong Li and Hidemitsu Ogawa, *Pseudo-duals of frames with applications*, Appl. Comput. Harmon. Anal. **11** (2001), no. 2, 289–304. MR 1848709 (2002f:46014)
36. R. Lyons and Y. Peres, *Probability on trees and networks*, Cambridge Series in Statistical and Probabilistic Mathematics, Cambridge University Press, 2017.
37. Frank Natterer, *The mathematics of computerized tomography*, Teubner, Stuttgart, 1986.

38. M. A. Naĭmark, *The direct integral of pairs of dual spaces*, Dokl. Akad. Nauk SSSR **217** (1974), 762–765. MR 0365164
39. Ole A. Nielsen, *Direct integral theory*, Lecture Notes in Pure and Applied Mathematics, vol. 61, Marcel Dekker, Inc., New York, 1980. MR 591683
40. Kasso A. Okoudjou, Robert S. Strichartz, and Elizabeth K. Tuley, *Orthogonal polynomials on the Sierpinski gasket*, Constr. Approx. **37** (2013), no. 3, 311–340. MR 3054607
41. Yuval Peres and Boris Solomyak, *Self-similar measures and intersections of Cantor sets*, Trans. Amer. Math. Soc. **350** (1998), no. 10, 4065–4087. MR 1491873
42. A. D. Pollington, *The Hausdorff dimension of a set of normal numbers. II*, J. Austral. Math. Soc. Ser. A **44** (1988), no. 2, 259–264. MR 922610
43. A. G. Poltoratskiĭ, *Boundary behavior of pseudocontinuable functions*, Algebra i Analiz **5** (1993), no. 2, 189–210, English translation in St. Petersburg Math. 5:2 (1994): 389–406. MR 1223178 (94k:30090)
44. Robert J. Ravier and Robert S. Strichartz, *Sampling theory with average values on the Sierpinski gasket*, Constr. Approx. **44** (2016), no. 2, 159–194. MR 3543997
45. V. A. Rohlin, *On the decomposition of a dynamical system into transitive components*, Mat. Sbornik N.S. **25(67)** (1949), 235–249. MR 0032958
46. ——, *On the fundamental ideas of measure theory*, Mat. Sbornik N.S. **25(67)** (1949), 107–150. MR 0030584
47. Reinhard Schaflitzel, *Direct integrals of unitarily equivalent representations of nonseparable C^*-algebras*, J. Funct. Anal. **111** (1993), no. 1, 62–75. MR 1200636
48. Wolfgang M. Schmidt, *Über die Normalität von Zahlen zu verschiedenen Basen*, Acta Arith. **7** (1961/1962), 299–309. MR 0140482
49. Robert S. Strichartz, *Remarks on: "Dense analytic subspaces in fractal L^2-spaces" [J. Anal. Math. **75** (1998), 185–228; MR1655831 (2000a:46045)] by P. E. T. Jorgensen and S. Pedersen*, J. Anal. Math. **75** (1998), 229–231. MR 1655832
50. ——, *Analysis on fractals*, Notices Amer. Math. Soc. **46** (1999), no. 10, 1199–1208. MR MR1715511 (2000i:58035)
51. ——, *Mock Fourier series and transforms associated with certain Cantor measures*, J. Anal. Math. **81** (2000), 209–238. MR MR1785282 (2001i:42009)
52. ——, *Convergence of mock Fourier series*, J. Anal. Math. **99** (2006), 333–353. MR 2279556
53. Boming Yu, *Fractal dimensions for multiphase fractal media*, Fractals **14** (2006), no. 02, 111–118.
54. Boming Yu, Mingqing Zou, and Yongjin Feng, *Permeability of fractal porous media by monte carlo simulations*, International Journal of Heat and Mass Transfer **48** (2005), 2787–2794.

Chapter 10
Blowups and Tops of Overlapping Iterated Function Systems

Louisa F. Barnsley and Michael F. Barnsley

Dedicated to Robert Strichartz

10.1 Introduction

In "Fractals in the large" [11], Robert Strichartz observes that fractal structure is characterized by repetition of detail at all small scales. He asks "Why not large scales as well?" He proposes two ways to study large scaling structures using developments of iterated function systems. Here, we review geometrical aspects of his paper and make a contribution in the area of tiling theory (Fig. 10.1).

In his first approach, Strichartz defines a *reverse iterated function system* (r.i.f.s.) to be a set of $m > 1$ expansive maps

$$T = \{t_i : M \to M | i = 1, 2, \ldots, m\}$$

acting on a locally compact discrete metric space M, where every point of M is isolated. Here, the large scaling structures are the invariant sets of T, sets $S \subset M$ which obey

$$S = \bigcup_{i=1}^{m} t_i (S) .$$

Why does Strichartz restrict his definition to functions acting on *discrete* metric spaces? (i) He establishes that there are interesting nontrivial examples. (ii) He shows that such objects (act as a kind of skeleton to) play a role in his second kind of large-scale fractal structure that he calls a *fractal blowup*. Probably he had

---

L. F. Barnsley · M. F. Barnsley (✉)
Australian National University, Canberra, ACT, Australia
e-mail: michael.barnsley@anu.edu.au

P. Alonso Ruiz et al. (eds.), *From Classical Analysis to Analysis on Fractals*, Applied and Numerical Harmonic Analysis, https://doi.org/10.1007/978-3-031-37800-3_10

Fig. 10.1 Small ferns growing wildly: can fractal geometry model such images?

other reasons related to situations where his approach to analysis on fractals could be explored.

In Sect. 10.2, we present the notation for iterated function systems (i.f.s.) acting on $\mathbb{R}^n$. We are particularly concerned with notation for chains of compositions of functions and properties of addresses of points on fractals. In Sect. 10.3, we review Strichartz's definition and the basic theorem concerning invariant sets of reverse iterated function systems, and we compare them to the corresponding situation for contractive i.f.s. We describe some kinds of invariant sets of contractive i.f.s. and consider how they compare to Strichartz's large scaling structures. It is a notable feature of Strichartz's definition that he restricts attention to functions acting on compact discrete metric spaces. We mention that, if this restriction is lifted, sometimes interesting structures, characterized by repetition of structure at large scales, may be obtained. See, for example, Fig. 10.2.

In Sect. 10.4, we define fractal blowups, Strichartz's second kind of large-scale fractal structure, and present his characterization of them, when the open set condition (OSC) is obeyed, as unions of scaled copies of an i.f.s. attractor, with the scaling restricted to a finite range. We outline the proof of his characterization theorem using different notation, anticipating fractal tops. We recall Strichartz's theorem, where he restricts attention to blowups of an i.f.s. all of the same scaling factor. Here, he combines his two ideas: he reveals that the fractal blowup is in fact a copy of the original fractal translated by all the points on an invariant set of an r.i.f.s.

In Sect. 10.5, we discuss how tilings of blowups can be extended to overlapping i.f.s. In [3], it was shown how, in the overlapping (OSC not obeyed) case, tilings

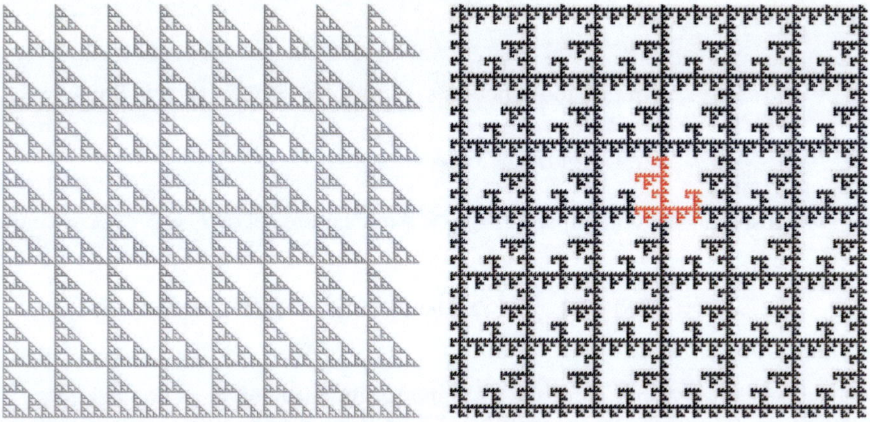

Fig. 10.2 Two examples of invariant sets of inverse iterated function systems. The left image illustrates part of the fast basin of a Sierpinski triangle i.f.s. The right image illustrates the fast basin of an i.f.s. whose attractor is illustrated in red. These unbounded sets are "invariant in the large" but are not discrete

of blowups can be defined using an artificial recursive system of "masks." Here, the approach is more natural, but we pay a price: namely, sequences of tilings are not necessarily nested. Here, tilings are defined by using "fractal tops," namely attractors with their points labeled by lexicographically highest addresses. The needed theory of fractal tops is developed in Sect. 10.5.1. Then, in Sect. 10.5.2, we use these top addresses to define and establish the existence of tilings of some blowups for overlapping i.f.s. The main theorem concerns the relationship between the successive tilings that may be used to define a tiling of a blowup. In Sect. 10.5.3, we present an example involving a tile that resembles a leaf.

Strichartz's paper has overlap with [1], published about the same time by Christoph Bandt. Both papers consider the relationship between i.f.s. theory and self-similar tiling theory. Most current work in tiling theory does not use the mapping point of view, but both Bandt and Stricharz do. Bandt is not only particularly focused on the open set condition and the algebraic structure of tilings but also has a clear understanding of tilings of blowups when the OSC is obeyed.

Strichartz's paper also contains measure theory aspects that we do not discuss. But from the little we have focused on here, much has been learned concerning the subtlety, the depth, and the elegant simplicity of the mathematical thinking of Robert Strichartz.

Examples of one-dimensional "fractal top" address systems are related to beta-expansions introduced by Renyi [10]; see, for example, Parry [9], and Frougny and Solomyak [7]. A specific relationship between beta-expansions and self-similar tilings is mentioned in Thurston's lecture notes [14].

10.2 Preliminaries

Let $\mathbb{N} = \{1, 2, \dots\}$. An iterated function system (i.f.s.) is a set of functions

$$F = \{f_i : \mathbb{X} \to \mathbb{X} | i = 1, 2, \dots, m\}$$

mapping a space $\mathbb{X}$ into itself, with $m \in \mathbb{N}$. An invariant set of F is $S \subset \mathbb{X}$ such that

$$S = F(S) := \bigcup_{i=1}^{m} f_i(S) \text{ where } f_i(S) = \{f_i(s) | s \in S\}.$$

We use the same symbol F for the i.f.s. and for its action on S, as defined here.

The i.f.s. F is said to be contractive when $\mathbb{X}$ is equipped with a metric d such that $d(f_i x, f_i y) \leq \lambda d(x, y)$ for some $0 < \lambda < 1$ and all $x, y \in \mathbb{X}$. If $\mathbb{X} = \mathbb{R}^n$, we take d to be the Euclidean metric. A contractive i.f.s. on $\mathbb{R}^n$ is associated with its attractor A, the unique nonempty closed and bounded invariant set of F [8]. But Strichartz is also interested in the case where the underlying space is discrete and the maps are expansive.

We use *addresses* to describe compositions of maps. Addresses are defined in terms of the indices of the maps of F. Let $\Sigma = \{1, 2, \dots, m\}^{\mathbb{N}}$, the set of strings of the form $\mathbf{j} = j_1 j_2 \dots$ where each j_i belongs to $\{1, 2, \dots, m\}$. We write $\Sigma_n = \{1, 2, \dots, m\}^n$ and let $\Sigma_{\mathbb{N}} = \cup_{n=1}^{\infty} \Sigma_n$. The address $\mathbf{j} \in \Sigma$ truncated to length n is denoted by $\mathbf{j}|n = j_1 j_2 \dots j_n \in \Sigma_{\mathbb{N}}$, and we define

$$f_{\mathbf{j}|n} = f_{j_1} f_{j_2} \dots f_{j_n} = f_{j_1} \circ f_{j_2} \circ \dots \circ f_{j_n},$$
$$f_{-\mathbf{j}|n} = f_{j_1}^{-1} f_{j_2}^{-1} \dots f_{j_n}^{-1} = f_{j_1}^{-1} \circ f_{j_2}^{-1} \circ \dots \circ f_{j_n}^{-1}.$$

Define a metric d on Σ by $d(\mathbf{j}, \mathbf{k}) = 2^{-\max\{n | j_m = k_m, m=1,2,\dots,n\}}$ for $\mathbf{j} \neq \mathbf{k}$, so that (Σ, d) is a compact metric space.

The *forward orbit* of a point x under (the semigroup generated by) F is

$$\{f_{\mathbf{j}|n}(x) | \mathbf{j} \in \Sigma, n \in \mathbb{N}\}.$$

Here, we do not allow $\mathbf{j}|n$ to be the empty set, so x is not necessarily an element of its forward orbit under the i.f.s. Indeed, x is a member of its forward orbit if and only if x is a fixed point of one of the composite maps $f_{\mathbf{j}|n}$.

Now, let F be a contractive IFS of invertible maps on $\mathbb{R}^n$. Then, a continuous surjection $\pi : \Sigma \to A$ is defined by

$$\pi(\mathbf{j}) = \lim_{N \to \infty} f_{\mathbf{j}|N}(x) = \lim_{N \to \infty} f_{j_1} f_{j_2} \dots f_{j_N}(x).$$

The limit is independent of x. The convergence is uniform in $\mathbf{j}$ over Σ and uniform in x over any compact subset of $\mathbb{R}^n$. We say $\mathbf{j} \in \Sigma$ is an *address* of the point $\pi(\mathbf{j}) \in A$.

We define $i : \Sigma \to \Sigma$ by $i(\mathbf{j}) = ij_1j_2\ldots$ But we may also write $k_1k_2\ldots k_l\mathbf{j}$ to mean the address $k_1k_2\ldots k_lj_1j_2\cdots \in \Sigma$. Let $\sigma : \Sigma \to \Sigma$ be the shift operator defined by $\sigma(\mathbf{j}) = j_2j_3\ldots$. It is well known that

$$f_i \circ \pi = \pi \circ i \text{ and } \pi \circ \sigma(\mathbf{j}) = f_{j_1}^{-1} \circ \pi(\mathbf{j})$$

for all $i \in \{1, 2, \ldots m\}, \mathbf{j} \in \Sigma$.

A notable shift invariant subset of Σ is the set of disjunctive addresses Σ_{dis}. An address $\mathbf{j} \in \Sigma$ is *disjunctive* when, for each finite address $i_1i_2i_3\ldots i_k \in \{1, 2, \ldots, m\}^k$, there is $l \in \mathbb{N}$ so that $j_{l+1}\ldots j_{l+k} = i_1i_2i_3\ldots i_k$. The set of disjunctive addresses $\Sigma_{dis} \subset \Sigma$ is totally invariant under the shift, according to $\sigma(\Sigma_{dis}) = \Sigma_{dis}$. A point $a \in A$ is disjunctive if there is a disjunctive address $\mathbf{j} \in \Sigma$ such that $\pi(\mathbf{j}) = a$. Disjunctive points play a role in the structure of attractors. For example, if the i.f.s. obeys the open set condition (OSC) and its attractor has nonempty interior, then all the disjunctive points belong to the interior of the attractor [2]. Recall that F obeys the OSC when there exists a nonempty open set O such that $\cup f_i(O) \subset O$ and $f_i(O) \cap f_j(O) = \emptyset$ whenever $i \neq j$.

10.3 Reverse Iterated Function Systems

In his first approach to large scaling structures, Strichartz defines a *reverse iterated function system* (r.i.f.s.) to be a set of $m > 1$ expansive maps

$$T := \{t_i : M \to M | i = 1, 2, \ldots, m\}$$

acting on a locally compact discrete (i.e., every point is isolated) metric space M. We write T and t_i in place of F and f_i to distinguish this special kind of i.f.s. A mapping $t_i : M \to M$ is said to be expansive if there is a constant $r > 1$ such that $d(t_ix, t_iy) \geq rd(x, y)$ for all $x \neq y$ in M. An expansive mapping is necessarily one to one and has at most one fixed point.

In this case, Strichartz's large scaling structures are the invariant sets of r.i.f.s.; that is, sets $S \subset M$ which obey

$$S = T(S) = \bigcup_{i=1}^{m} t_i(S).$$

By requiring that M is discrete, Strichartz restricts the possible invariant sets to be discrete.

Let P be the fixed points of $\{t_{\mathbf{i}|k} : M \to M | k \in \mathbb{N}, \mathbf{i} \in \Sigma\}$. Contrast Theorem 10.1 with Theorem 10.2.

Theorem 10.1 (Strichartz) *A set is invariant for an r.i.f.s. if and only if it is a finite union of forward orbits of points in P. In particular, invariant sets exist if and only if P is nonempty, and there are at most a finite number of invariant sets.*

Example 1 Let $M = \mathbb{Z}$, $T = \{t_i : M \to M; t_1(x) = 2x, t_2(x) = 2x - 1\}$. It is readily verified that M is invariant for this r.i.f.s, T. It consists of the forward orbits of the fixed points of t_1 and t_2.

Example 2 Strichartz presents the following example of an r.i.f.s. Let M be the set of integer lattice points $\mathbb{Z}^2$ in the plane, lying between or on the lines $y = \rho x$ and $y = \rho x + 1$ where $\rho + \rho^2 = 1$, $\rho = \left(\sqrt{5} - 1\right)/2$. The r.i.f.s. comprises the two maps

$$t_1(x, y) = (-x - y, -x), t_2(x, y) = (1 - x - y, 1 - x).$$

These maps are expansive on M, even though when viewed as transformations acting on $\mathbb{R}^2$, they contract pairs of points that lie on any straight line with slope $-1/\rho$. The fixed point of t_1 is $(0, 0)$ and of t_2 is $(0, 1)$, both of which lie in M. The union of the forward orbits of these two points is M. So, this unlikely looking set of discrete points is invariant under the r.i.f.s.

This example yields, by projection onto the line $y = \rho x$, an example of a quasi-periodic linear tiling using tiles of lengths ρ and $1+\rho$. Strichartz also points out that by projection onto the perpendicular line $y = -x/\rho$ of a natural measure on M, one obtains, after renormalizing, the unique self-similar measure on $[0, 1]$ associated with the overlapping i.f.s. $f_1(x) = \rho x$, $f_2(x) = \rho x + 1$ with equal probabilities.

Theorem 10.1 leads one to wonder: What are the invariant sets of an i.f.s.? Usually, the focus is on compact invariant sets, namely attractors. The following theorem is simply a list of some of the invariant sets of a contractive i.f.s. The wealth of such invariants here stands in sharp contrast to Theorem 10.1.

Theorem 10.2 (Some Invariant Sets of an i.f.s.) *Let F be a contractive i.f.s. of invertible maps on $\mathbb{R}^n$. If $S \subset \mathbb{R}^n$ is invariant and bounded, then either $\overline{S} = \emptyset$ or $\overline{S} = A$. The following sets are invariant:*

1. *The attractor A and the whole space $\mathbb{R}^n$.*
2. *The forward orbit under F of any periodic point $p \in P$.*
3. *The set of disjunctive points of A.*
4. *The orbit of any $x \in \mathbb{R}^n$ under the free group generated by the maps of F and their inverses.*
5. *The union of any collection of invariant sets.*

There are other invariant sets. For example, let A be a Sierpinski triangle, the attractor of an i.f.s. F_{sierp} in the usual way. Let B be the union of the sides of all triangles in A. Then, B is an invariant set for F_{sierp}. It is not covered by Theorem 10.2.

We note that the invariant set in (4) is also invariant under the *inverse i.f.s.*

$$F^{-1} := \left\{ f_i^{-1} : \mathbb{R}^n \to \mathbb{R}^n | i = 1, \ldots, m \right\}.$$

The orbit under F^{-1} of the attractor A is invariant under F^{-1}. This set may be referred to as the *fast basin* of A with respect to F, see [5]. It is an example of a set which is "invariant in the large," admitted when Strichartz's constraint, that the underlying space is discrete and locally compact, is lifted.

Figure 10.2 illustrates the fast basin associated with (left) a Sierpinski triangle i.f.s. and (right) a different i.f.s. of three similitudes of scaling factor $1/2$. Fast basins are interesting from a computational point of view because they comprise the points x in $\mathbb{R}^n$ for which there is an address $\mathbf{j} \in \Sigma_{\mathbb{N}}$ such that $f_{\mathbf{j}}(x) \in A$.

10.4 Strichartz's Fractal Blowups

Strichartz uses r.i.f.s. to analyze the structure of what he christened "fractal blowups." These structures have been used to develop differential operators on unbounded fractals, see, for example, [12, 13].

Let F be an i.f.s. of similitudes. The maps take the form

$$f_j\left(x\right) = r_j U_j x + b_j,$$

where $0 < r_j < 1$, $b_j \in \mathbb{R}^n$, and U_j is an orthogonal transformation. It is convenient to write $r_j = r^{a_j}$, where $r = \max\left\{r_j\right\}$, so that $1 \leq a_j < a_{\max}$. A *blowup* $\mathcal{A}$ of A is the union of an increasing sequence of sets

$$A = A_0 \subset A_1 \subset A_2 \subset \ldots, \tag{10.1}$$

where $A_j = f_{-\mathbf{k}|j}\left(A\right)$ for some fixed $\mathbf{k} \in \Sigma$ and all $j \in \mathbb{N}$. We have

$$\mathcal{A} = \mathcal{A}\left(\mathbf{k}\right) = \bigcup_{j=1}^{\infty} f_{-\mathbf{k}|j}\left(A\right). \tag{10.2}$$

Strichartz starts with a more general definition of a blowup but restricts consideration to the one given here.

Theorem 10.3 (Strichartz) *Let $\mathcal{A}\left(\mathbf{k}\right)$ be a blowup of A of the form in Eq. (10.2) and assume F satisfies the OSC. Then, $\mathcal{A}\left(\mathbf{k}\right)$ is the union of sets $\mathcal{G}_n$ which are similar to A with the contraction ratios bounded from above and below, and the number of sets $\mathcal{G}_n$ that intersect any ball of radius R is at most a multiple of R^n. In particular, the union $\mathcal{A}\left(\mathbf{k}\right) = \cup_{n=1}^{\infty}\mathcal{G}_n$ is locally finite, and the intersection of $\mathcal{A}\left(\mathbf{k}\right)$ with any compact set is equal to the intersection of $\cup_{n=1}^{N}\mathcal{G}_n$ with that compact set for large N.*

Proof We outline a proof for the case of a single scaling factor $0 < r < 1$ with $f_i(x) = rU_i x + b_i$. At heart, our proof is the same as Strichartz, but we introduce the notation that helps with our generalization to overlapping i.f.s. in Sect. 10.5.

Since F satisfies the OSC, there is a bounded open set O such that $A \subset \overline{O}$, $f_i(O) \subset O$ for all i, $f_i(O) \cap f_j(O) = \emptyset$ for all $i \neq j$. Note that the latter condition implies that the sets in $\left\{ f_{j_1 j_2 \ldots j_l}(O) | j_1 j_2 \ldots j_l \in \Sigma_l \right\}$ are disjoint.

Define a collection of sets

$$\Pi_S(\mathbf{k}|n) := \left\{ f_{-\mathbf{k}|n} f_{\mathbf{m}|n}(S) | \mathbf{m} \in \Sigma \right\},$$

where S may be O, $\overline{O}$, or A. Observe that

$$\Pi_A(\mathbf{k}|1) \subset \Pi_A(\mathbf{k}|2) \subset \ldots$$

and

$$f_{-\mathbf{k}|l}(A) = \bigcup_{n=1}^{l} \Pi_A(\mathbf{k}|n).$$

Also,

$$\Pi_O(\mathbf{k}|\,(n+1)) \backslash \Pi_O(\mathbf{k}|n) = \left\{ f_{-\mathbf{k}|n+1} f_{\mathbf{m}|(n+1)}(O) | \mathbf{m} \in \Sigma_{n+1}, k_{n+1} \neq m_1 \right\}$$

consists of $m^n(m-1)$ disjoint open sets. It follows that $\{\Pi_O(\mathbf{k}|n) | n = 1, 2, \ldots\}$ is a nested increasing sequence of disjoint open sets, whose closed union contains $\mathcal{A}(\mathbf{k})$. The closure of each open set contains a copy of A. Since each open set has volume bounded below by a positive constant, local finiteness is assured.

A general case of a Strichartz style blowup is captured by defining

$$\Pi_S(\mathbf{k}|j) = f_{-\mathbf{k}|l}(\{f_{\mathbf{m}}(S) | \eta^{-}(\mathbf{m}) < \eta(\mathbf{k}|l) \leq \eta(\mathbf{m}), \mathbf{m} \in \Sigma_{\mathbb{N}}\})$$
$$\Pi_S(\mathbf{k}) = \bigcup_{j \in \mathbb{N}} \Pi_S(\mathbf{k}|j),$$

where

$$\eta^{-}(m_1 m_2 \ldots m_n) = a_{m_1} + a_{m_2} + \cdots + a_{n-1}$$
$$\eta(m_1 m_2 \ldots m_n) = a_{m_1} + a_{m_2} + \cdots + a_n.$$

These formulas provide a specific form to Strichartz's stopping time argument. Using these more general expressions, one obtains, for fixed $\mathbf{k}$, an increasing union of copies of A scaled by factors that lie between $r^{a_{\max}}$ and r. See, for example, [3, 4]. The argument concerning local finiteness is essentially the same as above. □

Strichartz unites his two ideas, reverse i.f.s. and blowups, by considering the case where $f_j(x) = rx + b_j$ for all j and studying the blowup $\mathcal{A}\left(\overline{1}\right)$ where $\overline{1} = 111\ldots$, that is,

$$\mathcal{A}\left(\overline{1}\right) = \cup_{n=1}^{\infty}(f_1^{-1})^n A.$$

Theorem 10.4 (Strichartz Combines r.i.f.s. and Blowups) *Let $f_j x = rx + b_j$. Then, $\mathcal{A}\left(\overline{1}\right) = A + D$, where D is an invariant set for the r.i.f.s.*

$$t_j(x) = r^{-1}(x + b_j - b_1),\ j = 1, 2, \ldots, m.$$

Specifically, D is the forward orbit of 0, the fixed point of t_1.

That is, $\mathcal{A}\left(\overline{1}\right)$ is the Minkowski sum of the attractor of the i.f.s. and an invariant set of an r.i.f.s.

10.5 Tops Tilings

In this section, we study tilings of fractal blowups in the case of overlapping i.f.s. attractors. First, in Sect. 10.5.1, we give relevant theory of fractal tops. In Sect. 10.5.2, we show how fractal tops may be used to generate tilings of fractal blowups for overlapping i.f.s. The approach here is distinct from the one in [3]. In Sect. 10.5.3, we illustrate fractal tops for an i.f.s. of two maps, with overlapping attractor that looks like a leaf, suggesting applications to modeling of complicated real-world images.

10.5.1 Fractal Tops

Let F be a strictly contractive i.f.s. acting on a complete metric space $\mathbb{X}$, with maps f_i and attractor A. We assume that there are two or more maps, at least two of which have different fixed points. Also, all of the maps are invertible.

Lemma 10.1 *Let C be a closed subset of Σ. Let $\mathbf{j} = \max\{\mathbf{k} \in C\}$. Then, $\mathbf{j} = j_1 \max\{\mathbf{m} \in \Sigma | (j_1\mathbf{m}) \in C\}$.*

Proof C is the union of the three closed sets $\{\mathbf{k} \in C | k_1 > j_1\}$, $\{\mathbf{k} \in C | k_1 = j_1\}$, and $\{\mathbf{k} \in C | k_1 < j_1\}$. The maximum over C is the maximum of the maxima over these three sets. But the set $\{\mathbf{k} \in C | k_1 > j_1\}$ is empty, because if not then $\max\{\mathbf{k} \in C\} \geq \max\{\mathbf{k} \in C | k_1 > j_1\} > \mathbf{j}$, which is a contradiction. If $\max\{\mathbf{k} \in C\} = \max\{\mathbf{k} \in C | k_1 < j_1\}$, then $\mathbf{j} > \mathbf{j}$, again a contradiction. □

Since $\pi : \Sigma \to A$ is continuous and onto, it follows that $\pi^{-1}(x)$ is closed for all $x \in A$. Lemma 10.1 tells us that a map $\tau : A \to \Sigma$ and subset $\Sigma_{top} \subset \Sigma$ are well defined by

$$\tau(x) := \max\{\mathbf{k} \in \Sigma | \pi(\mathbf{k}) = x\},\ \Sigma_{top} := \tau(A).$$

Conventionally, the maximum here is with respect to lexicographical ordering. We refer loosely to these objects and the ideas around them as relating to the *top* of A. Formally, the top of A is the graph of τ, namely $\{(x, \tau(x)) | x \in A\}$.

Top addresses of points in A, namely points in Σ_{top}, can be calculated by following the orbits of the shift map $\sigma : \Sigma_{top} \to \Sigma_{top}$. Simply partition A into $A_1 = f_1(A)$, $A_2 = f_2(A)\backslash A_1$, $A_3 = f_3(A)\backslash(A_1 \cup A_2), \ldots, A_m = f_m(A)\backslash \cup_{n \neq m} A_n$. Define the orbit $\{x_n\}_{n=1}^{\infty}$ of $x = x_1 \in A$, under the tops dynamical system, by $x_{n+1} = f_{i_n}^{-1}(x_n)$, where i_n is the unique index such that $x_n \in A_{i_n}$.

A version of the following observation can be found in [6]. See also [2].

Theorem 10.5 *The set of top addresses is shift invariant, according to* $\Sigma_{top} = \sigma\left(\Sigma_{top}\right)$*, where* σ *is the left shift.*

Proof First, we show that $\sigma(\tau(A)) \subset \tau(A)$. If $\mathbf{j} \in \tau(A)$, then

$$\begin{aligned}
\mathbf{j} &= \max\{\mathbf{k} \in \Sigma | \pi(\mathbf{k}) = \pi(\mathbf{j})\} \text{(by definition)} \\
&= \max\{j_1\mathbf{l} \in \Sigma | \pi(j_1\mathbf{l}) = \pi(\mathbf{j})\} \text{ (by Lemma 10.1)} \\
&= j_1 \max\{\mathbf{l} \in \Sigma | f_{j_1}(\pi(\mathbf{l})) = f_{j_1}(\pi(\sigma\mathbf{j}))\} \\
&= j_1 \max\{\mathbf{l} \in \Sigma | \pi(\mathbf{l}) = \pi(\sigma(\mathbf{j}))\} \text{ (since } f_{j_1} \text{ is invertible)} \\
&= j_1\tau(\pi(\sigma(\mathbf{j}))).
\end{aligned}$$

Hence, $\sigma(\mathbf{j}) = \tau(\pi(\sigma(\mathbf{j})))$. Hence, $\{\sigma(\mathbf{j}) | \mathbf{j} \in \tau(A)\} = \{\tau(\pi(\sigma(\mathbf{j}))) | \mathbf{j} \in \tau(A)\}$, which implies $\sigma(\tau(A)) \subset \tau(A)$.

We also have $1(\Sigma) \subset \Sigma$ so $\tau(\pi(1(\Sigma))) \subset \tau(\pi(\Sigma)) = \tau(A)$. But $\tau(\pi(1(\Sigma)))) = 1(\tau(\pi(\Sigma)))$ by a similar argument to the proof of Lemma 10.1, so $1\tau(A) \subset \tau(A)$. Applying σ to both sides, we obtain $\tau(A) \subset \sigma(\tau(A))$. □

It appears that the shift space Σ_{top} is not of finite type in general, and graph directed constructions cannot be used in general. This is a topic of ongoing research.

Define $\Sigma_{top,n}$ to be the elements of Σ_{top} truncated to the first n elements. That is,

$$\Sigma_{top,n} = \{(\mathbf{k}|n) | \mathbf{k} \in \Sigma_{top}\}.$$

Define $\pi_{top} : \Sigma_{top} \to A$ to be the restriction of $\pi : \Sigma \to A$ to Σ_{top}. Extend the definition of π_{top} so that it acts on truncated top addresses according to

$$\pi_{top}(\mathbf{k}|n) = \{x \in f_{\mathbf{k}|n}(A) | x \notin f_{\mathbf{c}|n}(A) \text{ for all } \mathbf{c}|n > \mathbf{k}|n\}$$

for all $\mathbf{k} \in \Sigma_{top}$ and all $n \in \mathbb{N}$. We will make use of the following observation.

Lemma 10.2 *If* $\mathbf{k} \in \Sigma_{top}$*, then* $f_{k_1}\left(\pi_{top,n-1}(\sigma(\mathbf{k}|n))\right) \supset \pi_{top,n}(\mathbf{k}|n)$ *for all* $n \in \mathbb{N}$.

Proof We need to compare the sets

$$\{f_{k_1\ldots k_n}(x)|\, f_{k_1\ldots k_n}(x) \notin f_{l_1 l_2\ldots l_n}(A) \text{ for all } l_1\ldots l_n > k_1\ldots k_n\}$$

and

$$\{f_{k_1\ldots k_n}(x)|\, f_{k_1} f_{k_2\ldots k_n}(x) \notin f_{k_1} f_{l_2\ldots l_n}(A) \text{ for all } l_2\ldots l_n > k_2\ldots k_n\}.$$

The condition in the latter expression is less restrictive. □

The sets of truncated top addresses $\Sigma_{top,n}$ have an interesting structure. Any addresses in $\Sigma_{top,n}$ can be truncated on the left or on the right to obtain an address in $\Sigma_{top,n-1}$. The following lemma is readily verified.

Lemma 10.3 *Let $n > 1$. If $i_1 i_2 \ldots i_{n-1} i_n \in \Sigma_{top,n}$, then both $i_2 \ldots i_{n-1} i_n$ and $i_1 i_2 \ldots i_{n-1}$ belong to $\Sigma_{top,n-1}$.*

10.5.2 Top Blowups and Tilings

Here, we are particularly interested in the overlapping case, where the OSC does not hold. We show that natural partitions of fractal blowups, which we call tilings, may still be obtained.

Throughout this subsection, F is an i.f.s. with

$$f_j(x) = rU_j x + b_j, \tag{10.3}$$

where $b_j \in \mathbb{R}^n$ and U_j is an orthogonal transformation. We assume that there are two or more maps, at least two of which have distinct fixed points. We have in mind the situation where A is homeomorphic to a ball, although this is not required by Theorems 10.6 and 10.7.

As in Sect. 10.4, but restricted to $\mathbf{i} \in \Sigma_{top}$, fractal blowups are well defined by

$$\mathcal{A}_n = \mathcal{A}(\mathbf{i}|n) = \bigcup_{l=1}^{n} f_{-\mathbf{i}|l}(A) \text{ and } \mathcal{A} = \mathcal{A}(\mathbf{i}) = \bigcup_{l=1}^{\infty} f_{-\mathbf{i}|l}(A).$$

The unions are of increasing nested sequences of sets, so $\mathcal{A}_n = f_{\mathbf{i}|n}^{-1}(A)$ and $\mathcal{A} = \cup \mathcal{A}_n$. Note that $\mathcal{A}(\mathbf{i}|n)$ is related to $\mathcal{A}(\mathbf{j}|n)$ by the isometry $\left(f_{-\mathbf{j}|n}\right)\left(f_{-\mathbf{i}|n}\right)^{-1}$. But possible relationships between $\mathcal{A}(\mathbf{i})$ and $\mathcal{A}(\mathbf{j})$ are quite subtle because inverse limits are involved.

Under conditions on F and $\mathbf{i}$, stated in Theorems 10.6 and 10.7, we can define generalized tilings of $\mathcal{A}(\mathbf{i})$ with the aid of the following two definitions:

$$\Pi_{top}(\mathbf{i}|k) := \left\{ f_{-\mathbf{i}|k}\left(\left\{ x \in \pi_{top}\left(\mathbf{t}|\left(k+1\right)\right)\right\}\right) | \mathbf{t} \in \Sigma_{top} \right\},$$
$$\Pi_{top}(\mathbf{i}) := \lim_{k\to\infty} \Pi_{top}(\mathbf{i}|k), \text{ when this limit is well defined.}$$

For example, the limit is well defined when $\Pi_{top}(\mathbf{i}|k) \subset \Pi_{top}(\mathbf{i}|k+1)$ for all k, as occurs when the OSC holds. As we will show, it is also well defined in some more complicated situations.

We call each set $f_{-\mathbf{i}|k}\left(\left\{x \in \pi_{top}\left(\mathbf{t}|k\right)\right\}\right)$ a *tile*, and we call the collection of disjoint sets $\left\{f_{-\mathbf{i}|k}\left(\left\{x \in \pi_{top}\left(\mathbf{t}|k\right)\right\}\right) |\mathbf{t} \in \Sigma_{top}\right\}$ a *partial tiling*. The partial tilings $\left\{f_{-\mathbf{i}|k}\left(\left\{x \in \pi_{top}\left(\mathbf{t}|k\right)\right\}\right) |\mathbf{t} \in \Sigma_{top}\right\}$ are well defined. However, $\Pi_{top}(\mathbf{i})$ may not be well defined because there may not be any simple relationship between successive partial tilings. But when it is well defined, we call it a tiling.

The tiles in the partial tiling $\left\{f_{-\mathbf{i}|k}\left(\left\{x \in \pi_{top}\left(\mathbf{t}|k\right)\right\}\right) |\mathbf{t} \in \Sigma_{top}\right\}$ may be referred to by their addresses. It is convenient to define

$$tile(i_1 i_2 \ldots i_k . t_1 t_2 \ldots t_k) = f_{-\mathbf{i}|k}\left(\left\{x \in \pi_{top}\left(\mathbf{t}|k\right)\right\}\right)$$

for all $\mathbf{i}|k$ and all $\mathbf{t}|k \in \Sigma_{top}$. We also define $tile(\varnothing) = A$, corresponding to $k = 0$.

Lemma 10.4 *This concerns the sequence of tilings $\Pi_{top}(\mathbf{i}|n)$. If*

$$i_n p_1 p_2 \ldots p_{n-1} \in \Sigma_{top,n},$$

then

$$tile(i_1 i_2 \ldots i_{n-1} . p_1 p_2 \ldots p_{n-1}) \subset tile(i_1 i_2 \ldots i_n . j_1 j_2 \ldots j_n)$$

implies $i_n p_1 p_2 \ldots p_{n-1} = j_1 j_2 j_3 .. j_n$.

Proof Suppose $tile(i_1 i_2 \ldots i_{n-1} . p_1 p_2 \ldots p_{n-1}) \subset tile(i_1 i_2 \ldots i_n . j_1 j_2 \ldots j_n)$. Then, applying $\left(f_{-\mathbf{i}|(n-1)}\right)^{-1}$ to both sides, we obtain

$$\pi_{top,n-1}(p_1 p_2 \ldots p_{n-1}) \subset f_{i_n}^{-1}(\pi_{top,n}(j_1 j_2 \ldots j_n)),$$

which is equivalent to

$$f_{i_n} \pi_{top,n-1}(p_1 p_2 \ldots p_{n-1}) \subset \pi_{top,n}(j_1 j_2 \ldots j_n).$$

But $\pi_{top,n}(i_n p_1 p_2 \ldots p_{n-1}) \subset f_{i_n} \pi_{top,n-1}(p_1 p_2 \ldots p_{n-1})$ by Lemma 10.2, so

$$\pi_{top,n}(i_n p_1 p_2 \ldots p_{n-1}) \subset \pi_{top,n}(j_1 j_2 \ldots j_n).$$

This implies $i_n p_1 p_2 \ldots p_{n-1} = j_1 j_2 j_3 .. j_n$ because otherwise $\pi_{top,n}(i_n p_1 p_2 \ldots p_{n-1})$ and $\pi_{top,n}(j_1 j_2 \ldots j_n)$ are disjoint subsets of A. □

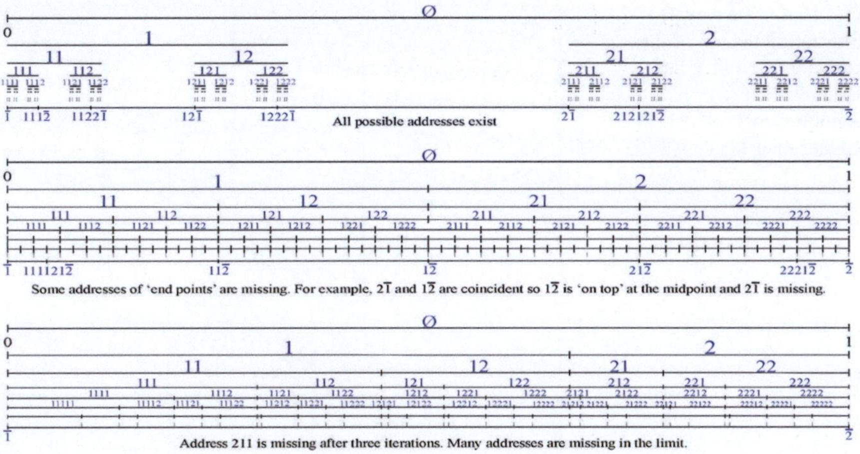

Fig. 10.3 This compares the development of the top addresses for an i.f.s. of two maps in the cases (**a**) where each scaling is 1/3, (**b**) each scaling is 1/2, and (**c**) each scaling is 2/3

We say that $\mathbf{i} \in \Sigma_{top}$ is *reversible* when, for each $n \in \mathbb{N}$, there exists $\mathbf{j} = \mathbf{j}_n \in \Sigma_{top}$ such that $j_1 = i_n, j_2 = i_{n-1}, \ldots, j_n = i_1$. Note that $\mathbf{j}$ depends on n. The address $\overline{1} = 11111\ldots$ is reversible and belongs to Σ_{top} in all cases.

Example 3 For the i.f.s. $\{\mathbb{R}; f_1(x) = 2x/3; f_2(x) = 2x/3 + 1/3\}$, the strings $\overline{1}$ and $\overline{2}$ both belong to Σ_{top} and are reversible. Figures 10.3 and 10.4 illustrate two ways of looking at the development of top addresses. Figure 10.5a illustrates the sets in $\Sigma_{top,n}$ for $n = 0, 1, 2, 3, 4, 5$. We usually use lexicographic ordering to define top addresses, but Fig. 10.4 uses standard ordering.

Example 4 For the i.f.s. $\{\mathbb{R}; f_1(x) = 2x/3; f_2(x) = 1 - 2x/3\}$, each of the strings $\overline{1}, \overline{2}, \overline{12}, \overline{21}$ belongs to Σ_{top} and is reversible. Figure 10.5b illustrates the sets in $\Sigma_{top,n}$ for $n = 0, 1, 2, 3, 4, 5$. Here, it appears that all addresses are reversible.

Let us define a *new tile* to be a tile at level $n+1$ that is not contained in any tile at level n. Also, a *child* or *child tile* is a tile at level $n+1$ that is contained in a tile and its *parent* at level n.

Theorem 10.6 *Let F be an invertible contractive i.f.s. on $\mathbb{R}^n$, as defined in Eqs. 10.3. Let $\mathbf{i} \in \Sigma_{top}$ be reversible. Each tile in $\Pi_{top}(\mathbf{i}|k+1)$ is either (i) a nonempty subset, the child of a tile in $\Pi_{top}(\mathbf{i}|k)$, of the form $tile(i_1 \ldots i_{k+1}.i_{k+1}p_1 \ldots p_k)$ or (ii) a nonempty set of the form $tile(i_1 \ldots i_{k+1}.q_1q_2 \ldots q_kq_{k+1})$, where $q_1 \neq i_{k+1}$, a new tile. Each tile in $\Pi_{top}(\mathbf{i}|k)$ contains exactly one child in $\Pi_{top}(\mathbf{i}|k+1)$.*

Proof We can write

$$\Pi_{top}(\mathbf{i}|k+1) = \left\{tile(i_1i_2 \ldots i_{k+1}.i_{k+1}p_2 \ldots p_{k+1}) | i_{k+1}p_2 \ldots p_{k+1} \in \Sigma_{top,k+1}\right\}$$

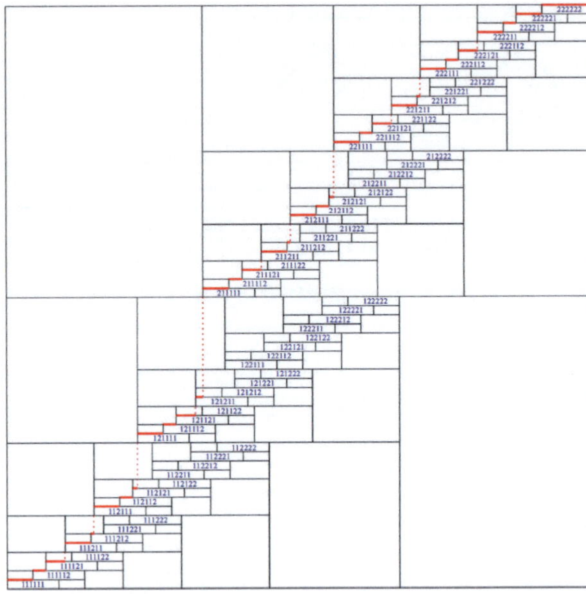

Fig. 10.4 One way of illustrating the top of the attractor of an i.f.s. See Example 3. The ordering here is not lexicographical, so 2 is greater than 1

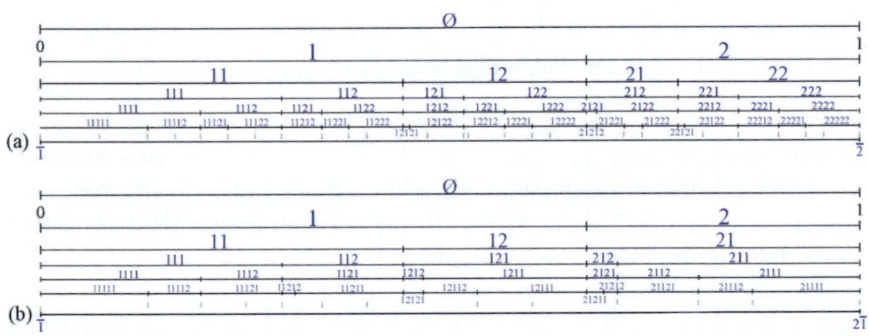

Fig. 10.5 See Examples 3 and 4

$$\cup \left\{tile(i_1 i_2 \ldots i_{k+1}.j_1 j_2 \ldots j_{k+1}) | j_1 j_2 \ldots j_{k+1} \in \Sigma_{top,k+1}, \, j_1 \neq i_{k+1}\right\}$$

Each tile in the first set is a subset of a tile in $\Pi_{top}(\mathbf{i}|k)$, and it is nonempty because $\mathbf{i}$ is reversible. (By reversibility, the set of top addresses $\{i_{k+1} p_1 p_2 \ldots p_k \in \Sigma_{top,k+1} | p_1 p_2 \ldots p_k \in \Sigma_{top,k}\}$ is nonempty.)

Consider any tile $tile(i_1 i_2 \ldots i_{k+1}.p_1 p_2 \ldots p_{k+1})$ in the second set. By Lemma 10.4, if $tile(i_1 i_2 \ldots i_k.p_1 p_2 \ldots p_k) \subset tile(i_1 i_2 \ldots i_{k+1}.i_{k+1} p_2 \ldots p_{k+1})$, then $i_{k+1} p_1 p_2 \ldots p_k = j_1 j_2 j_3 .. j_{k+1}$, which is not possible because $j_1 \neq i_{k+1}$. So no tile in the second set is contained in a tile in the first set. That is to say, the tiles in the second set, which have non-canceling addresses, are "new" and do not contain any tile in the first set.

This says that every tile at level k has a unique child at level $k+1$, either equal to its parent or smaller but not empty; also, there are new tiles at level $k+1$ which do not have predecessors at level k because $\mathcal{A}_{k+1} \neq \mathcal{A}_k$. Each tile in $\Pi_{top}(\mathbf{i}|k)$ contains a child in $\Pi_{top}(\mathbf{i}|k+1)$. One deduces that $\mathcal{A}_{k+1} \backslash \cup$ {children of tiles at level k} is tiled by new tiles. □

In the special case $\mathbf{i} = \overline{\mathbf{1}}$, also considered by Strichartz in Theorem 10.4, we have the theorem:

Theorem 10.7 *Let F be an invertible contractive i.f.s. on $\mathbb{R}^n$, as defined in Eqs. 10.3. Then, $\Pi_{top}(\overline{\mathbf{1}})$ is a well-defined tiling of $\mathcal{A}(\overline{\mathbf{1}})$: specifically, $\Pi_{top}(\overline{\mathbf{1}}|k) \subset \Pi_{top}(\overline{\mathbf{1}}|k+1)$, and*

$$\Pi_{top}(\overline{\mathbf{1}}) = \bigcup_{k=1}^{\infty} \Pi_{top}(\overline{\mathbf{1}}|k).$$

Each tile $\Pi_{top}(\overline{\mathbf{1}}|k)$ (for all $k \in \mathbb{N}$) in $\Pi_{top}(\overline{\mathbf{1}})$ can be written $tile((\overline{1}|k)|t_1t_2\ldots t_k)$ for some $t_1t_2\ldots t_k \in \Sigma_{top,k}$ for some k, with $t_1 \neq 1$. The tile A corresponds to $k=0$.

Proof The result follows from the observation that in this case all children are exact copies of their parents. To see this, simply note that $f_1^{-1}\pi_{top}(1t_1t_2\ldots t_k) = \pi_{top}(t_1t_2\ldots t_k)$ for all $1t_1t_2\ldots t_k \in \Sigma_{top,k+1}$. □

For future work, one can consider the case where A is homeomorphic to a ball. By introducing a stronger notion of reversibility (see also [3]), which requires the top dynamical system orbit of a reversible point $\mathbf{i} \in \Sigma_{top}$ to be contained in a compact set A' contained in the interior of A, one can ensure that new tiles are located further and further away from A. This means that new tiles have only finitely many successive generations of children (one child at each subsequent generation) before children are identical to their parents. Hence, given any ball B of finite radius, the set of tiles in $\Pi_{top}(\mathbf{i}|k)$ that have nonempty intersection with B remains constant for all large enough k. In such cases, one $\Pi_{top}(\mathbf{i}|k) \cap B$ is constant for all k sufficiently large and so the tiling $\Pi_{top}(\mathbf{i})$ is well defined. We note that if $\mathbf{i}$ is disjunctive, then $\mathcal{A}(\mathbf{i}) = \mathbb{R}^n$, see [3].

We conjecture that if A is homeomorphic to a ball and if $\mathbf{i} \in \mathbf{\Sigma}_{top}$ is both reversible and disjunctive (relative to the top), then $\Pi_{top}(\mathbf{i})$ is a well-defined tiling of $\mathbb{R}^n$.

10.5.3 A Leafy Example of a Two-Dimensional Top Tiling

For a two-dimensional affine transformation $f : \mathbb{R}^2 \to \mathbb{R}^2$, we write

$$f = \begin{bmatrix} a & b & e \\ c & d & g \end{bmatrix} \text{ for } f(x, y) = (ax + by + e, cx + dy + g),$$

where $a, b, c, d, e, g \in \mathbb{R}$. We consider the i.f.s. defined by the two similitudes

$$f_1 = \begin{bmatrix} 0.7526 & -.2190 & .2474 \\ 0.2190 & 0.7526 & -.0726 \end{bmatrix}, f_2 = \begin{bmatrix} -0.7526 & 0.2190 & 1.0349 \\ 0.2190 & 0.7526 & 0.0678 \end{bmatrix}. \tag{10.4}$$

The attractor, $L =$ leaf, illustrated in Fig. 10.6, is made of two overlapping copies of itself. The copy illustrated in black is associated with f_1. The point with top address $\overline{\mathbf{1}} = 111\ldots$ is represented by the tip of the stem of the leaf. The stem is actually arranged in an infinite spiral, not visible in the picture. In all tiling pictures, the colors of the tiles were obtained by overlaying the tiling on a colorful photograph: the color of each tile is the color of a point beneath it. In this way, if the tiles were very small, the tiling would look like a mosaic representation of the underlying picture.

Figure 10.7 illustrates the top of L at depths $n \in \{1, 2, \ldots, 6\}$ labeled by the addresses in $\Sigma_{n,top}$.

Figure 10.8 illustrates the successive blowups $\Pi_{top,n}(\overline{1}|n)$ for $n = 1, 2, \ldots, 6$ for the i.f.s. in Eq. (5.2). See also Fig. 10.9 where the successive images are illustrated in their correct relative positions.

Figure 10.10 shows a patch of a leaf tiling, illustrating its complexity. Figure 10.11 illustrates a patch of a top tiling obtained using an i.f.s. of four maps.

Fig. 10.6 The overlapping attractor of an i.f.s. of two similitudes, each with the same scaling factor

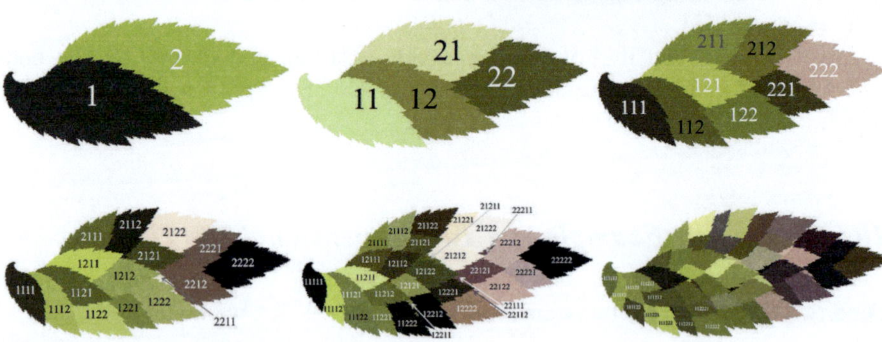

Fig. 10.7 Successive fractal tops

Fig. 10.8 This shows the sequence of tops $\Pi(111\ldots|n)$ for $n = 0, 1, \ldots, 6$ for the leaf i.f.s. In each case, the tip of the stem is at the origin

Fig. 10.9 This illustrates the relationship between the successive partial tilings in Fig. 10.8

Fig. 10.10 Patch of a leaf tiling

Fig. 10.11 Fractal top for an i.f.s of three maps looks both random and somewhat natural, but is not the real thing, compare with Fig. 10.1

Acknowledgments We thank both Brendan Harding and Giorgio Mantica for careful reading and corrections.

References

1. C. Bandt, Self-similar tilings and patterns described by mappings, *Mathematics of Aperiodic Order* (ed. R. Moody) Proc. NATO Advanced Study Institute C489, Kluwer, (1997) 45–83.
2. C. Bandt, M. F. Barnsley, M. Hegland, A. Vince, Old wine in fractal bottles I: Orthogonal expansions on self-referential spaces via fractal transformations, *Chaos, Solitons and Fractals,* **91** (2016), 478–489.
3. M. F. Barnsley, A. Vince, Fractal tilings from iterated function systems, *Discrete and Computational Geometry,* **51** (2014), 729–752.
4. M. F. Barnsley, A. Vince, Self-similar polygonal tilings, *Amer. Math. Monthly,* **124** (2017), 905–921.
5. M. F. Barnsley, K. Leśniak, M. Rypka, Basic topological structure of fast basins, *Fractals* **26** (2018), no. 1, 1850011.
6. M. F. Barnsley, M. F., *Theory and Applications of Fractal Tops,* in: Levy-Vehel, J., Lutton, E., (eds.)Fractals in Engineering: New Trends in Theory and Applications. Springer, London (2005).
7. C. Frougny, B. Solomyak, Finite beta-expansions, *Ergodic Theory and Dyn. Systems*, **12** (1992) 713–723.
8. J. Hutchinson, Fractals and self-similarity, *Indiana Univ. Math. J.* **30** (1981), 713–747.
9. W. Parry, On beta-expansions of real numbers, *Acta Math. Acad. Sci. Hungary* **11** (1960) 401–416.
10. A. Reyni, Representation for real numbers and their ergodic properties, *Aca Math. Acad. Sci. Hungary* **8** (1957) 477–493.
11. R. S. Strichartz, Fractals in the large, *Canad. J. Math.*, **50** (1998), 638–657.
12. R. S. Strichartz, *Differential Equations on Fractals,* Princeton University Press, Princeton, New Jersey (2006).
13. A. Teplyaev, Spectral analysis on infinite Sierpinski gaskets, *J. Funct. Anal.* **129** (1998), 357–567.
14. W. Thurston, *Groups, tilings, and finite state automata*, A.M.S. Colloquium Lecture Notes, 1989.

Chapter 11
Estimates of the Local Spectral Dimension of the Sierpinski Gasket

Masanori Hino

11.1 Introduction

Let us recall how to construct the two-dimensional Sierpinski gasket and the associated Dirichlet form. We take three points p_1, p_2, and p_3 in $\mathbb{R}^2$ that are the vertices of an equilateral triangle. Let ψ_i $(i = 1, 2, 3)$ be a contraction map from $\mathbb{R}^2$ to itself that is defined by $\psi_i(x) = (x + p_i)/2$, $x \in \mathbb{R}^2$. Denoted herein by K, the two-dimensional Sierpinski gasket is a unique nonempty compact subset of $\mathbb{R}^2$ such that $K = \bigcup_{i=1}^{3} \psi_i(K)$.

Let $V_0 = \{p_1, p_2, p_3\}$ and $V_n = \bigcup_{i=1}^{3} \psi_i(V_{n-1})$ for $n \geq 1$ inductively. Then, $\{V_n\}_{n=0}^{\infty}$ is an increasing sequence, and the closure of $V_* := \bigcup_{n=0}^{\infty} V_n$ is equal to K. Let $S = \{1, 2, 3\}$, and $W_n = S^n$ for $n \in \mathbb{Z}_{\geq 0}$. For each $w = w_1 w_2 \cdots w_n \in W_n$, we define a map $\psi_w\colon K \to K$ by $\psi_w = \psi_{w_1} \circ \cdots \circ \psi_{w_n}$ and a compact set K_w by $K_w = \psi_w(K)$. Note that for $w = \emptyset \in W_0$, ψ_w is defined as the identity map. Let W_* denote $\bigcup_{n \in \mathbb{Z}_{\geq 0}} W_n$. For $w = w_1 w_2 \cdots w_m \in W_m$ and $w' = w'_1 w'_2 \cdots w'_n \in W_n$, we write ww' for $w_1 w_2 \cdots w_m w'_1 w'_2 \cdots w'_n \in W_{m+n}$.

We write $p \sim q$ for distinct $p, q \in V_n$ if there exist $p', q' \in V_0$ and $w \in W_n$ such that $p = \psi_w(p')$ and $q = \psi_w(q')$. The relation $\sim$ associates V with a graph structure by setting $\{(p, q) \in V_n \times V_n \mid p \sim q\}$ as the set of edges. In general, let $l(X)$ denote the space of all real-valued functions on a countable set X. For $n \in \mathbb{Z}_{\geq 0}$ and $f, g \in l(V_n)$, let

$$Q_n(f, g) = \frac{1}{2} \sum_{x, y \in V_n,\ x \sim y} (f(x) - f(y))(g(x) - g(y))$$

M. Hino (✉)
Department of Mathematics, Kyoto University, Kyoto, Japan
e-mail: hino@math.kyoto-u.ac.jp

P. Alonso Ruiz et al. (eds.), *From Classical Analysis to Analysis on Fractals*, Applied and Numerical Harmonic Analysis, https://doi.org/10.1007/978-3-031-37800-3_11

and $Q_n(f) = Q_n(f, f)$. We regard $Q_n(f)$ as the total energy of the function f. The sequence $\{(5/3)^n Q_n(f|_{V_n})\}_{n=0}^{\infty}$ is proved to be nondecreasing for any function f in $l(V_*)$. For each $g \in l(V_0)$, there exists a unique $f \in l(V_*)$ such that $f|_{V_0} = g$ and this sequence is a constant one. In this sense, $5/3$ is the correct scaling factor for K. Let $C(K)$ denote the space of all continuous real-valued functions on K. For $f \in C(K)$, define $\mathcal{E}(f) = \lim_{n\to\infty}(5/3)^n Q_n(f|_{V_n})$ ($\le +\infty$) and

$$\mathcal{F} = \{f \in C(K) \mid \mathcal{E}(f) < \infty\}.$$

For $f, g \in \mathcal{F}$, let

$$\mathcal{E}(f, g) = \frac{1}{2}\{\mathcal{E}(f + g) - \mathcal{E}(f) - \mathcal{E}(g)\}.$$

Then, for any finite Borel measure κ on K with full support, $(\mathcal{E}, \mathcal{F})$ is a strongly local regular Dirichlet form on $L^2(K, \kappa)$. Here, $C(K)$ is identified with a subspace of $L^2(K, \kappa)$. This Dirichlet form has the following self-similarity: for $f \in \mathcal{F}$ and $n \in \mathbb{N}$, $\psi_w^* f := f \circ \psi_w$ belongs to $\mathcal{F}$ for all $w \in W_n$ and it holds that

$$\mathcal{E}(f, f) = \sum_{w\in W_n} \left(\frac{5}{3}\right)^n \mathcal{E}(\psi_w^* f, \psi_w^* f). \tag{11.1}$$

By invoking the general theory of Dirichlet forms, the energy measure ν_f of $f \in \mathcal{F}$ is characterized by a unique finite Borel measure on K such that

$$\int_K g(x)\, \nu_f(dx) = 2\mathcal{E}(f, fg) - \mathcal{E}(f^2, g) \quad \text{for all } g \in \mathcal{F} \cap C(K).$$

(Note that the above definition is simpler than usual because K is compact and $C(K)$ is continuously embedded in $L^2(K, \kappa)$.) The measure ν_f does not have mass on any one-point sets. From the self-similarity (11.1) of $(\mathcal{E}, \mathcal{F})$, it holds for all $f \in \mathcal{F}$ and $n \in \mathbb{N}$ that

$$\nu_f = \sum_{w\in W_n} \left(\frac{5}{3}\right)^n \nu_{\psi_w^* f}.$$

In particular, we have the following identity: for $f \in \mathcal{F}$ and $w \in W_n$,

$$\nu_f(K_w) = 2\left(\frac{5}{3}\right)^n \mathcal{E}(\psi_w^* f, \psi_w^* f).$$

Unlike those on differentiable spaces, energy measures on fractals generally have no simple expressions that reveal their distributions. In this respect, Bell, Ho, and Strichartz [3] studied the infinitesimal behaviors of energy measures. To introduce their study, we state several further notations and their properties.

For each $g \in l(V_0)$, there exists a unique $f \in \mathcal{F}$ such that $f|_{V_0} = g$ and the sequence $\{(5/3)^n Q_n(f|_{V_n})\}_{n=0}^{\infty}$ is a constant one. Such f is called harmonic, and the totality of harmonic functions will be denoted by $\mathcal{H}$. This is three-dimensional as a real vector space. We can take functions h_1 and h_2 from $\mathcal{H}$ such that

$$2\mathcal{E}(h_i, h_j) = \begin{cases} 1 & (i = j) \\ 0 & (i \neq j). \end{cases}$$

Define $\nu = (\nu_{h_1} + \nu_{h_2})/2$. This measure does not depend on the choice of h_1 and h_2 and is sometimes called the Kusuoka measure after Kusuoka [11].[1] For all $f \in \mathcal{F}$, ν_f is absolutely continuous with respect to ν. The measure ν is singular with respect to not only the Hausdorff measure on K [11] but also any self-similar measures on K [8]. For $w \in W_*$, define

$$c^{(w)} = \left(c_j^{(w)}\right)_{j \in S} = \left(\frac{\nu(K_{wj})}{\nu(K_w)}\right)_{j \in S} \in \mathbb{R}^3.$$

Clearly, $c^{(w)}$ lies in the plane $H = \{{}^t(x_1, x_2, x_3) \in \mathbb{R}^3 \mid x_1 + x_2 + x_3 = 1\}$. This vector describes the ratio of the distribution of $\nu|_{K_w}$ to one-step smaller similarities. We are interested in how $\{c^{(w)}\}_{w \in W_n}$ are distributed in H. Let

$$\mathbb{D} = \left\{ {}^t(x_1, x_2, x_3) \in H \;\middle|\; \sum_{j=1}^{3} \left(x_j - \frac{1}{3}\right)^2 < \frac{8}{75} \right\},$$

and let $\overline{\mathbb{D}}$ (respectively, $\partial\mathbb{D}$) be defined similarly as above by replacing $<$ by $\leq$ (respectively, $=$). Let (r, θ) be the polar coordinates of $\mathbb{D}$ with center ${}^t(1/3, 1/3, 1/3)$. More specifically, $(r, \theta) \in [0, \sqrt{8/75}) \times (-\pi, \pi]$ corresponds to

$$\begin{pmatrix} 1/3 \\ 1/3 \\ 1/3 \end{pmatrix} + \frac{r\cos\theta}{\sqrt{6}} \begin{pmatrix} -1 \\ 2 \\ -1 \end{pmatrix} + \frac{r\sin\theta}{\sqrt{2}} \begin{pmatrix} 1 \\ 0 \\ -1 \end{pmatrix} \in \mathbb{D}.$$

We regard r and θ as maps $\mathbb{D} \to [0, \sqrt{8/75})$ and $\mathbb{D} \to (-\pi, \pi]$, respectively. Here, we set $\theta(1/3, 1/3, 1/3) = 0$ by convention, which does not affect later discussions. Bell, Ho, and Strichartz [3] obtained the following result and posed conjectures.[2]

[1] Note that more general situations are considered in [11].

[2] In fact, $b^{(w)} := \frac{1}{3} + \frac{5}{4}(c^{(w)} - \frac{1}{3}) = \frac{5}{4}c^{(w)} - \frac{1}{12}$ is treated in [3, 7] in place of $c^{(w)}$ (for this relation, see also [3, Theorem 6.3]). Theorem 11.1, Conjecture 11.1, Theorems 11.2 and 11.4 below are translations of their descriptions in terms of $c^{(w)}$.

Theorem 11.1 ([3, Theorem 6.5], see also [7, Theorem 3.2]) *For all $w \in W_*$, $c^{(w)} \in \mathbb{D}$. Moreover, $c^{(w)}$ can be arbitrarily close to $\partial\mathbb{D}$.*

Conjecture 11.1 (see [3, Conjectures 7.1 and 7.2]) Let λ_m be the uniform probability distribution on W_m.

(i) The law of $r \circ c^{(w)}$ under λ_m converges to the Dirac measure at $\sqrt{8/75}$ as $m \to \infty$.
(ii) The law of $\theta \circ c^{(w)}$ under λ_m converges to an absolutely continuous measure on the interval $(-\pi, \pi]$.

Although Conjecture (ii) remains unsolved, Conjecture (i) has been solved affirmatively in a stronger sense as follows.

Theorem 11.2 (See [7, Theorem 3.5]) *Let κ be either the normalized Hausdorff measure λ on K or the Kusuoka measure ν on K. For $x \in K \setminus V_*$ and $m \in \nabla$, let $[x]_m$ denote the unique element in W_m such that $x \in K_{[x]_m}$. Then,*

$$\lim_{m\to\infty} \sum_{j=1}^{3} \left(c_j^{([x]_m)} - \frac{1}{3} \right)^2 = \frac{8}{75}, \qquad \kappa\text{-a.e. } x.^3$$

The result for $\kappa = \lambda$ implies Conjecture (i) because almost everywhere convergence implies convergence in law. For $\kappa = \lambda$, a key to the proof is the general theory of products of random matrices (Furstenberg's theorem). For $\kappa = \nu$, a key to the proof is the fact that the martingale dimension is 1, which was first proved by Kusuoka [11] for Sierpinski gaskets of arbitrary dimension; see also [5, 6] for more general fractals.

In the next section, we discuss an application of Theorem 11.2 for $\kappa = \nu$ to quantitative estimates of the local spectral dimension of the Sierpinski gasket with respect to the Kusuoka measure ν.

11.2 Quantitative estimates of local spectral dimension

The transition density $p_t(x, y)$ of Brownian motion on Sierpinski gasket K—which is associated with the Dirichlet form $(\mathcal{E}, \mathcal{F})$ on $L^2(K, \lambda)$ in our context—was extensively studied by Barlow and Perkins [2]. In particular, the following sub-Gaussian estimate is known:

$$c_1 t^{-d_s/2} \exp\left(-c_2 \left(\frac{|x-y|_{\mathbb{R}^2}^{d_w}}{t} \right)^{-1/(d_w - 1)} \right) \le p_t(x, y)$$

[3] Since $\kappa(V_*) = 0$, it is sufficient to define $[x]_m$ for only $x \in K \setminus V_*$.

$$\le c_3 t^{-d_s/2} \exp\left(-c_4\left(\frac{|x-y|_{\mathbb{R}^2}^{d_w}}{t}\right)^{-1/(d_w-1)}\right), \qquad x, y \in K,\ t \in (0, 1],$$

where c_j $(j = 1, 2, 3, 4)$ are positive constants, $d_s = 2\log_5 3 = 1.36521\cdots$ is the *spectral dimension*, and $d_w = \log_2 5 = 2.32192\cdots > 2$ is the *walk dimension*. On the other hand, the transition density of the singular time-changed Brownian motion with symmetrizing measure, say μ—which is associated with the Dirichlet form $(\mathcal{E}, \mathcal{F})$ on $L^2(K, \mu)$—was studied in several cases. The case when μ is a self-similar measure was studied in [1, 4], and in particular, the multifractal properties of the (local) spectral dimension and walk dimension were observed. The case when μ is equal to the Kusuoka measure ν was treated in [9, 10, 12]. We will focus on such a case here. The behavior of the transition density $q_t(x, y)$ is somewhat Gaussian-like. Concerning the short-time asymptotics of the on-diagonal $q_t(x, x)$, in particular, the following result is known.

Theorem 11.3 ([9, Theorem 1.3 (2) and Proposition 6.6]) *There exists a constant $d_s^{\mathrm{loc}} \in (1, 2\log_{25/3} 5]$ such that*

$$\lim_{t\downarrow 0} \frac{2\log q_t(x, x)}{-\log t} = d_s^{\mathrm{loc}}, \qquad \nu\text{-a.e. } x.$$

Moreover, d_s^{loc} is described as

$$d_s^{\mathrm{loc}} = 2 - \frac{2\log(5/3)}{\log(5/3) - \rho}, \tag{11.2}$$

where $\rho = \lim_{m\to\infty} \rho_m = \inf_{m\in\mathbb{N}} \rho_m$ with

$$\rho_m = \frac{1}{m} \sum_{w\in W_m} \nu(K_w) \log \nu(K_w). \tag{11.3}$$

We call d_s^{loc} the *local spectral dimension* of K with respect to the Kusuoka measure ν. From the numerical computation of ρ_m with $m = 16$, a quantitative estimate of d_s^{loc} is given in [9, Remark 6.7 (1)] as

$$\left(2 - \frac{2\log(5/3)}{\log(5/3) - \rho_{16}} =\right) 1.27874\cdots \le d_s^{\mathrm{loc}} \le 1.51814\cdots \left(= 2\log_{25/3} 5\right).$$

It seems difficult to obtain a substantially sharper estimate of d_s^{loc} by using only the above Eqs. (11.2) and (11.3). The main objective of this chapter is to discuss quantitative estimates of d_s^{loc} by another approach using Theorem 11.2 with $\kappa = \nu$. Theorem 11.5, which is stated later, provides an estimate of d_s^{loc}; by using this, we will give a rigorous proof of the estimate

$$(1.271650\cdots=)\ \frac{15\log 3+15\log 5-14\log 7}{15\log 5-7\log 7}\le d_{\mathrm{s}}^{\mathrm{loc}} \le \frac{5\log 5-3\log 3}{5\log 5-4\log 3}\ (=1.300763\cdots) \tag{11.4}$$

(see Theorem 11.6). We will also explain that the numerical calculation by *Mathematica* [13] suggests the estimate

$$1.291008\cdots\le d_{\mathrm{s}}^{\mathrm{loc}}\le 1.291026\cdots. \tag{11.5}$$

The first ingredient for the arguments is the following.

Theorem 11.4 (see [3, Theorem 6.2]) *The correspondence* $c^{(w)}\mapsto {}^t(c^{(w1)},c^{(w2)},c^{(w3)})$ *for* $w\in W_*$ *is given by* $c^{(w)}\mapsto\Psi(c^{(w)})$*, where* $\Psi={}^t(\Psi_1,\Psi_2,\Psi_3)\colon\mathbb{D}\to\mathbb{D}\times\mathbb{D}\times\mathbb{D}$ *is defined as*

$$\Psi_1\begin{pmatrix}x_1\\x_2\\x_3\end{pmatrix}=\frac{1}{15x_1}\begin{pmatrix}10x_1\\4x_1+3x_2\\4x_1+3x_3\end{pmatrix}-\frac{1}{25x_1}\begin{pmatrix}1\\2\\2\end{pmatrix},$$

$$\Psi_2=R^{-1}\circ\Psi_1\circ R,\quad \Psi_3=R\circ\Psi_1\circ R^{-1},$$

$$R\begin{pmatrix}x_1\\x_2\\x_3\end{pmatrix}=\begin{pmatrix}x_2\\x_3\\x_1\end{pmatrix}.$$

Each Ψ_j extends continuously to the map from $\overline{\mathbb{D}}$ to itself. We remark that the restriction map $\Psi_j|_{\partial\mathbb{D}}$ provides a homeomorphism from $\partial\mathbb{D}$ to itself for each $j\in S$.

We define a Markov chain $\{X_m\}_{m=0}^{\infty}$ on $\overline{\mathbb{D}}$ as follows. We set $X_0=\begin{pmatrix}1/3\\1/3\\1/3\end{pmatrix}$, and for $m\ge 0$,

$$\mathbb{P}\left(X_{m+1}=\Psi_j(X_m)\,\middle|\,X_m=\begin{pmatrix}x_1\\x_2\\x_3\end{pmatrix}\right)=x_j,\quad j\in S.$$

Proposition 11.1 *For all* $m\ge 0$*, the law* P^{X_m} *of* X_m *is equal to* $\sum_{w\in W_m}\nu(K_w)\delta_{c^{(w)}}$*, where* δ_z *denotes the Dirac measure at* z*. In other words,* P^{X_m} *coincides with the image measure of* ν *by the map* $x\mapsto c^{([x]_m)}$*, where* $[x]_m$ *is provided in Theorem 11.2.*

Proof The claim is true for $m=0$ by noting that $c^{(\emptyset)}={}^t(1/3,1/3,1/3)$ from the symmetry of the Kusuoka measure ν. Let us assume that the claim is true for $m=n$. Then, $P^{X_{n+1}}$ is equal to

$$\sum_{w\in W_n} \nu(K_w)\left(\sum_{j\in S} c_j^{(w)}\delta_{\Psi_j(c^{(w)})}\right) = \sum_{w\in W_n,\ j\in S} \nu(K_{wj})\delta_{c^{(wj)}}.$$

Therefore, the claim is true for $m = n + 1$. □

The Markov chain $\{X_m\}_{m=0}^{\infty}$ is Feller, that is, its transition operator $\mathcal{P}$ defined as

$$\mathcal{P}f(x) = \sum_{j=1}^{3} f(\Psi_j(x))x_j, \quad x = \begin{pmatrix} x_1 \\ x_2 \\ x_3 \end{pmatrix} \in \overline{\mathbb{D}}, \quad f \in C(\overline{\mathbb{D}})$$

satisfies that $\mathcal{P}(C(\overline{\mathbb{D}})) \subset C(\overline{\mathbb{D}})$.

We define a function g on $\overline{\mathbb{D}}$ by

$$g\begin{pmatrix} x_1 \\ x_2 \\ x_3 \end{pmatrix} = \sum_{j\in S} x_j \log x_j, \tag{11.6}$$

where $0 \log 0 := 0$. For $m \in \mathbb{N}$, let

$$\xi_m = \frac{1}{m}\sum_{k=0}^{m-1} P^{X_k}.$$

The following proposition describes the connection between $\{X_m\}_{m=0}^{\infty}$ and ρ_m, which was introduced in (11.3).

Proposition 11.2 *For each $m \in \mathbb{N}$,*

$$\rho_m = \int_{\overline{\mathbb{D}}} g(x)\, \xi_m(dx). \tag{11.7}$$

Proof From Proposition 11.1, for $k \geq 0$,

$$\begin{aligned}
\mathbb{E}[g(X_k)] &= \sum_{w\in W_k} \nu(K_w) g(c^{(w)}) \\
&= \sum_{w\in W_k} \nu(K_w) \sum_{j\in S} \frac{\nu(K_{wj})}{\nu(K_w)} \log \frac{\nu(K_{wj})}{\nu(K_w)} \\
&= \sum_{w\in W_k} \sum_{j\in S} \nu(K_{wj}) \log \frac{\nu(K_{wj})}{\nu(K_w)} \\
&= \sum_{w'\in W_{k+1}} \nu(K_{w'}) \log \nu(K_{w'}) - \sum_{w\in W_k} \nu(K_w) \log \nu(K_w).
\end{aligned}$$

Therefore,

$$
\begin{aligned}
\int_{\overline{\mathbb{D}}} g(x)\,\xi_m(dx) &= \frac{1}{m}\sum_{k=0}^{m-1}\mathbb{E}[g(X_k)] \\
&= \frac{1}{m}\left(\sum_{w\in W_m}\nu(K_w)\log\nu(K_w) - \nu(K_\emptyset)\log\nu(K_\emptyset)\right) \\
&= \rho_m,
\end{aligned}
$$

since $\nu(K_\emptyset) = \nu(K) = 1$. □

Since $\overline{\mathbb{D}}$ is compact, there exists a subsequence $\{\xi_{m_l}\}$ of $\{\xi_m\}$ converging weakly to a probability measure ξ. By letting $m \to \infty$ along $\{m_l\}$ in (11.7),

$$
\rho = \lim_{l\to\infty}\rho_{m_l} = \int_{\overline{\mathbb{D}}} g(x)\,\xi(dx).
$$

It is a standard fact that ξ is an invariant measure. Indeed, for any $f \in C(\overline{\mathbb{D}})$, by letting $l \to \infty$ in the equation

$$
\begin{aligned}
&\left|\int_{\overline{\mathbb{D}}} \mathcal{P}f(x)\,\xi_{m_l}(dx) - \int_{\overline{\mathbb{D}}} f(x)\,\xi_{m_l}(dx)\right| \\
&= \left|\frac{1}{m_l}\sum_{k=0}^{m_l-1}\mathbb{E}[f(X_{k+1})] - \frac{1}{m_l}\sum_{k=0}^{m_l-1}\mathbb{E}[f(X_k)]\right| \\
&= \left|\frac{1}{m_l}(\mathbb{E}[f(X_{m_l})] - \mathbb{E}[f(X_0)])\right| \\
&\le \frac{2}{m_l}\sup_{x\in\overline{\mathbb{D}}}|f(x)|,
\end{aligned}
$$

we have

$$
\int_{\overline{\mathbb{D}}} \mathcal{P}f(x)\,\xi(dx) - \int_{\overline{\mathbb{D}}} f(x)\,\xi(dx) = 0.
$$

Therefore, for all $n \in \mathbb{Z}_{\ge 0}$,

$$
\rho = \int_{\overline{\mathbb{D}}} \mathcal{P}^n g(x)\,\xi(dx). \tag{11.8}
$$

Since $P^{X_m} \circ r^{-1}$ converges to the Dirac measure at $\sqrt{8/75}$ as $m \to \infty$ from Proposition 11.1 and Theorem 11.2 with $\kappa = \nu$, $\xi \circ r^{-1}$ is the Dirac measure at

$\sqrt{8/75}$. That is, ξ concentrates on $\partial\mathbb{D}$. We can then rewrite (11.8) as

$$\rho = \int_{\partial\mathbb{D}} \mathcal{P}^n g(x)\,\xi(dx). \tag{11.9}$$

Thus, we obtain the following estimate.

Theorem 11.5 *For all $n \in \mathbb{Z}_{\geq 0}$, it holds that*

$$\min_{x\in\partial\mathbb{D}} \mathcal{P}^n g(x) \leq \rho \leq \max_{x\in\partial\mathbb{D}} \mathcal{P}^n g(x) \tag{11.10}$$

and

$$2 - \frac{2\log(5/3)}{\log(5/3) - \max_{x\in\partial\mathbb{D}} \mathcal{P}^n g(x)} \leq d_{\mathrm{s}}^{\mathrm{loc}} \leq 2 - \frac{2\log(5/3)}{\log(5/3) - \min_{x\in\partial\mathbb{D}} \mathcal{P}^n g(x)}. \tag{11.11}$$

Proof Eq. (11.10) follows from (11.9). Eq. (11.11) follows from (11.10) and (11.2). □

Remark 11.1 Since $\mathcal{P}$ is positivity-preserving on $C(\partial\mathbb{D})$ and $\mathcal{P}1 = 1$, inequality (11.10) provides a finer estimate as n increases. It is expected that $\min_{x\in\partial\mathbb{D}} \mathcal{P}^n g(x)$ and $\max_{x\in\partial\mathbb{D}} \mathcal{P}^n g(x)$ have the same limit as $n \to \infty$, but this remains to be proved.

The functions $\mathcal{P}^n g$ are explicitly described in theory. Fig. 11.1 shows graphs of $\mathcal{P}^n g$ on $\partial\mathbb{D}$ for $0 \leq n \leq 5$, where $\partial\mathbb{D}$ is identified with the interval $(-\pi, \pi]$ via the map

$$\phi\colon (-\pi, \pi] \ni \theta \mapsto \begin{pmatrix} 1/3 \\ 1/3 \\ 1/3 \end{pmatrix} + \frac{2\cos\theta}{15}\begin{pmatrix} -1 \\ 2 \\ -1 \end{pmatrix} + \frac{2\sqrt{3}\sin\theta}{15}\begin{pmatrix} 1 \\ 0 \\ -1 \end{pmatrix} \in \partial\mathbb{D}. \tag{11.12}$$

Table 11.1 gives the results of some numerical calculations by *Mathematica*.[4] According to these computations, Eq. (11.5) holds numerically; in particular, the first few digits of $d_{\mathrm{s}}^{\mathrm{loc}}$ are $1.2910\cdots$, a value that happens to be close to $\sqrt{5/3} = 1.290994\cdots$.

For reference, we provide a rigorous proof for the estimate of $\mathcal{P}^0 g\ (= g)$, which implies Eq. (11.4). Even such an estimate ensures that $d_{\mathrm{s}}^{\mathrm{loc}}$ is less than $d_{\mathrm{s}} = 1.36521\cdots$ (see Corollary 11.1 below), which was previously unconfirmed.

Theorem 11.6 *It holds that*

[4] We used the command `NMaxValue` to obtain the maximum and minimum of $\mathcal{P}^n g$.

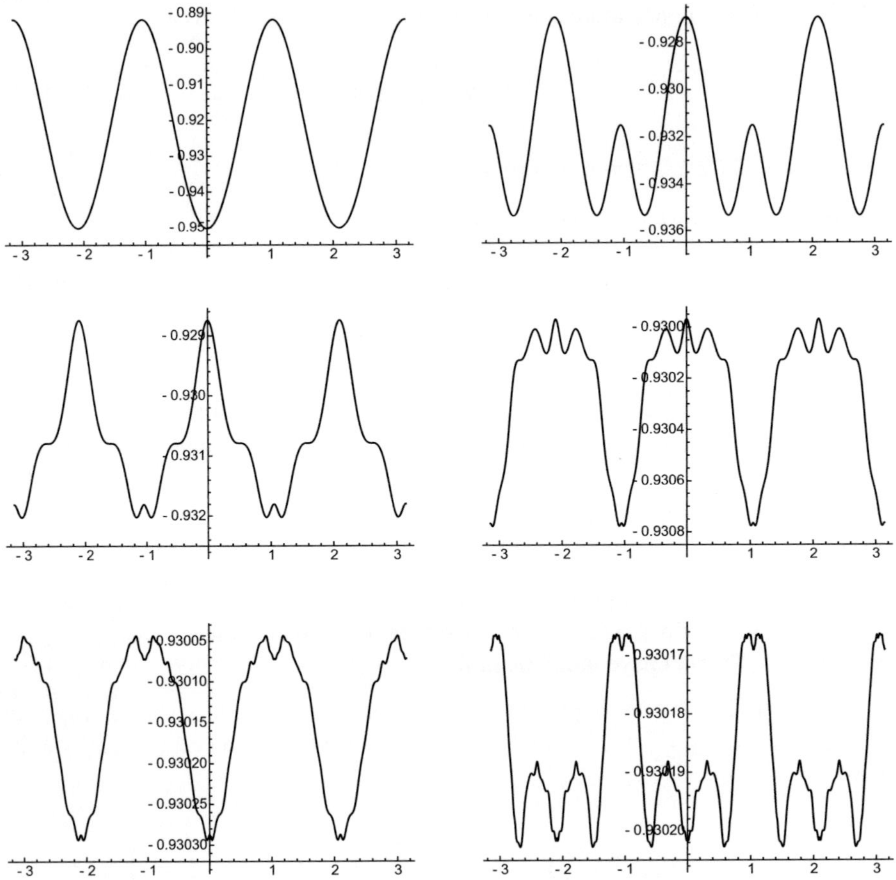

Fig. 11.1 Graphs of $\mathcal{P}^n g$, where the horizontal axis represents the argument $\theta \in (-\pi, \pi]$

Table 11.1 Upper and lower estimates of ρ and d_s^{loc} based on Theorem 11.5

n	Estimates of ρ	Estimates of d_s^{loc}
0	$-0.9502705\cdots \le \rho \le -0.8918673\cdots$	$1.271650\cdots \le d_s^{\mathrm{loc}} \le 1.300763\cdots$
1	$-0.9353387\cdots \le \rho \le -0.9269092\cdots$	$1.289402\cdots \le d_s^{\mathrm{loc}} \le 1.293544\cdots$
2	$-0.9320224\cdots \le \rho \le -0.9287450\cdots$	$1.290308\cdots \le d_s^{\mathrm{loc}} \le 1.291920\cdots$
3	$-0.9307764\cdots \le \rho \le -0.9299684\cdots$	$1.290911\cdots \le d_s^{\mathrm{loc}} \le 1.291308\cdots$
4	$-0.9302937\cdots \le \rho \le -0.9300433\cdots$	$1.290947\cdots \le d_s^{\mathrm{loc}} \le 1.291071\cdots$
5	$-0.9302027\cdots \le \rho \le -0.9301663\cdots$	$1.291008\cdots \le d_s^{\mathrm{loc}} \le 1.291026\cdots$

$$\min_{x \in \partial \mathbb{D}} g(x) = g(\phi(0)) = \frac{3}{5} \log 3 - \log 5 \tag{11.13}$$

and

$$\max_{x \in \partial \mathbb{D}} g(x) = g\left(\phi\left(\frac{\pi}{3}\right)\right) = \frac{14}{15}\log 7 - \log 15. \tag{11.14}$$

Consequently, we have

$$2 - \frac{2\log(5/3)}{\log(5/3) - g(\phi(\pi/3))} \le d_{\mathrm{s}}^{\mathrm{loc}} \le 2 - \frac{2\log(5/3)}{\log(5/3) - g(\phi(0))},$$

that is, Eq. (11.4) *holds.*

Proof First, we note from (11.6) and (11.12) that

$$\begin{aligned} g(\phi(\theta)) = &\left(\frac{1}{3} - \frac{2}{15}\cos\theta + \frac{2\sqrt{3}}{15}\sin\theta\right)\log\left(\frac{1}{3} - \frac{2}{15}\cos\theta + \frac{2\sqrt{3}}{15}\sin\theta\right) \\ &+ \left(\frac{1}{3} + \frac{4}{15}\cos\theta\right)\log\left(\frac{1}{3} + \frac{4}{15}\cos\theta\right) \\ &+ \left(\frac{1}{3} - \frac{2}{15}\cos\theta - \frac{2\sqrt{3}}{15}\sin\theta\right)\log\left(\frac{1}{3} - \frac{2}{15}\cos\theta - \frac{2\sqrt{3}}{15}\sin\theta\right). \end{aligned}$$

Because we can easily check the periodicity and symmetry of $g(\phi(\theta))$:

$$g(\phi(\theta)) = g\left(\phi\left(\frac{2\pi}{3} + \theta\right)\right) = g\left(\phi\left(\frac{2\pi}{3} - \theta\right)\right),$$

it suffices to prove that $\frac{d}{d\theta}(g(\phi(\theta))) \ge 0$ for $\theta \in [0, \pi/3]$ for the validity of (11.13) and (11.14). From direct computation, we have

$$\begin{aligned} \frac{d}{d\theta}(g(\phi(\theta))) = &\frac{1}{\sqrt{3}}(-x+y)\log\left(\frac{1}{3} - \frac{x}{3} - y\right) - \frac{2y}{\sqrt{3}}\log\left(\frac{1}{3} + \frac{2x}{3}\right) \\ &+ \frac{1}{\sqrt{3}}(x+y)\log\left(\frac{1}{3} - \frac{x}{3} + y\right), \end{aligned}$$

where

$$x = \frac{2}{5}\cos\theta \quad \text{and} \quad y = \frac{2\sqrt{3}}{15}\sin\theta.$$

Note that $0 \le y \le 1/5 \le x \le 2/5$ for $\theta \in [0, \pi/3]$. By letting

$$\alpha = \frac{3y}{1-x} \quad \text{and} \quad \beta = \frac{3(x-y)}{1-x+3y},$$

it holds that

$$\frac{d}{d\theta}(g(\phi(\theta))) = \frac{1}{\sqrt{3}}(x-y)\log\frac{1+\alpha}{1-\alpha} - \frac{2y}{\sqrt{3}}\log(1+\beta).$$

We now use the general inequalities

$$\log\frac{1+\alpha}{1-\alpha} \geq 2\alpha \quad \text{and} \quad \log(1+\beta) \leq \beta$$

for $\alpha \in [0, 1)$ and $\beta \geq 0$ to obtain that

$$\begin{aligned}\frac{d}{d\theta}(g(\phi(\theta))) &\geq \frac{2}{\sqrt{3}}(x-y)\alpha - \frac{2y}{\sqrt{3}}\beta \\ &= 2\sqrt{3}(x-y)y\left(\frac{1}{1-x} - \frac{1}{1-x+3y}\right) \\ &\geq 0.\end{aligned}$$

Note that the last inequality becomes equality only if $y = 0$ or $x = y$, that is, when $\theta = 0$ or $\pi/3$. We can confirm that $\frac{d}{d\theta}(g \circ \phi)(0) = \frac{d}{d\theta}(g \circ \phi)(\pi/3) = 0$, and the remaining claims follow from Theorem 11.5. □

Corollary 11.1 $d_s^{\mathrm{loc}} < d_s$.

Proof In view of (11.4), it suffices to prove

$$\frac{5\log 5 - 3\log 3}{5\log 5 - 4\log 3} < 2\log_5 3.$$

By letting $a = \log_5 3 < 1$, this inequality is equivalent to $(5-3a)/(5-4a) < 2a$, that is, $8a > 5$. This is equivalent to $3^8 > 5^5$, which is true because $3^8 = 6561$ and $5^5 = 3125$. □

Acknowledgments This work was supported by JSPS KAKENHI Grant Numbers 19H00643 and 19K21833.

References

1. M. T. Barlow and T. Kumagai, Transition density asymptotics for some diffusion processes with multi-fractal structures, *Electron. J. Probab.* **6** (2001), no. 9, 23 pp.
2. M. T. Barlow and E. A. Perkins, Brownian motion on the Sierpiński gasket, *Probab. Theory Related Fields* **79** (1988), 543–623.
3. R. Bell, C.-W. Ho, and R. S. Strichartz, Energy measures of harmonic functions on the Sierpiński gasket, *Indiana Univ. Math. J.* **63** (2014), 831–868.
4. B. M. Hambly, J. Kigami, and T. Kumagai, Multifractal formalisms for the local spectral and walk dimensions, *Math. Proc. Cambridge Philos. Soc.* **132** (2002), 555–571.

5. M. Hino, Martingale dimensions for fractals, *Ann. Probab.* **36** (2008), 971–991.
6. M. Hino, Upper estimate of martingale dimension for self-similar fractals, *Probab. Theory Related Fields* **156** (2013), 739–793.
7. M. Hino, Some properties of energy measures on Sierpinski gasket type fractals, *J. Fractal Geom.* **3** (2016), 245–263.
8. M. Hino and K. Nakahara, On singularity of energy measures on self-similar sets II, *Bull. Lond. Math. Soc.* **38** (2006), 1019–1032.
9. N. Kajino, Heat kernel asymptotics for the measurable Riemannian structure on the Sierpinski gasket, *Potential Anal.* **36** (2012), 67–115.
10. J. Kigami, Measurable Riemannian geometry on the Sierpinski gasket: the Kusuoka measure and the Gaussian heat kernel estimate, *Math. Ann.* **340** (2008), 781–804.
11. S. Kusuoka, Dirichlet forms on fractals and products of random matrices, *Publ. Res. Inst. Math. Sci.* **25** (1989), 659–680.
12. V. Metz and K.-T. Sturm, Gaussian and non-Gaussian estimates for heat kernels on the Sierpiński gasket, *Dirichlet forms and stochastic processes (Beijing, 1993)*, 283–289, de Gruyter, Berlin, 1995.
13. Wolfram Research, Inc., Mathematica, Ver. 13.0, Champaign, IL (2021).

Chapter 12
Heat Kernel Fluctuations for Stochastic Processes on Fractals and Random Media

Sebastian Andres, David Croydon, and Takashi Kumagai

In memory of Professor Robert Strichartz

12.1 Introduction

The past 35 years have witnessed an extensive study of heat kernels for stochastic processes on fractals and associated graphs. (See, for instance, [3, 34] and also [31, 38] for textbooks related to analysis on fractals.) As a rough illustration of the kind of result that has been proved in this area, one has that if K is a suitably nice non-compact fractal, such as an infinite version of the Sierpinski gasket or Sierpinski carpet, then the corresponding 'Brownian motion' $X = \{X_t\}_{t\geq 0}$ on K exhibits sub-Gaussian heat kernel behaviour. More precisely, writing $(p_t(x, y))_{x,y\in K,\, t>0}$ for the transition density of X (typically with respect to a natural choice of Hausdorff measure on K), it holds that

$$c_1 t^{-\frac{d_f}{d_w}} \exp\left(-c_2\left(\frac{d(x, y)^{d_w}}{t}\right)^{\frac{1}{d_w-1}}\right) \leq p_t(x, y)$$

$$\leq c_3 t^{-\frac{d_f}{d_w}} \exp\left(-c_4\left(\frac{d(x, y)^{d_w}}{t}\right)^{\frac{1}{d_w-1}}\right), \qquad (12.1)$$

S. Andres
Department of Mathematics, The University of Manchester, Manchester, UK
e-mail: sebastian.andres@manchester.ac.uk

D. Croydon
Research Institute for Mathematical Sciences, Kyoto University, Kyoto, Japan
e-mail: croydon@kurims.kyoto-u.ac.jp

T. Kumagai (✉)
Department of Mathematics, Waseda University, Tokyo, Japan
e-mail: t-kumagai@waseda.jp

P. Alonso Ruiz et al. (eds.), *From Classical Analysis to Analysis on Fractals*, Applied and Numerical Harmonic Analysis, https://doi.org/10.1007/978-3-031-37800-3_12

for all $x, y \in K, t > 0$, where $d(\cdot, \cdot)$ is a geodesic distance on K, $d_f > 0$ is the Hausdorff dimension of K (with respect to d), $d_w \geq 2$ is a constant called the walk dimension of the process, and c_i, $i = 1, \ldots, 4$, are constants. (Similar estimates are known to hold for simple random walks on graphical versions of such fractals, with the time parameter being restricted to those $t \in 2\mathbb{N}$ satisfying $d(x, y) \leq t$.) Crucially, for many fractals, it is known that $d_w > 2$, which means the process in question admits anomalous, sub-diffusive space–time scaling. In particular, by integrating (12.1), one obtains that

$$c_5 t^{1/d_w} \leq E_x[d(x, X_t)] \leq c_6 t^{1/d_w},$$

where E_x represents the expectation under the law of X started from x, which, when $d_w > 2$, is a clear departure from the usual spatial scale of $t^{1/2}$ seen for Brownian motion in Euclidean spaces. (Note that when $d_w = 2$, the inequalities at (12.1) represent classical Gaussian estimates.)

For diffusions on domains within Euclidean spaces or on some nice manifolds for which Gaussian estimates hold, more precise asymptotic behaviour of the heat kernel and other objects associated with the spectrum of the diffusions' generators are known, and these are often related to geometric properties of the space. For instance, consider Brownian motion on a bounded domain $\Omega \subset \mathbb{R}^d$, killed upon hitting $\partial\Omega$, and write Δ for the corresponding Dirichlet Laplacian. Then, the well-known result of Weyl gives that the eigenvalue counting function $\rho(x) := \sharp\{\lambda \leq x : \lambda \text{ is an eigenvalue of } -\Delta\}$, the eigenvalue counting function, satisfies

$$\lim_{x \to \infty} \frac{\rho(x)}{x^{d/2}} = c_d \, m(\Omega), \tag{12.2}$$

where m is the Lebesgue measure on $\mathbb{R}^d$, and $c_d > 0$ is a constant depending only upon the dimension of the space. It turns out that on fractals the situation can be quite different, with fluctuations being seen to occur in various places.

It has recently been revealed that fluctuations also occur for stochastic processes on random media, in particular for processes in low dimensions (including certain fractals) and models at criticality. Indeed, in low dimensions, local irregularities in the random medium, in combination with the basic geometry of the space, may affect the process in a non-negligible way that leads to anomalous heat kernel behaviour. In this chapter, we will summarize some results in this area. Specifically, following a discussion of the heat kernel behaviour for two classes of random fractals, we will review heat kernel fluctuations on random trees, namely for random walks on the incipient infinite cluster of a particular critical Galton–Watson tree (i.e. the critical percolation cluster on a regular tree conditioned to be infinite) and for random walks on low-dimensional uniform spanning trees. For brevity, we will mainly discuss fluctuations of the on-diagonal parts of the heat kernels on the relevant media. We will also announce some new results concerning fluctuations of the heat kernels in the one-dimensional Bouchaud trap model.

The remainder of the chapter is organised as follows. In Sect. 12.2, we review results concerning fluctuations for Brownian motion on deterministic fractals. Sections 12.3 and 12.4 summarize fluctuations for processes on random fractals and random media, respectively. In Sect. 12.5, we announce our recent results concerning the fluctuations of the on-diagonal heat kernel for the one-dimensional Bouchaud trap model.

12.2 Fluctuations for Brownian Motion on Fractals

12.2.1 Fluctuations of the On-Diagonal Heat Kernel on Fractals

Unlike heat kernels on manifolds and regular graphs, heat kernels on fractals exhibit fluctuations. Indeed, in [30], N. Kajino shows on-diagonal short-time fluctuations for Brownian motion on various fractals. We next give a precise statement in the case of the Sierpinski gasket, as shown in Fig. 12.1. Note that, in the statement of the result, we write $d_s/2$ in place of the exponent d_f/d_w that appeared in (12.1); the constant d_s governing the short-time behaviour of the on-diagonal part of the heat kernel is often called the spectral dimension since it also determines the growth of the eigenvalue counting function as in (12.2) (cf. (12.3) below), while in the case of Euclidean domains it matches the actual dimension of the space. For the Sierpinski gasket itself, we have $d_f = \log_2 3$, $d_w = \log_2 5$ and $d_s = 2\log_5 3$.

Theorem 12.1 ([30, Theorem 1.1]) *Let $(p_t(x, y))_{x,y \in K,\, t>0}$ be the heat kernel of Brownian motion on the two-dimensional Sierpinski gasket K. It is then the case that the limit $\lim_{t\to 0} t^{d_s/2} p_t(x, x)$ does not exist for any $x \in K$.*

Fig. 12.1 The (two-dimensional) Sierpinski gasket

Remark 12.1

(i) In [30], Kajino obtained the same result for Brownian motion on the d-dimensional standard Sierpinski gasket for any $d \geq 2$ and on the N-polygasket with odd $N \geq 3$ (the latter class including the pentagasket, for example). He also obtained a slightly weaker statement (namely, the non-existence of the limit for ν-a.e. x, for any self-similar measure ν on the fractal) for Brownian motion on affine nested fractals with at least three boundary points (in the sense of cell boundaries appearing in the construction of the fractal, rather than the topological boundary). See [30, Theorem 1.2].

(ii) In [29], Kajino continued to study the finer behaviour of $t^{d_s/2}p_t(x,x)$ as $t \to 0$. For a certain class of self-similar fractals K, he proves that $\lim_{t\to 0} t^{d_s/2}p_t(x,x)$ does not exist for generic $x \in K$ if there exists $y, z \in K$ which are not contained in the closure of the boundary of K such that $\limsup_{t\to 0} p_t(y,y)/p_t(z,z) > 1$. (Again, boundary should be interpreted in a suitable sense.)

(iii) For random walk on the Sierpinski gasket graph, [24] established on-diagonal fluctuations for $t \to \infty$. (See also [39].)

12.2.2 Other Fluctuations

For Laplace operators and diffusions on fractals, other fluctuations are known, as we will briefly outline next.

12.2.2.1 Spectral Properties

Let Δ be the Laplace operator on the two-dimensional (compact) Sierpinski gasket K. It is known that $-\Delta$ has a compact resolvent and thus a discrete spectrum. Moreover, defining ρ to be the associated eigenvalue counting function as we did above (12.2),

$$0 < \liminf_{x\to\infty} \frac{\rho(x)}{x^{d_s/2}} < \limsup_{x\to\infty} \frac{\rho(x)}{x^{d_s/2}} < \infty, \qquad (12.3)$$

where $d_s = 2\log_5 3$ is the spectral dimension mentioned above [23]. Notably, unlike (12.2), the limit superior and the limit inferior do not coincide in (12.3). (The periodic nature of the fluctuations is described in [32].)

Later, in [7], (12.3) was extended to the Laplace operators on self-similar fractal spaces within quite a general class, including nested fractals. Moreover, it was shown in [7] that the non-existence of the limit is caused by the presence of 'many' localized eigenfunctions that produce eigenvalues with high multiplicities. We note that an eigenfunction u of $-\Delta$ is called a localized eigenfunction if u takes the value

0 outside some open subset $O \subset K$ that does not intersect an appropriately defined fractal boundary. By contrast, as we already noted for Euclidean domains at (12.2), the limit at (12.3) exists with $d_s = d$; no localized eigenfunctions exist in this case.

12.2.2.2 Off-diagonal Heat Kernels and Large Deviations

For Brownian motion $X = \{X_t\}_{t \geq 0}$ on the two-dimensional Sierpinski gasket K, fluctuations occur in the Varadhan-type short-time off-diagonal heat kernel asymptotics and in the Schilder-type large deviations principle. Precisely, the following results are known to hold.

Theorem 12.2 *Let $t_0 = 2^{1-d_w} = 2/5$, and set $\varepsilon_{n,u} = t_0^n u$ for any $u \in [t_0, 1)$.*

(i) ([33]) *It holds that*

$$-\lim_{n\to\infty} \varepsilon_{n,u}^{1/(d_w-1)} \log p_{\varepsilon_{n,u}}(x, y) = d(x, y)^{\frac{d_w}{d_w-1}} F\left(\frac{u}{d(x, y)}\right), \quad \forall x, y \in K, \tag{12.4}$$

where F is a continuous positive non-constant periodic function with period t_0^{-1}.

(ii) ([11]) *Let P_x^ε be the law of $B_{\varepsilon\cdot}$ starting at x. For each $u \in [t_0, 1)$ and Borel measurable $A \subseteq \Omega_x$, where we suppose that $\Omega_x := \{f \in C([0, T], K) : f(0) = x\}$ is endowed with the distance induced by the supremum norm, it holds that*

$$\begin{aligned} -\inf_{\phi \in Int(A)} I_x^u(\phi) &\leq \liminf_{n\to\infty} \varepsilon_{n,u}^{1/(d_w-1)} \log P_x^{\varepsilon_{n,u}}(A) \\ &\leq \limsup_{n\to\infty} \varepsilon_{n,u}^{1/(d_w-1)} \log P_x^{\varepsilon_{n,u}}(A) \leq -\inf_{\phi \in Cl(A)} I_x^u(\phi), \end{aligned} \tag{12.5}$$

where $Int(A)$ and $Cl(A)$ are the interior and closure of A, respectively, and $\{I_x^u\}_{u\in[t_0,1)}$ is a collection of rate functions, defined as follows for each $\phi \in \Omega_x$:

$$I_x^u(\phi) := \begin{cases} \int_0^T \left(\dot{\phi}(t)\right)^{d_w/(d_w-1)} F\left(\frac{u}{\dot{\phi}(t)}\right) dt, & \text{if } \phi \text{ is absolutely continuous}, \\ \infty, & \text{otherwise}, \end{cases}$$

with F being the same periodic function as in (i) and $\dot{\phi}(t) := \lim_{s\to t} \frac{d(\phi(s),\phi(t))}{|s-t|}$ for $t \in [0, T]$.

For comparison, we note that in the case of Brownian motion on $\mathbb{R}^n$, (12.4) and (12.5) hold independently of u, i.e. with F being a constant function and with $d_w = 2$. (In fact, there is no need to restrict to a subsequence when considering the

Euclidean case.) Hence, I_u^x is independent of u and, indeed, one may compute that $I_x(\phi) := \frac{1}{2}\int_0^T \dot{\phi}(t)^2 dt$.

In [11], it was further shown that when $A = \{f \in \Omega_x : f(T) = y\}$, then $\inf\{I_x^u(\phi) : \phi \in A\}$ is attained, independently of u, by any path which moves along a geodesic between x and y homogeneously. Thus, regardless of the choice of u, geodesics are always 'the most probable paths', but the energies (action functionals) of the paths in question depend on time sequences determined by u.

12.3 Fluctuations of Quenched On-Diagonal Heat Kernels on Random Fractals

Two classes of random Sierpinski gaskets, along with properties of Brownian motions on them, have been considered by B. M. Hambly, see [25, 26]. Both families are based on choosing randomly from the maps generating the generalised Sierpinski gaskets SG(ν), $\nu \geq 2$, which are constructed by replacing at each stage a single equilateral triangle with $\nu(\nu + 1)/2$ appropriately arranged smaller ones. See Fig. 12.2 for the first stage of the iterative construction of SG(2) and SG(3); note that SG(2) is the standard Sierpinski gasket.

In the model of [25], one starts with an i.i.d. sequence of integers $\{\nu_i\}_{i\in\mathbb{N}}$, $\nu_i \geq 2$, and then, beginning with a single equilateral triangle, at the i-th stage in the construction, divides each remaining equilateral triangle into $\nu_i(\nu_i + 1)/2$ sub-triangles, as per the arrangement in the construction of SG(ν_i). Via this procedure, one can define a limiting random Sierpinski gasket, which is called the homogeneous random gasket (or the random scale-irregular gasket); see Fig. 12.3 for an illustration and [25] for details. For the Brownian motion on this set, in [25] and [6, Section 6], there are derivations of transition density estimates, which show a sub-Gaussian form as at (12.1) up to some small order error terms, with the leading order exponents being deterministic. See [6, Theorem 6.1 and Corollary 6.3]

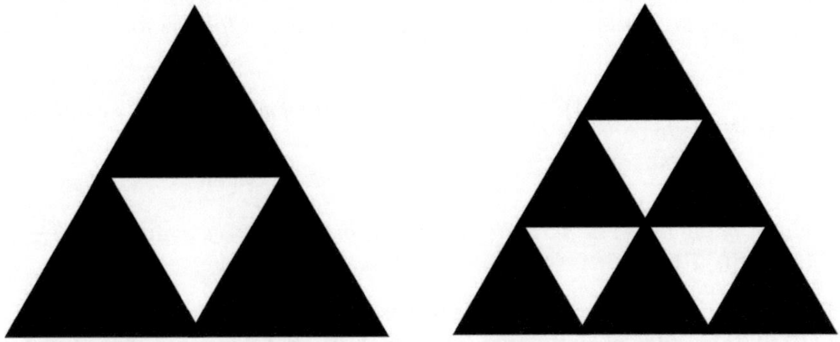

Fig. 12.2 The first stage in the iterative construction of SG(2) and SG(3)

Fig. 12.3 Example realisations of a random scale-irregular gasket (left, shown to the fourth level of construction) and a random recursive gasket (right, shown to the third level of construction)

for details and [6, Theorem 6.2] for sufficient conditions on the random sequence $\{\nu_i\}_{i\in\mathbb{N}}$ (not necessarily restricting to an i.i.d. one) for on-diagonal heat kernel fluctuations to occur. Moreover, concerning the associated eigenvalue counting function, [6, Corollary 7.2] establishes that this almost surely satisfies

$$0 < \limsup_{x\to\infty} \frac{\rho(x)}{x^{d_s/2} e^{c_1\phi(x)}} < \liminf_{x\to\infty} \frac{\rho(x) e^{c_2\phi(x)}}{x^{d_s/2}} < \infty,$$

where d_s is the appropriately defined spectral dimension, as appears in the on-diagonal part of the heat kernel, $\phi(x) := (\log x \log\log\log x)^{1/2}$, and c_1 and c_2 are constants. We highlight that d_s is deterministic, meaning that, on a set of probability one, it is independent of $\{\nu_i\}_{i\in\mathbb{N}}$. In keeping with the content of the present work on heat kernel fluctuations, we further note that the results of [6, 25] are detailed enough to yield that, by replacing the random sequence of integers $\{\nu_i\}_{i\in\mathbb{N}}$ with a suitably chosen deterministic sequence, one can exhibit examples of scale-irregular gaskets for which the leading polynomial order of the on-diagonal part of the heat kernel is not captured by a single exponent, i.e. the spectral dimension does not exist, see [6] for details.

Other random gaskets that are perhaps more closely related to examples of random fractals appearing organically as scaling limits of random graphs (such as those appearing in the next section) are the random recursive gaskets of [26]. In order to understand the construction, it is helpful to first consider the Sierpinski gasket SG(ν) as (the boundary at infinity of) a $\nu(\nu+1)/2$-ary tree, with vertices in the n-th generation of the tree corresponding to the equilateral triangles appearing in the n-th level of gasket construction. Generalising this picture, the random recursive gasket is constructed using a random tree corresponding to a simple Galton–Watson tree, whose offspring distribution is supported on integers greater than 2. Vertices in the n-th generation of the tree correspond to n-th level subsets of the gasket. See Fig. 12.3 for an illustration of the initial part of the construction of a random recursive gasket. Notice that, in contrast to the random scale-irregular gasket, the sizes of the different parts of the fractal in the n-th stage of the construction can vary.

In [28], it was shown that there are fluctuations in the on-diagonal part of the heat kernel and no fluctuations in the leading order of the eigenvalue counting function for random recursive gaskets. For simplicity in presenting these results, consider the case where the random recursive gasket is constructed from only SG(2)- and SG(3)-type replacements at each stage, i.e. the offspring distribution of the associated Galton–Watson branching process is supported on $\{2, 3\}$, and let $(\Omega, \mathcal{F}, \mathbf{P})$ be the probability space that governs the randomness of the fractal. Furthermore, write K_ω for a realisation of the random fractal and μ_ω for a suitably defined statistically self-similar measure on K_ω, where the suffix ω refers to the underlying element of Ω. In [28, Theorem 1.1], it is shown that there exist $\alpha > 0$, $\beta > \beta' > 0$, $\beta'' > 0$ and $c_1, c_2, c_3, c_4 \in (0, \infty)$ such that the following holds $\mathbf{P}$-a.s.: for μ_ω-a.e. $x \in K_\omega$,

$$c_1 \le \limsup_{t \to 0} \frac{t^{\alpha/(\alpha+1)} p_t^\omega(x, x)}{(\log|\log t|)^{\beta/(\alpha+1)}} \le c_2,$$

$$c_3 \le \liminf_{t \to 0} t^{\alpha/(\alpha+1)} (|\log t|)^{\beta''/(\alpha+1)} p_t^\omega(x, x), \quad \liminf_{t \to 0} \frac{t^{\alpha/(\alpha+1)} p_t^\omega(x, x)}{(\log|\log t|)^{\beta'/(\alpha+1)}} \le c_4.$$

(The lower bound for the lim inf here was obtained in [26] and is not expected to be optimal, rather one might conjecture a log log term would be sufficient.) Notably, the upper bound for the lim inf and the lower bound for the lim sup establish the on-diagonal heat kernel exhibits log-logarithmic fluctuations, which is a departure from the result for the standard Sierpinski gasket, where bounds of the form (12.1) are seen and so the fluctuations are only of constant order. We remark that [28] also considers global fluctuations, i.e. the behaviour of $\inf_{x \in K_\omega} p_t^\omega(x, x)$ and $\sup_{x \in K_\omega} p_t^\omega(x, x)$, and here logarithmic corrections are seen. Next, we discuss the eigenvalue counting function of the associated Laplacian (Dirichlet or Neumann). In [27, Theorem 1.1], it is shown that there exist a deterministic constant $c > 0$ and a strictly positive random variable W with mean one such that

$$\lim_{x \to \infty} \frac{\rho(x)}{x^{\alpha/(\alpha+1)}} = c W^{1/(\alpha+1)}, \qquad \mathbf{P}\text{-a.s.}$$

(In [27], the results are stated under a more general setting with a certain non-lattice assumption.) In this case, the randomness smooths the object of interest. Indeed, as we described in the preceding section, the eigenvalue counting function of the standard Sierpinski gasket displays periodic fluctuations.

12.4 Fluctuations of Quenched On-Diagonal Heat Kernels in Random Media 1: Random Trees

The behaviour of the heat kernel on random recursive gaskets is indicative of the on-diagonal heat kernel fluctuations that occur for other kinds of random media and in particular for various natural classes of random graphs and their scaling limits. (See [34] for a survey of random walks on disordered media, including a discussion of anomalous heat kernel estimates.) In this section, we discuss two examples of random media for which heat kernel fluctuations have been proved rigorously.

As before, we will write **P** for the probability measure determining the law of the random medium and **E** for the corresponding expectation. Given a particular realisation ω of the random medium, in the context of this section a random (locally finite) graph tree, we write P_x^ω for the law of the associated (continuous-time) simple random walk X (with unit mean holding times) started from x; this is the quenched law of X. The quenched transition density is given by

$$p_t^\omega(x, y) := \frac{P_x^\omega(X_t = y)}{\mu_y^\omega},$$

where μ_y^ω denotes the number of neighbours of y in the tree associated with the realisation ω. (We note that, for a given realisation ω, μ_y^ω gives the invariant measure of X.)

12.4.1 Incipient Infinite Cluster on Trees

Let $\{Z_n\}_{n\geq 0}$ be a critical Galton–Watson branching process with offspring distribution $\mathtt{Bin}(n_0, p_c)$, where $p_c = 1/n_0$, for some $n_0 \in \mathbb{N}$. Then, the family tree of the process can be understood in terms of a critical percolation process on a regular rooted n_0-ary tree $\mathbb{B}$. Indeed, write $\mathbb{B}_N$ for the N-th level of $\mathbb{B}$ and $\mathbb{B}_{\leq N}$ for the union of the first N levels of $\mathbb{B}$. Then, $Z_n = |C(0) \cap \mathbb{B}_n|$, where 0 is the root of $\mathbb{B}$ and

$$C(0) := \{x \in \mathbb{B} : \text{ there exists an open path from 0 to } x\},$$

see Fig. 12.4. Since this random tree is almost surely finite, we consider a conditioning procedure that results in a modification of this random tree that extends to infinity. Namely, for $A \subseteq \mathbb{B}_{\leq k}$, set

$$P_0(A) := \lim_{n\to\infty} \mathbf{P}_{p_c}\left(C(0) \cap \mathbb{B}_{\leq k} = A \mid Z_n \neq 0\right),$$

where $\mathbf{P}_{p_c}$ is the law of the critical percolation process. It may be checked that the right-hand side is equal to $|A \cap \mathbb{B}_k| \, \mathbf{P}_{p_c}(C(0) \cap \mathbb{B}_{\leq k} = A)$, so the limit exists. Moreover, P_0 has a unique extension to a probability measure on the set of infinite

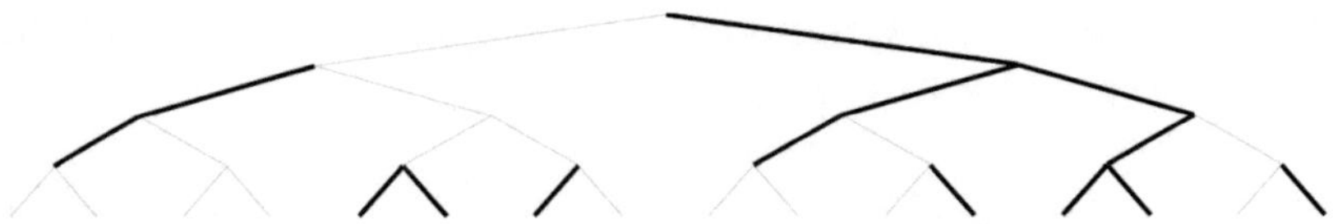

Fig. 12.4 Critical percolation on a binary tree. The first elements of the corresponding Galton–Watson branching process are given by $Z_0 = Z_1 = 1$, $Z_2 = Z_3 = Z_4 = 2$

connected subsets of $\mathbb{B}$ containing 0. We denote this measure by $\mathbf{P}$, since it will be this probability measure that governs the randomness of the medium in the following. We say that the infinite rooted tree $\mathcal{G}$ chosen according to the distribution $\mathbf{P}$ is called the incipient infinite cluster (IIC). Now, consider a simple random walk on the IIC $\mathcal{G}$. For such a process, the following on-diagonal heat kernel estimates are proved in [8], with $d_f = 2$ and $d_w = 3$.

Theorem 12.3

(i) There exist deterministic constants $c_1, c_2 \in (0, \infty)$ *and* $\alpha > 0$ *such that,* $\mathbf{P}$*-a.s.,*

$$c_1 t^{-d_f/d_w} (\log \log t)^{-\alpha} \leq p_t^{\omega}(0, 0) \leq c_2 t^{-d_f/d_w} (\log \log t)^{\alpha}$$

for large t.

(ii) There exists $\beta > 0$ *such that,* $\mathbf{P}$*-a.s.,*

$$\liminf_{t \to \infty} (\log \log t)^{\beta}\, t^{d_f/d_w}\, p_t^{\omega}(0, 0) = 0.$$

(iii) There exist constants $c_3, c_4 \in (0, \infty)$ *such that*

$$c_3 t^{-d_f/d_w} \leq \mathbf{E}[p_t^{\omega}(0, 0)] \leq c_4 t^{-d_f/d_w}.$$

We note that whilst part (i) and (ii) do not establish that heat kernel fluctuations occur $\mathbf{P}$-a.s., they do show that any fluctuations can be at most log-logarithmic in order. Moreover, in combination with the argument used to establish part (iii), they show that log-logarithmic fluctuations occur with positive probability. Heat kernel estimates for critical Galton–Watson trees with more general finite variance offspring distributions are obtained in [22]. Moreover, in the infinite variance case, similar results to the above theorem are derived in [19], with the distinction that the fluctuations are logarithmic in order in this case.

Finally, we also remark that, for a compact version of the scaling limit of critical Galton–Watson trees in the finite variance case, the continuum random tree, analogous fluctuation results for the heat kernel were obtained in [14]. In fact, the continuum random tree is known to be an example of a random recursive fractal, see [16], which goes some way to explaining why the results described above mirror those for the random recursive gaskets of the previous section.

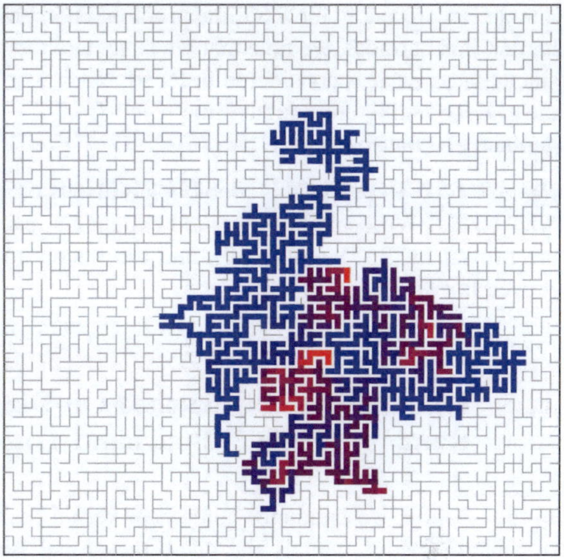

Fig. 12.5 The range of a realisation of the simple random walk on uniform spanning tree on a 60×60 box (with wired boundary conditions), shown after 50,000 steps. From most to least crossed edges, colours blend from red to blue. Originally presented in [4], based on code written by Sunil Chhita

12.4.2 Low-Dimensional Uniform Spanning Tree

Let $\Lambda_N := [-N, N]^2 \cap \mathbb{Z}^2$. A loopless connected graph whose vertex set consists of all the elements of Λ_N is called a spanning tree of Λ_N. Let $\mathcal{U}^{(N)}$ be the random graph obtained by picking one of the spanning trees from amongst all possible choices on Λ_N uniformly at random, see Fig. 12.5. The uniform spanning tree (UST) on $\mathbb{Z}^2$, $\mathcal{U}$ say, is the local limit of $\mathcal{U}^{(N)}$ as $N \to \infty$, see [35] for details. Note that it was to describe the potential scaling limit of this model, and the closely connected loop-erased random walk, that O. Schramm introduced the Schramm-Loewner evolution (SLE) in his celebrated paper [36].

Now, consider a (discrete-time) simple random walk on $\mathcal{U}$. As for critical Galton–Watson trees, the behaviour of the random walk is anomalous and the quenched on-diagonal heat kernel admits log-logarithmic fluctuations, namely the discrete-time analogue of Theorem 12.3 holds with $d_f = 8/5$ and $d_w = 13/5$. (Note that $8/5 = 2/(5/4)$; here, 2 is the dimension of the underlying space with respect to the Euclidean metric and $5/4$ is the exponent describing the scaling of the intrinsic metric on $\mathcal{U}$, also with respect to the Euclidean metric.) Parts (i) and (iii) are proved in [9, Theorems 4.4 and 4.5], and part (ii) is proved in [5, Corollary 1.2], together with the following estimate that establishes the fluctuations occur almost surely in this case: there exists a deterministic constant $\beta \in (0, \infty)$ such that, $\mathbf{P}$-a.s.,

$$\limsup_{n\to\infty} (\log\log n)^{-\beta}\, n^{d_f/d_w}\, p^{\omega}_{2n}(0,0) = \infty.$$

Although the estimates and the basic strategies used for proving them are similar to those for the random walk discussed in Sect. 12.4.1, due to the long-range interaction present in the UST, the proof of Theorem 12.3 (in particular (ii)) is much harder. Furthermore, we note that we state here the result for the discrete-time walk since that is what is considered in [5, 9], but it would be straightforward to deduce the corresponding continuous-time result from the same essential estimates on the geometry of the space.

Remark 12.2 Recently, this problem has also been studied for random walks on the UST on $\mathbb{Z}^3$. With suitably modified exponents, parts (i) and (iii) of Theorem 12.3 are proved in [2, Theorem 1.9], and the fluctuation result of Theorem 12.3(ii) is proved in [37, Theorem 1.2].

12.5 Fluctuations of Quenched On-Diagonal Heat Kernels in Random Media 2: Bouchaud Trap Models

In this section, we announce some heat kernel fluctuation results for another kind of random media, namely the one-dimensional Bouchaud trap model, which are obtained in the ongoing work [1].

To introduce the model, let $\tau = \{\tau_x\}_{x\in\mathbb{Z}}$ be i.i.d. random variables with distribution given by

$$\mathbf{P}(\tau_x > u) = u^{-\alpha}$$

for $u \geq 1$, where $\alpha > 0$ is some fixed constant; this will represent the random environment in the current setting. Given a realisation of τ, let $X = (X_t)_{t\geq 0}$ be the continuous-time Markov chain with generator

$$\mathcal{L}f(x) = \frac{1}{2\tau_x}\sum_{y\sim x}(f(y) - f(x));$$

this is the one-dimensional (symmetric) Bouchaud trap model. The dynamics of X are as follows: X waits at a vertex x an exponentially distributed time with mean τ_x, and, at this time, it jumps to one of the two neighbours of x, chosen uniformly at random. It is easy to see that the measure $\boldsymbol{\tau} := \sum_{x\in\mathbb{Z}}\tau_x\delta_x$ is invariant for the reversible Markov chain X. As before, we will write P_x^ω for the quenched law of X started from x, where again we suppose ω is a variable that determines the environment, so in particular we have that $\tau = \tau(\omega)$. Moreover, for this model, we will also consider

$$\mathbb{P}(\cdot) = \int P_0^\omega(\cdot)\,\mathbf{P}(d\omega),$$

which is typically called the annealed law of X started from 0.

We first recall the known scaling limits for the process X. The topology considered in the following theorem is the usual Skorohod J_1 topology on the space of càdlàg paths $D([0,\infty),\mathbb{R})$, and we omit to mention a (non-trivial) deterministic constant in the time scaling of the limiting process.

Theorem 12.4

(i) If $\alpha \in (0,1)$, then $(\varepsilon X_{t/\varepsilon^{1+1/\alpha}})_{t\geq 0}$ converges in distribution under the annealed law $\mathbb{P}$ to the FIN diffusion (for Fontes–Isopi–Newman, see [21, Definition 1.2]) with parameter α.

(ii) If $\alpha = 1$, then $(\varepsilon X_{t\log(\varepsilon^{-1})/\varepsilon^2})_{t\geq 0}$ converges in distribution under the annealed law $\mathbb{P}$ to Brownian motion.

(iii) If $\alpha > 1$, then, for **P***-a.e. realisation of the environment τ, $(\varepsilon X_{t/\varepsilon^2})_{t\geq 0}$ converges in distribution under the quenched law P_0^ω to Brownian motion.*

We note that part (i) is contained within [10, Theorem 3.2(ii)]. We also refer to [21] for the introduction of the limiting process and a scaling limit for a single marginal of the process and [12] for a related result in the non-symmetric case. Part (iii) is contained within [10, Theorem 3.2(ii)]. Part (ii) is described in [10, Remark 3.3] and can also be checked by applying the resistance scaling techniques of [15, 17]. Indeed, via such an approach, a similar result (i.e. convergence to Brownian motion under the annealed law, with logarithmic terms in the scaling) has been obtained in the more challenging case of the random walk on the range of four-dimensional random walks in [20].

To capture the leading order term in the on-diagonal part of the heat kernel of X, we define

$$\phi_\alpha(t) := \begin{cases} t^{-\frac{1}{1+\alpha}}, & \text{if } \alpha \in (0,1); \\ t^{-\frac{1}{2}}(\log t)^{-\frac{1}{2}}, & \text{if } \alpha = 1; \\ t^{-\frac{1}{2}}, & \text{if } \alpha > 1. \end{cases}$$

The following theorem, which shows quenched fluctuations of the heat kernel about ϕ_α when $\alpha \leq 1$, is due to [1]. We note that for the short-time heat kernel asymptotics of the limiting FIN diffusion, similar results to those contained in part (i) of the result were obtained in [18].

Theorem 12.5

(i) If $\alpha \in (0,1)$, then it **P***-a.s. holds that*

$$\limsup_{t\to\infty} \frac{p_t^\omega(0,0)}{\phi_\alpha(t)(\log\log t)^{\frac{1-\alpha}{1+\alpha}}} \in (0,\infty),$$

and

$$\liminf_{t\to\infty} \frac{p_t^\omega(0,0)\,(\log t)^\theta}{\phi_\alpha(t)} = \begin{cases} 0, & \text{if } \theta = \frac{1}{1+\alpha}, \\ \infty, & \text{if } \theta > \frac{3}{\alpha}. \end{cases}$$

(ii) If $\alpha = 1$, then it **P***-a.s. holds that*

$$\limsup_{t\to\infty} \frac{p_t^\omega(0,0)}{\phi_\alpha(t)} \in (0,\infty),$$

and

$$\liminf_{t\to\infty} \frac{p_t^\omega(0,0)\,(\log\log t)^\theta}{\phi_\alpha(t)} = \begin{cases} 0, & \text{if } \theta = \frac{1}{2}, \\ \infty, & \text{if } \theta > 3. \end{cases}$$

(iii) If $\alpha > 1$, then there exists a constant $\sigma^2 > 0$ such that, **P***-a.s., for every $x_0 \in [0,\infty)$ and compact interval $I \subset (0,\infty)$,*

$$\lim_{\lambda\to\infty} \sup_{|x|\le x_0} \sup_{t\in I} \left| \lambda p^\omega_{\lambda^2 t}(0, \lfloor \lambda x \rfloor) - \frac{1}{\sqrt{2\pi\sigma^2 t}} e^{-\frac{x^2}{2\sigma^2 t}} \right| = 0.$$

In particular,

$$\lim_{t\to\infty} \phi_\alpha(t)^{-1} p_t^\omega(0,0) = \frac{1}{\sqrt{2\pi\sigma^2}}.$$

Let us very briefly mention the idea of the proof, which is based on standard ideas in the area that relate heat kernel estimates to estimates on the resistance and volume growth of a space. (A framework for deriving such estimates that is broad enough to include random recursive gaskets, the continuum random tree and the FIN diffusion is presented in [13].) In the setting of the one-dimensional Bouchaud trap model, the resistance metric is easy to understand since it simply coincides with the Euclidean metric. As for the volume growth of the invariant measure, consider

$$V(x,n) := \sum_{y=x-n}^{x+n} \tau_y, \qquad x \in \mathbb{Z},\ n \in \mathbb{N}.$$

In particular, as a sum of i.i.d. random variables, the asymptotic behaviour of $V(0,n)$ can be understood in terms of classical results of probability theory. When $\alpha > 1$, so the random variables in the sequence τ have a mean, the law of large numbers immediately implies that, **P**-a.s., $V(0,n) \sim cn$, for some deterministic constant n. Indeed, with a little more work, one can check that a suitably scaled version of the invariant measure converges to Lebesgue measure on the line. Since there are no fluctuations of the volume in this case, there are no fluctuations of the heat kernel, as seen in part (iii) of the above theorem. On the other hand, when

$\alpha < 1$, the random variables in the sequence τ fall into the domain of attraction of an α-stable law and so we see subordinators in the distributional scaling limit of $V(0, n)$. Again, with more work, one can describe the distributional scaling limit of the invariant measure, this time in terms of a Poisson random measure. Now, in such a regime, there is no quenched convergence of the volume, but rather we see **P**-a.s. fluctuations about the leading order scaling factor. It is this that leads to the fluctuations of part (i) of Theorem 12.5. Note that, in this case, we see asymmetric fluctuations, with upper fluctuations of log-logarithmic order, but lower ones of logarithmic order. As for $\alpha = 1$, this falls somewhere between the two other cases, with a deterministic distributional limit for the volume, but quenched fluctuations about this. Thus, apart from the intermediate case of $\alpha = 1$, this example captures the basic heuristic that one might expect to see heat kernel fluctuations when the scaling limit of the random media is itself random.

Finally, we note that, as for the volume measure, the heat kernel of X shows smoother behaviour in terms of its distributional scaling [1]; this is reminiscent of the result for critical Galton–Watson trees that was presented above as Theorem 12.3(iii).

Theorem 12.6

(i) For every $\alpha > 0$,

$$\lim_{\lambda\to\infty} \liminf_{t\to\infty} \mathbf{P}\left(\lambda^{-1} \leq \phi_\alpha(t)^{-1} p_t^\omega(0,0) \leq \lambda\right) = 1.$$

(ii) For every $\alpha > 0$, there exists an $\varepsilon > 0$ such that

$$0 < \liminf_{t\to\infty} \phi_\alpha(t)^{-\varepsilon}\mathbf{E}\left(p_t^\omega(0,0)^\varepsilon\right) \leq \limsup_{t\to\infty} \phi_\alpha(t)^{-\varepsilon}\mathbf{E}\left(p_t^\omega(0,0)^\varepsilon\right) < \infty.$$

Moreover, if $\alpha > \frac{3}{2}$, then we may take $\varepsilon = 1$.

The restriction on the moments in part (ii) above relates to the integrability of certain powers of the volume that appear in the heat kernel estimates. Similar bounds were obtained in [19] for random walks on critical Galton–Watson trees with infinite variance offspring distributions.

Acknowledgments This research was supported by JSPS Grant-in-Aid for Scientific Research (A) 22H00099, JSPS Grant-in-Aid for Scientific Research (C) 19K03540, and the Research Institute for Mathematical Sciences, an International Joint Usage/Research Center located in Kyoto University.

References

1. S. Andres, D. A. Croydon, and T. Kumagai. Heat kernel fluctuations and quantitative homogenization for the one-dimensional Bouchaud trap model. *In preparation*, 2022.

2. O. Angel, D. A. Croydon, S. Hernandez-Torres, and D. Shiraishi. Scaling limits of the three-dimensional uniform spanning tree and associated random walk. *Ann. Probab.*, 49(6):3032–3105, 2021.
3. M. T. Barlow. Diffusions on fractals. In *Lectures on probability theory and statistics (Saint-Flour, 1995)*, volume 1690 of *Lecture Notes in Math.*, pages 1–121. Springer, Berlin, 1998.
4. M. T. Barlow, D. A. Croydon, and T. Kumagai. Subsequential scaling limits of simple random walk on the two-dimensional uniform spanning tree. *Ann. Probab.*, 45(1):4–55, 2017.
5. M. T. Barlow, D. A. Croydon, and T. Kumagai. Quenched and averaged tails of the heat kernel of the two-dimensional uniform spanning tree. *Probab. Theory Related Fields*, 181(1–3):57–111, 2021.
6. M. T. Barlow and B. M. Hambly. Transition density estimates for Brownian motion on scale irregular Sierpinski gaskets. *Ann. Inst. H. Poincaré Probab. Statist.*, 33(5):531–557, 1997.
7. M. T. Barlow and J. Kigami. Localized eigenfunctions of the Laplacian on p.c.f. self-similar sets. *J. London Math. Soc. (2)*, 56(2):320–332, 1997.
8. M. T. Barlow and T. Kumagai. Random walk on the incipient infinite cluster on trees. *Illinois J. Math.*, 50(1–4):33–65, 2006.
9. M. T. Barlow and R. Masson. Spectral dimension and random walks on the two dimensional uniform spanning tree. *Comm. Math. Phys.*, 305(1):23–57, 2011.
10. G. Ben Arous, M. Cabezas, J. Černý, and R. Royfman. Randomly trapped random walks. *Ann. Probab.*, 43(5):2405–2457, 2015.
11. G. Ben Arous and T. Kumagai. Large deviations of Brownian motion on the Sierpinski gasket. *Stochastic Process. Appl.*, 85(2):225–235, 2000.
12. G. Ben Arous and J. Černý. Bouchaud's model exhibits two different aging regimes in dimension one. *Ann. Appl. Probab.*, 15(2):1161–1192, 2005.
13. D. A. Croydon. Heat kernel fluctuations for a resistance form with non-uniform volume growth. *Proc. Lond. Math. Soc. (3)*, 94(3):672–694, 2007.
14. D. A. Croydon. Volume growth and heat kernel estimates for the continuum random tree. *Probab. Theory Related Fields*, 140(1–2):207–238, 2008.
15. D. A. Croydon. Scaling limits of stochastic processes associated with resistance forms. *Ann. Inst. Henri Poincaré Probab. Stat.*, 54(4):1939–1968, 2018.
16. D. A. Croydon and B. Hambly. Self-similarity and spectral asymptotics for the continuum random tree. *Stochastic Process. Appl.*, 118(5):730–754, 2008.
17. D. A. Croydon, B. Hambly, and T. Kumagai. Time-changes of stochastic processes associated with resistance forms. *Electron. J. Probab.*, 22:Paper No. 82, 41, 2017.
18. D. A. Croydon, B. M. Hambly, and T. Kumagai. Heat kernel estimates for FIN processes associated with resistance forms. *Stochastic Process. Appl.*, 129(9):2991–3017, 2019.
19. D. A. Croydon and T. Kumagai. Random walks on Galton-Watson trees with infinite variance offspring distribution conditioned to survive. *Electron. J. Probab.*, 13:no. 51, 1419–1441, 2008.
20. D. A. Croydon and D. Shiraishi. Scaling limit for random walk on the range of random walk in four dimensions. *Ann. Inst. Henri Poincaré Probab. Stat.*, 59(1):166–184, 2023.
21. L. R. G. Fontes, M. Isopi, and C. M. Newman. Random walks with strongly inhomogeneous rates and singular diffusions: convergence, localization and aging in one dimension. *Ann. Probab.*, 30(2):579–604, 2002.
22. I. Fujii and T. Kumagai. Heat kernel estimates on the incipient infinite cluster for critical branching processes. In *Proceedings of RIMS Workshop on Stochastic Analysis and Applications*, RIMS Kôkyûroku Bessatsu, B6, pages 85–95. Res. Inst. Math. Sci. (RIMS), Kyoto, 2008.
23. M. Fukushima and T. Shima. On a spectral analysis for the Sierpiński gasket. *Potential Anal.*, 1(1):1–35, 1992.
24. P. J. Grabner and W. Woess. Functional iterations and periodic oscillations for simple random walk on the Sierpiński graph. *Stochastic Process. Appl.*, 69(1):127–138, 1997.
25. B. M. Hambly. Brownian motion on a homogeneous random fractal. *Probab. Theory Related Fields*, 94(1):1–38, 1992.

26. B. M. Hambly. Brownian motion on a random recursive Sierpinski gasket. *Ann. Probab.*, 25(3):1059–1102, 1997.
27. B. M. Hambly. On the asymptotics of the eigenvalue counting function for random recursive Sierpinski gaskets. *Probab. Theory Related Fields*, 117(2):221–247, 2000.
28. B. M. Hambly and T. Kumagai. Fluctuation of the transition density for Brownian motion on random recursive Sierpinski gaskets. *Stochastic Process. Appl.*, 92(1):61–85, 2001.
29. N. Kajino. Non-regularly varying and non-periodic oscillation of the on-diagonal heat kernels on self-similar fractals. In *Fractal geometry and dynamical systems in pure and applied mathematics. II. Fractals in applied mathematics*, volume 601 of *Contemp. Math.*, pages 165–194. Amer. Math. Soc., Providence, RI, 2013.
30. N. Kajino. On-diagonal oscillation of the heat kernels on post-critically finite self-similar fractals. *Probab. Theory Related Fields*, 156(1–2):51–74, 2013.
31. J. Kigami. *Analysis on fractals*, volume 143 of *Cambridge Tracts in Mathematics*. Cambridge University Press, Cambridge, 2001.
32. J. Kigami and M. L. Lapidus. Weyl's problem for the spectral distribution of Laplacians on p.c.f. self-similar fractals. *Comm. Math. Phys.*, 158(1):93–125, 1993.
33. T. Kumagai. Short time asymptotic behaviour and large deviation for Brownian motion on some affine nested fractals. *Publ. Res. Inst. Math. Sci.*, 33(2):223–240, 1997.
34. T. Kumagai. *Random walks on disordered media and their scaling limits*, volume 2101 of *Lecture Notes in Mathematics*. Springer, Cham, 2014. Lecture notes from the 40th Probability Summer School held in Saint-Flour, 2010, École d'Été de Probabilités de Saint-Flour. [Saint-Flour Probability Summer School].
35. R. Pemantle. Choosing a spanning tree for the integer lattice uniformly. *Ann. Probab.*, 19(4):1559–1574, 1991.
36. O. Schramm. Scaling limits of loop-erased random walks and uniform spanning trees. *Israel J. Math.*, 118:221–288, 2000.
37. D. Shiraishi and S. Watanabe. Volume and heat kernel fluctuations for the three-dimensional uniform spanning tree. Preprint available at arXiv:2211.15031, 2022.
38. R. S. Strichartz. *Differential equations on fractals*. Princeton University Press, Princeton, NJ, 2006. A tutorial.
39. W. Woess. *Random walks on infinite graphs and groups*, volume 138 of *Cambridge Tracts in Mathematics*. Cambridge University Press, Cambridge, 2000.

Index

P. Alonso Ruiz et al. (eds.), *From Classical Analysis to Analysis on Fractals*, Applied and Numerical Harmonic Analysis, https://doi.org/10.1007/978-3-031-37800-3

Applied and Numerical Harmonic Analysis (107 volumes)

1. A. I. Saichev and W. A. Woyczyñski: *Distributions in the Physical and Engineering Sciences* (ISBN: 978-0-8176-3924-2)
2. C. E. D'Attellis and E. M. Fernandez-Berdaguer: *Wavelet Theory and Harmonic Analysis in Applied Sciences* (ISBN: 978-0-8176-3953-2)
3. H. G. Feichtinger and T. Strohmer: *Gabor Analysis and Algorithms* (ISBN: 978-0-8176-3959-4)
4. R. Tolimieri and M. An: *Time-Frequency Representations* (ISBN: 978-0-8176-3918-1)
5. T. M. Peters and J. C. Williams: *The Fourier Transform in Biomedical Engineering* (ISBN: 978-0-8176-3941-9)
6. G. T. Herman: *Geometry of Digital Spaces* (ISBN: 978-0-8176-3897-9)
7. A. Teolis: *Computational Signal Processing with Wavelets* (ISBN: 978-0-8176-3909-9)
8. J. Ramanathan: *Methods of Applied Fourier Analysis* (ISBN: 978-0-8176-3963-1)
9. J. M. Cooper: *Introduction to Partial Differential Equations with MATLAB* (ISBN: 978-0-8176-3967-9)
10. Procházka, N. G. Kingsbury, P. J. Payner, and J. Uhlir: *Signal Analysis and Prediction* (ISBN: 978-0-8176-4042-2)
11. W. Bray and C. Stanojevic: *Analysis of Divergence* (ISBN: 978-1-4612-7467-4)
12. G. T. Herman and A. Kuba: *Discrete Tomography* (ISBN: 978-0-8176-4101-6)
13. K. Gröchenig: *Foundations of Time-Frequency Analysis* (ISBN: 978-0-8176-4022-4)
14. L. Debnath: *Wavelet Transforms and Time-Frequency Signal Analysis* (ISBN: 978-0-8176-4104-7)
15. J. J. Benedetto and P. J. S. G. Ferreira: *Modern Sampling Theory* (ISBN: ISBN: 978-0-8176-4023-1)
16. D. F. Walnut: *An Introduction to Wavelet Analysis* (ISBN: 978-0-8176-3962-4)

P. Alonso Ruiz et al. (eds.), *From Classical Analysis to Analysis on Fractals*, Applied and Numerical Harmonic Analysis, https://doi.org/10.1007/978-3-031-37800-3

17. A. Abbate, C. DeCusatis, and P. K. Das: *Wavelets and Subbands* (ISBN: ISBN: 978-0-8176-4136-8)
18. O. Bratteli, P. Jorgensen, and B. Treadway: *Wavelets Through a Looking Glass* (ISBN: 978-0-8176-4280-80
19. H. G. Feichtinger and T. Strohmer: *Advances in Gabor Analysis* (ISBN: ISBN: 978-0-8176-4239-6)
20. O. Christensen: *An Introduction to Frames and Riesz Bases* (ISBN: 978-0-8176-4295-2)
21. L. Debnath: *Wavelets and Signal Processing* (ISBN: 978-0-8176-4235-8)
22. G. Bi and Y. Zeng: *Transforms and Fast Algorithms for Signal Analysis and Representations* (ISBN: 978-0-8176-4279-2)
23. J. H. Davis: *Methods of Applied Mathematics with a MATLAB Overview* (ISBN: 978-0-8176-4331-7)
24. J. J. Benedetto and A. I. Zayed: *Sampling, Wavelets, and Tomography* (ISBN: 978-0-8176-4304-1)
25. E. Prestini: *The Evolution of Applied Harmonic Analysis* (ISBN: 978-0-8176-4125-2)
26. L. Brandolini, L. Colzani, A. Iosevich, and G. Travaglini: *Fourier Analysis and Convexity* (ISBN: 978-0-8176-3263-2)
27. W. Freeden and V. Michel: *Multiscale Potential Theory* (ISBN: 978-0-8176-4105-4)
28. O. Christensen and K. L. Christensen: *Approximation Theory* (ISBN: 978-0-8176-3600-5)
29. O. Calin and D.-C. Chang: *Geometric Mechanics on Riemannian Manifolds* (ISBN: 978-0-8176-4354-6)
30. J. A. Hogan: *Time?Frequency and Time?Scale Methods* (ISBN: 978-0-8176-4276-1)
31. C. Heil: *Harmonic Analysis and Applications* (ISBN: 978-0-8176-3778-1)
32. K. Borre, D. M. Akos, N. Bertelsen, P. Rinder, and S. H. Jensen: *A Software-Defined GPS and Galileo Receiver* (ISBN: 978-0-8176-4390-4)
33. T. Qian, M. I. Vai, and Y. Xu: *Wavelet Analysis and Applications* (ISBN: ISBN: 978-3-7643-7777-9)
34. G. T. Herman and A. Kuba: *Advances in Discrete Tomography and Its Applications* (ISBN: 978-0-8176-3614-2)
35. M. C. Fu, R. A. Jarrow, J.-Y. Yen, and R. J. Elliott: *Advances in Mathematical Finance* (ISBN: 978-0-8176-4544-1)
36. O. Christensen: *Frames and Bases* (ISBN: 978-0-8176-4677-6)
37. P. E. T. Jorgensen, J. D. Merrill, and J. A. Packer: *Representations, Wavelets, and Frames* (ISBN: 978-0-8176-4682-0)
38. M. An, A. K. Brodzik, and R. Tolimieri: *Ideal Sequence Design in Time-Frequency Space* (ISBN: 978-0-8176-4737-7)
39. S. G. Krantz: *Explorations in Harmonic Analysis* (ISBN: 978-0-8176-4668-4)
40. B. Luong: *Fourier Analysis on Finite Abelian Groups* (ISBN: 978-0-8176-4915-9)

41. G. S. Chirikjian: *Stochastic Models, Information Theory, and Lie Groups, Volume 1* (ISBN: 978-0-8176-4802-2)
42. C. Cabrelli and J. L. Torrea: *Recent Developments in Real and Harmonic Analysis* (ISBN: 978-0-8176-4531-1)
43. M. V. Wickerhauser: *Mathematics for Multimedia* (ISBN: 978-0-8176-4879-4)
44. B. Forster, P. Massopust, O. Christensen, K. Gröchenig, D. Labate, P. Vandergheynst, G. Weiss, and Y. Wiaux: *Four Short Courses on Harmonic Analysis* (ISBN: 978-0-8176-4890-9)
45. O. Christensen: *Functions, Spaces, and Expansions* (ISBN: 978-0-8176-4979-1)
46. J. Barral and S. Seuret: *Recent Developments in Fractals and Related Fields* (ISBN: 978-0-8176-4887-9)
47. O. Calin, D.-C. Chang, and K. Furutani, and C. Iwasaki: *Heat Kernels for Elliptic and Sub-elliptic Operators* (ISBN: 978-0-8176-4994-4)
48. C. Heil: *A Basis Theory Primer* (ISBN: 978-0-8176-4686-8)
49. J. R. Klauder: *A Modern Approach to Functional Integration* (ISBN: 978-0-8176-4790-2)
50. J. Cohen and A. I. Zayed: *Wavelets and Multiscale Analysis* (ISBN: 978-0-8176-8094-7)
51. D. Joyner and J.-L. Kim: *Selected Unsolved Problems in Coding Theory* (ISBN: 978-0-8176-8255-2)
52. G. S. Chirikjian: *Stochastic Models, Information Theory, and Lie Groups, Volume 2* (ISBN: 978-0-8176-4943-2)
53. J. A. Hogan and J. D. Lakey: *Duration and Bandwidth Limiting* (ISBN: 978-0-8176-8306-1)
54. G. Kutyniok and D. Labate: *Shearlets* (ISBN: 978-0-8176-8315-3)
55. P. G. Casazza and P. Kutyniok: *Finite Frames* (ISBN: 978-0-8176-8372-6)
56. V. Michel: *Lectures on Constructive Approximation* (ISBN: 978-0-8176-8402-0)
57. D. Mitrea, I. Mitrea, M. Mitrea, and S. Monniaux: *Groupoid Metrization Theory* (ISBN: 978-0-8176-8396-2)
58. T. D. Andrews, R. Balan, J. J. Benedetto, W. Czaja, and K. A. Okoudjou: *Excursions in Harmonic Analysis, Volume 1* (ISBN: 978-0-8176-8375-7)
59. T. D. Andrews, R. Balan, J. J. Benedetto, W. Czaja, and K. A. Okoudjou: *Excursions in Harmonic Analysis, Volume 2* (ISBN: 978-0-8176-8378-8)
60. D. V. Cruz-Uribe and A. Fiorenza: *Variable Lebesgue Spaces* (ISBN: 978-3-0348-0547-6)
61. W. Freeden and M. Gutting: *Special Functions of Mathematical (Geo-)Physics* (ISBN: 978-3-0348-0562-9)
62. A. I. Saichev and W. A. Woyczyñski: *Distributions in the Physical and Engineering Sciences, Volume 2: Linear and Nonlinear Dynamics of Continuous Media* (ISBN: 978-0-8176-3942-6)
63. S. Foucart and H. Rauhut: *A Mathematical Introduction to Compressive Sensing* (ISBN: 978-0-8176-4947-0)

64. G. T. Herman and J. Frank: *Computational Methods for Three-Dimensional Microscopy Reconstruction* (ISBN: 978-1-4614-9520-8)
65. A. Paprotny and M. Thess: *Realtime Data Mining: Self-Learning Techniques for Recommendation Engines* (ISBN: 978-3-319-01320-6)
66. A. I. Zayed and G. Schmeisser: *New Perspectives on Approximation and Sampling Theory: Festschrift in Honor of Paul Butzer's 85th Birthday* (ISBN: 978-3-319-08800-6)
67. R. Balan, M. Begue, J. Benedetto, W. Czaja, and K. A. Okoudjou: *Excursions in Harmonic Analysis, Volume 3* (ISBN: 978-3-319-13229-7)
68. H. Boche, R. Calderbank, G. Kutyniok, and J. Vybiral: *Compressed Sensing and its Applications* (ISBN: 978-3-319-16041-2)
69. S. Dahlke, F. De Mari, P. Grohs, and D. Labate: *Harmonic and Applied Analysis: From Groups to Signals* (ISBN: 978-3-319-18862-1)
70. A. Aldroubi: *New Trends in Applied Harmonic Analysis* (ISBN: 978-3-319-27871-1)
71. M. Ruzhansky: *Methods of Fourier Analysis and Approximation Theory* (ISBN: 978-3-319-27465-2)
72. G. Pfander: *Sampling Theory, a Renaissance* (ISBN: 978-3-319-19748-7)
73. R. Balan, M. Begue, J. Benedetto, W. Czaja, and K. A. Okoudjou: *Excursions in Harmonic Analysis, Volume 4* (ISBN: 978-3-319-20187-0)
74. O. Christensen: *An Introduction to Frames and Riesz Bases, Second Edition* (ISBN: 978-3-319-25611-5)
75. E. Prestini: *The Evolution of Applied Harmonic Analysis: Models of the Real World, Second Edition* (ISBN: 978-1-4899-7987-2)
76. J. H. Davis: *Methods of Applied Mathematics with a Software Overview, Second Edition* (ISBN: 978-3-319-43369-1)
77. M. Gilman, E. M. Smith, and S. M. Tsynkov: *Transionospheric Synthetic Aperture Imaging* (ISBN: 978-3-319-52125-1)
78. S. Chanillo, B. Franchi, G. Lu, C. Perez, and E. T. Sawyer: *Harmonic Analysis, Partial Differential Equations and Applications* (ISBN: 978-3-319-52741-3)
79. R. Balan, J. Benedetto, W. Czaja, M. Dellatorre, and K. A. Okoudjou: *Excursions in Harmonic Analysis, Volume 5* (ISBN: 978-3-319-54710-7)
80. I. Pesenson, Q. T. Le Gia, A. Mayeli, H. Mhaskar, and D. X. Zhou: *Frames and Other Bases in Abstract and Function Spaces: Novel Methods in Harmonic Analysis, Volume 1* (ISBN: 978-3-319-55549-2)
81. I. Pesenson, Q. T. Le Gia, A. Mayeli, H. Mhaskar, and D. X. Zhou: *Recent Applications of Harmonic Analysis to Function Spaces, Differential Equations, and Data Science: Novel Methods in Harmonic Analysis, Volume 2* (ISBN: 978-3-319-55555-3)
82. F. Weisz: *Convergence and Summability of Fourier Transforms and Hardy Spaces* (ISBN: 978-3-319-56813-3)
83. C. Heil: *Metrics, Norms, Inner Products, and Operator Theory* (ISBN: 978-3-319-65321-1)
84. S. Waldron: *An Introduction to Finite Tight Frames: Theory and Applications.* (ISBN: 978-0-8176-4814-5)

85. D. Joyner and C. G. Melles: *Adventures in Graph Theory: A Bridge to Advanced Mathematics.* (ISBN: 978-3-319-68381-2)
86. B. Han: *Framelets and Wavelets: Algorithms, Analysis, and Applications* (ISBN: 978-3-319-68529-8)
87. H. Boche, G. Caire, R. Calderbank, M. März, G. Kutyniok, and R. Mathar: *Compressed Sensing and Its Applications* (ISBN: 978-3-319-69801-4)
88. A. I. Saichev and W. A. Woyczyñski: *Distributions in the Physical and Engineering Sciences, Volume 3: Random and Fractal Signals and Fields* (ISBN: 978-3-319-92584-4)
89. G. Plonka, D. Potts, G. Steidl, and M. Tasche: *Numerical Fourier Analysis* (ISBN: 978-3-030-04305-6)
90. K. Bredies and D. Lorenz: *Mathematical Image Processing* (ISBN: 978-3-030-01457-5)
91. H. G. Feichtinger, P. Boggiatto, E. Cordero, M. de Gosson, F. Nicola, A. Oliaro, and A. Tabacco: *Landscapes of Time-Frequency Analysis* (ISBN: 978-3-030-05209-6)
92. E. Liflyand: *Functions of Bounded Variation and Their Fourier Transforms* (ISBN: 978-3-030-04428-2)
93. R. Campos: *The XFT Quadrature in Discrete Fourier Analysis* (ISBN: 978-3-030-13422-8)
94. M. Abell, E. Iacob, A. Stokolos, S. Taylor, S. Tikhonov, J. Zhu: *Topics in Classical and Modern Analysis: In Memory of Yingkang Hu* (ISBN: 978-3-030-12276-8)
95. H. Boche, G. Caire, R. Calderbank, G. Kutyniok, R. Mathar, P. Petersen: *Compressed Sensing and its Applications: Third International MATHEON Conference 2017* (ISBN: 978-3-319-73073-8)
96. A. Aldroubi, C. Cabrelli, S. Jaffard, U. Molter: *New Trends in Applied Harmonic Analysis, Volume II: Harmonic Analysis, Geometric Measure Theory, and Applications* (ISBN: 978-3-030-32352-3)
97. S. Dos Santos, M. Maslouhi, K. Okoudjou: *Recent Advances in Mathematics and Technology: Proceedings of the First International Conference on Technology, Engineering, and Mathematics, Kenitra, Morocco, March 26-27, 2018* (ISBN: 978-3-030-35201-1)
98. Á. Bényi, K. Okoudjou: *Modulation Spaces: With Applications to Pseudodifferential Operators and Nonlinear Schrödinger Equations* (ISBN: 978-1-0716-0330-7)
99. P. Boggiato, M. Cappiello, E. Cordero, S. Coriasco, G. Garello, A. Oliaro, J. Seiler: *Advances in Microlocal and Time-Frequency Analysis* (ISBN: 978-3-030-36137-2)
100. S. Casey, K. Okoudjou, M. Robinson, B. Sadler: *Sampling: Theory and Applications* (ISBN: 978-3-030-36290-4)
101. P. Boggiatto, T. Bruno, E. Cordero, H. G. Feichtinger, F. Nicola, A. Oliaro, A. Tabacco, M. Vallarino: *Landscapes of Time-Frequency Analysis: ATFA 2019* (ISBN: 978-3-030-56004-1)

102. M. Hirn, S. Li, K. Okoudjou, S. Saliana, Ö. Yilmaz: *Excursions in Harmonic Analysis, Volume 6: In Honor of John Benedetto's 80th Birthday* (ISBN: 978-3-030-69636-8)
103. F. De Mari, E. De Vito: *Harmonic and Applied Analysis: From Radon Transforms to Machine Learning* (ISBN: 978-3-030-86663-1)
104. G. Kutyniok, H. Rauhut, R. J. Kunsch, *Compressed Sensing in Information Processing* (ISBN: 978-3-031-09744-7)
105. P. Flandrin, S. Jaffard, T. Paul, B. Torresani, *Theoretic Physics, Wavelets, Analysis, Genomics: An Indisciplinary Tribute to Alex Grossmann* (ISBN: 978-3-030-45846-1)
106. G. Plonka-Hoch, D. Potts, G. Steidl, M. Tasche, *Numerical Fourier Analysis, Second Edition* (ISBN: 978-XXX)
107. P. Alonso Ruiz, M. Hinz, K. Okoudjou, L. Rogers, A. Teplyaev, *From Classical Analysis to Analysis on Fractals: A Tribute to Robert Strichartz, Volume 1* (978-3-031-37799-0)

For an up-to-date list of ANHA titles, please visit http://www.springer.com/series/4968

MIX
Papier aus verantwortungsvollen Quellen
Paper from responsible sources
FSC® C105338

If you have any concerns about our products,
you can contact us on
ProductSafety@springernature.com

In case Publisher is established outside the EU,
the EU authorized representative is:
Springer Nature Customer Service Center GmbH
Europaplatz 3, 69115 Heidelberg, Germany

Printed by Libri Plureos GmbH
in Hamburg, Germany